Aridland Springs in North America

Arizona-Sonora Desert Museum Studies in Natural History

SERIES EDITORS

Richard C. Brusca
Thomas R. Van Devender
Mark A. Dimmitt

Aridland Springs in North America
Ecology and Conservation

Edited by
Lawrence E. Stevens and Vicky J. Meretsky

With a Foreword by Gary Paul Nabhan

The University of Arizona Press and
The Arizona-Sonora Desert Museum | Tucson

The University of Arizona Press
© 2008 The Arizona Board of Regents
All rights reserved
www.uapress.arizona.edu

Library of Congress Cataloging-in-Publication Data
Aridland springs in North America : ecology and conservation / edited
by Lawrence E. Stevens and Vicky J. Meretsky ; with a foreword by
Gary Paul Nabhan.
p. cm. — (Arizona-Sonora Desert Museum studies in natural history)
Includes bibliographical references and index.
ISBN 978-0-8165-2645-1 (hardcover : alk. paper)
1. Arid regions ecology—North America—Congresses. 2. Spring
ecology—North America—Congresses. I. Stevens, Lawrence E.
II. Meretsky, Vicky J., 1958–
QH102.A75 2008
577.609154—dc22   2008003985

Wendell Minckley, PhD
(1935–2001)

Scientist, savant, natural historian, professor, and friend, whose work provides much of the foundation for our understanding of southwestern springs ecology. Minck never failed to speak his mind and tell the truth. His wisdom and fervent concern for our desert waters and remarkable fish were inspiring to all who knew him and are sorely missed. His voice will not be forgotten. (Photo courtesy of the Arizona State University School of Life Sciences Visualization Lab)

Contents

Foreword

GARY PAUL NABHAN

Few of us would argue that anything but water has ever been the hottest topic discussed at gatherings among environmental activists, natural scientists, farmers, ranchers, developers, or politicians of the arid Southwest; neither sex nor violence come out as close seconds, although the frequency of both is probably positively correlated with the availability of water in any given locale. For this reason, a collection of writings about freshwater springs must be considered a sort of historical wonder: that we have paid immense attention to both river water and groundwater but not to one of the most vital linkages between the aquatic habitats, where underground aquifers literally bubble up to the surface, creating streams and wetlands.

Most states in the western United States have focused increasing attention on legal means to protect groundwater reserves and in-stream flows from depletion, but there remains a paucity of legal protection for springs and the biologically diverse habitats they create. Many geologists agree that springs occur in some of the most interesting geomorphic settings within our many geologic provinces, but few have tried to classify those settings or ask the big "why" and "how" questions associated with them. And the majority of biologists who have worked within the four North American deserts can immediately recall remarkable assemblages of species or unusually dynamic ecological processes that they have witnessed around springs, but hardly any of them have systematically articulated the patterns of productivity, diversity, and resilience that underlie their observations. The consequences of this historic oversight take the form of ecological, cultural, and economic costs for which our descendents will curse us.

Fortunately, there is a growing cadre of environmental scientists who are working to rid us of this curse, and some of the most dynamic and interdisciplinary among them are contributors to this book. At last, a theory and practice of springs appears to be on the horizon, just as naturalist Jim

Harrison has spoken of the theory and practice of rivers. We are beginning to fathom the geological, biological, and cultural significance of springs, and this understanding will, we hope, be inspiring enough to motivate us to protect and restore these habitats before they blink out, one by one, like so many isolated candles on a geological layer cake.

Read on, my friends, and when you are done reading, take some time to decide what is within your intellectual, physical, and financial means to help protect these most ecologically productive, geologically unique, biologically diverse, and culturally significant habitats in western North America. The clock is ticking; the faucet is running.

Preface and Acknowledgments

LAWRENCE E. STEVENS AND VICKY J. MERETSKY

This volume on the ecology and conservation of springs in the arid portions of North America arose out of two symposia, one at the Arizona-Sonora Desert Museum in 2000 and the other at the Ecological Society of America meetings in 2002, both in Tucson, Arizona. While the issues addressed herein pertain specifically to springs in the arid portions of North America, the scope of these issues and problems, and our concern regarding springs conservation, are global.

The Arizona-Sonora Desert Museum symposium provided a forum for numerous experts from the diverse fields of geology, hydrology, geomorphology, water quality, ecology, ecosystem dynamics, prehistoric and historical human uses, water law, and conservation biology to describe their work on springs ecosystems in the arid portions of North America. This volume includes an array of chapters presented at that symposium, as well as others that arose from the panel discussions there.

We begin this book with an introductory chapter by Larry Stevens and Vicky Meretsky that provides an overview of springs ecosystem distribution, ecology, and conservation challenges. Chapter 2 is by Peter Unmack and the late W. L. Minckley, who expand on lifelong concerns enunciated in their 2000 essay; this chapter illustrates both the biological uniqueness of springs in arid regions around the world and the imperiled state of the fish and other biota associated with these ecosystems. Minckley made enormous contributions to our understanding of fish ecology, southwestern springs and riparian ecology, and conservation of those ecosystems. He was one of the first to sound the alarm on the state of western North America's springs ecosystems, and it is fitting that this book is dedicated to him.

In chapter 3, David Kreamer and Abraham Springer describe the complex groundwater hydrology of springs on the southern Colorado Plateau. This vast, poorly understood system of aquifers is presently threatened by rapidly increasing groundwater extraction, a practice that is eliminating springs around the world at an alarming but largely undocumented rate.

The next chapter, by Abraham Springer and several colleagues, describes the history and the elements of a springs classification system, why such an undertaking is elusive, and the information needed to initiate that effort. Efforts to categorize springs have largely ignored ecological considerations, habitat, and biota, instead focusing primarily on springs-groundwater relationships, geomorphology, and other physical factors (e.g., Meinzer 1923; Van Everdingen 1991; but see Van Der Kamp 1995); moreover, no one has yet developed a comprehensive classification scheme that integrates these disciplines. In part, this is attributable to the daunting complexity of springs ecosystems; it also derives from the sometimes-great dissimilarities that exist between adjacent springs. Previous studies provide some guidance, and this classification scheme will require rigorous testing, but it takes a fundamental step toward springs classification.

In chapter 5, Vance Haynes provides insight into the paleontology and paleoecology of springs, with his descriptions of Quaternary soils, megafaunal remains, and early evidence of humans from study sites across North America. Springs were often focal spots at which Paleo-Indian hunters ambushed their prey. His findings support the contention that early humans appeared in North America at the conclusion of the last ice age and are likely to have exerted impacts on some springs during the entire Holocene epoch. Although not part of his central thesis, his results taken in concert with the findings of other contributors (chapters 12 and 13) suggest that no "pristine" (that is, prehuman) condition has existed for many large North American springs in the past 12,000 years. This is a substantial challenge for land managers attempting to manage or restore ecologically impaired springs ecosystems to an unaltered condition.

We next present a series of case studies of springs ecosystems and vegetation, including studies of three of arid North America's most remarkable springs systems. In chapter 6, Dean Blinn summarizes the results of his and his students' long-term studies on the ecology of Montezuma Well in central Arizona. In chapter 7, Dean Hendrickson and colleagues describe and summarize the ecology of Cuatro Ciénegas in Coahuila, Mexico. In chapter 8, James Cornett describes the ecology and distribution of palm oases throughout the lower deserts of the Southwest.

We then shift our emphasis to the ecological processes operating at springs. In chapter 9, John Spence describes plant biogeographic processes among the diverse array of springs on the southern Colorado Plateau. In chapter 10, Vicky Meretsky presents a study on the rate and mechanisms of vegetation change at springs in Zion and Grand Canyon national parks. Bianca Perla and Larry Stevens, in chapter 11, present some of the first data on the terrestrial productivity of springs in comparison with that of the surrounding uplands.

The fate of springs biodiversity under different patterns of human land use is explored in a pair of chapters by Gary Nabhan (chapter 12) and Amadeo Rea (chapter 13). They use data from Quitovac and Quitobaquito springs in northern Sonora and southern Arizona, respectively, to examine Holocene interactions of humans at springs. Although coming at the question from different angles, they both conclude that indigenous management of springs by intact cultures may be more effective at preserving natural biodiversity than strict preservationist management by the federal government.

Both biophysical and legal aspects of contemporary nonindigenous anthropogenic management of springs are considered in the next three chapters. Groundwater pumping is among the most immediate and important threats to springs ecosystem integrity. In chapter 14, Duncan Patten, Leigh Rouse, and Juliet Stromberg describe the flora of Great Basin springs, including Ash Meadows, and how groundwater withdrawal will likely affect the extent and composition of springs there. Fire is a less common disturbance at springs. In chapter 15, Tim Graham provides an in-depth examination of the recovery of a hanging garden in Glen Canyon National Recreation Area from a severe fire of human origin. In chapter 16, Nancy Nelson presents a thoughtful examination of the legal options for springs conservation. She concludes that aridland springs often fall between the (legal) cracks, as they are considered neither as groundwater nor as surface water.

Larry Stevens concludes this volume with a summary of the panel discussion that concluded the May 2000 Arizona-Sonora Desert Museum symposium. That discussion emphasized the need for improved inventory, management assessment, basic and applied research, conservation, and restoration of springs, as well as the challenge of engaging a presently polarized public in improving the care and restoration of springs ecosystems. The panel indicated that those issues were critical if springs ecosystem ecology and conservation are to be recognized as important by the scientific community, governmental land managers, and the public.

Acknowledgments

The symposium and this volume would not have been possible without the enthusiastic support of the Arizona-Sonora Desert Museum and its staff, particularly Thomas Van Devender, Robert Scarborough, Gary Nabhan, and Richard Brusca. We thank the many symposium contributors, both those represented in this volume and those not. We also thank the many anonymous peer reviewers whose expert opinions and editorial efforts made our jobs so much easier. Andrea Jaussi, Chad Runyon, Keri Stiverson, Kimberly Whitley, and especially Emily Omana helped bring this book to fruition by diligently compiling the chapters and illustrations, and we thank them warmly for their assistance. We thank Chris Brod of Flagstaff, Arizona, for assistance with GIS mapping of western North American springs. We also thank the staff of the University of Arizona Press for their fastidious care in publishing this volume.

Aridland Springs in North America

Chapter 1

Springs Ecosystem Ecology and Conservation

LAWRENCE E. STEVENS AND VICKY J. MERETSKY

Springs ecosystems are among the most structurally complicated, ecologically and biologically diverse, productive, evolutionarily provocative, and threatened ecosystems on earth. Springs are sacred places to almost all cultures, particularly those in arid regions. Humans have used springs for water, habitat, and hunting sites throughout our evolutionary existence. The prehistoric and modern technologies allowing us to capture or divert the flow of springs have a lengthy history, including the use of fire to facilitate hunting, irrigation, stock watering, and piping for potable water supplies and, in the case of geothermal springs, heating and extraction of rare minerals. The Siloam Tunnel in Israel has delivered water from Gihon Spring, the only perennial water source in the region, to Jerusalem since at least 700 BC (Frumkin et al. 2003); it is just one of many ancient water supply systems in the Middle East (Tsuk 2000). Although reverence for water miraculously emerging from the earth is deep and widely professed, it has done little to protect these remarkable ecosystems from the onslaught of human exploitation. The ancient Chinese saying that "the sweetest spring is the first consumed" aptly summarizes human treatment of these ecosystems, as each spring discovered is quickly altered to benefit human needs, and few of those in developed regions are preserved for their natural qualities.

Springs are ecosystems in which groundwater reaches the earth's surface in complex, sometimes lengthy, flow paths through subsurface structural, geochemical, and geomorphic environments. Prior to the point of emergence, groundwater pursues tortuous pathways, and springs may arise one or more times in caves, again at the surface, and then sink and rise one or more times in losing channels (e.g., Tintoretto 2004). At the point of emergence, the physical geomorphic setting allows some springs to support large arrays of aquatic, wetland, and terrestrial species and assemblages, occasionally including cave and hyporheic biota. In ecological time frames, springs in arid regions may be tremendously productive and may provide the only available water and habitat in the landscape for many plant and animal species, as well as humans. In the words of Bianca Perla, one of the chapter authors in this volume, aridlands springs function as "keystone ecosystems," exerting vastly disproportionate impacts on regional ecology, evolutionary processes, and sociocultural economics in relation to their size.

In evolutionary and biogeographic time frames, springs often function as isolated islands of habitat. Many springs serve as paleorefugia (*sensu* Nekola 1999): long-term stable habitats in which the evolutionary processes of natural selection, isolation, and adaptation (sometimes to extreme environmental conditions) result in restricted or endemic species. The greatest concentration of endemic species in North America occurs around desert springs at Cuatro Ciénegas in northern Mexico, Ash Meadows in southern Nevada, and Montezuma Well in central Arizona. Springs may contain paleontological evidence that reveals much about changing climates and landscape responses over time. Also, many springs emerge on the floors of rivers, lakes, and oceans; recent information on subaqueous springs in freshwater and marine settings demonstrates many ecological and evolutionary parallels with those of subaerial springs, including high levels of biodiversity, productivity, species packing, and endemism.

Springs ecosystems remain remarkably poorly studied. Although Odum's (1957) studies on Silver Springs in Florida laid the groundwork for much of the science of ecosystem ecology, his study remains one of the few comprehensive examples of a springs ecosystem. Dean Blinn's long-term study of Montezuma Well in Arizona (see chapter 6) and Riggs and Deacon's (2002) synopsis of Devils Hole in southern Nevada are among the few examples of detailed, long-term ecosystem analyses of individual springs. We suggest that there is an enormous amount to learn about springs ecosystems, both terrestrial and aquatic, but there is not much time left for such study, as underinformed land management practices and rampant groundwater depletion are rapidly eliminating springs (Grand Canyon Wildlands Council 2002).

Springs Distribution

The distribution of springs in North America is relatively (but far from perfectly) known, with the most reliable data available found on 7.5′

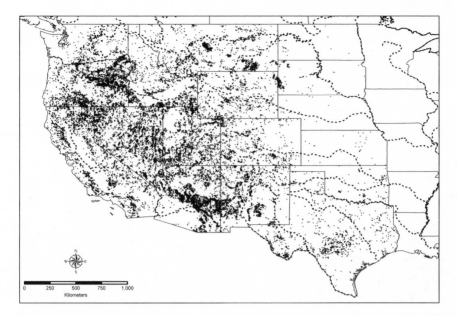

Figure 1.1. Springs in the western half of the United States. Note the abundance of springs in the topographically diverse terrain of the Rocky Mountain and Intermountain West, the low density of springs in the Great Plains, and the low density of springs in the lower reaches of large river basins.

U.S. Geological Survey maps. We searched USGS maps for named springs and developed a georeferenced database for those in the western United States. This is a crude measure of springs distribution, because state-level inventories have not been completed and fewer than half of the known springs are named, particularly in topographically diverse terrains. For example, springs tend to be rather thoroughly mapped and often are named in the flatlands terrain of northern Arizona, but the Grand Canyon Wildlands Council (2002) and Abraham E. Springer (pers. comm., 2005) estimate that fewer than 25 percent of the springs have been mapped in Grand Canyon and other cliff-dominated portions of the southern Colorado Plateau.

Nonetheless, a map of the named springs of the western United States provides insight into their distribution and potential biogeographic significance (fig. 1.1). The density of named springs per square kilometer varies tremendously among the western states (table 1.1). The lowest density of springs occurs on the Great Plains, while the highest density occurs in the topographically diverse terrain of the Rocky Mountains and the Intermoun-

TABLE 1.1. Abundance and density of named springs
by state in the western United States

State	Number (N)	Density (N/km²)
Arizona	4,812	0.0163
California	3,074	0.0076
Colorado	1,043	0.0039
Idaho	1,672	0.0078
Kansas	28	0.0001
Montana	2,860	0.0076
Nebraska	22	0.0001
Nevada	4,179	0.0147
New Mexico	2,293	0.0073
North Dakota	34	0.0002
Oklahoma	111	0.0006
Oregon	4,414	0.0178
South Dakota	283	0.0014
Texas	745	0.0011
Utah	2,540	0.0119
Washington	769	0.0045
Wyoming	983	0.0039

Note: Data are from springs mapped on 7.5′ U.S. Geological Survey
maps.

tain West. Kansas and Nebraska have the lowest density of springs, with
only 0.0001 spring per square kilometer. From a biogeographic perspective,
the extreme isolation of southern Great Plains springs suggests that they
may be paleorefugia (*sensu* Nekola 1999), potentially supporting relatively
high levels of endemism. As such, they are likely to be excellent candidate
sites for biological research and conservation. In contrast, the mountainous
state of Oregon has the highest density of named springs, with 0.018 springs
per square kilometer. Arizona, another mountainous state, has the second-
highest density, with 0.017 springs per square kilometer; however, we know
of hundreds of unmapped, unnamed springs in Arizona, which likely make
our driest state the one with the highest density of springs.

Additionally, figure 1.1 shows that the density of named springs
is remarkably low in the lower reaches of large river basins, including the
Columbia, Missouri, Colorado, and Sacramento rivers. The causes and implications of this pattern have yet to be explored but may indicate limitation

of refugial biogeographic dynamics in large river systems. Of course, the subaqueous springs in such river basins usually are not mapped or studied.

Springs Ecosystem Ecology

Although several symposia and survey studies of springs have been conducted in North America, springs generally remain rarely studied and poorly known. Recent texts on North American freshwater ecoregions and the legal protection of wetlands scarcely mention springs ecosystems (e.g., Shine and de Klemm 1999; Abell et al. 2000), and most inventories have been conducted at relatively small spatial scales or on a restricted suite of springs characteristics: flow and water quality (Meinzer 1923; Mundorff 1971); individual taxa or biota, in terms of vegetation (Bowers 1980; Malanson 1980; Welsh and Toft 1981; Malanson 1982; Welsh 1989b); Trichoptera (Erman 1992); aquatic snails (Hershler 1994; Hershler et al. 1999); aquatic invertebrates in general (Williams and Danks 1991b; Ferrington 1995); and springs biota in general (Botosaneanu 1998). Only rarely have comprehensive inventories of springs ecosystems studies been conducted (e.g., Odum 1957; Teal 1957; Tilly 1968; Riggs and Deacon 2002; also see chapters by Blinn and by Hendrickson et al., this vol.). Most studies conducted in recent decades have recognized the threatened ecological condition of many springs ecosystems and the imperiled state of their biota.

River ecology (but not necessarily riparian ecology) is conceptually well grounded in the river continuum concept (Vannote et al. 1980); however, springs ecosystem ecology has no such theoretical foundation. At present, springs ecosystem ecology is a poorly synthesized amalgam of other disciplines, including historical and structural geology and stratigraphy, geomorphology, groundwater and surface water hydrology, paleontology, microclimatology, cave biology, lentic and lotic limnology, as well as archaeology, anthropology, contemporary socioeconomics, water law, and conservation science. Controversial demands for water have retarded development of a comprehensive understanding of springs as ecosystems. This research area has been further obfuscated by the lack of basic information and inventory and mapping data. Indeed, many trees of knowledge and scientific controversies need integration into a coherent theoretical framework of springs ecosystem ecology, making this subject a difficult forest to see.

Springs ecosystem ecology is a compelling field for continuing basic and applied research, and exciting scientific topics await research. In addition to inventory and ecosystem health assessment, one of the more pressing scientific issues in this field is the need for a common lexicon or classification system with which we can categorize and quantify the various kinds of springs (see chapter 4). Beyond inventory and classification, several important scientific questions deserve attention:

1. Do springs ecosystems function in a "bottom-up" trophic fashion (as represented by the structure of this volume)? Water supplies and geomorphology are the primary controlling factors of springs biological assemblages. The geophysical setting provides the habitat template, on which life takes place. "Top-down" forces and processes, such as keystone predators, competition and predation, may influence ecosystem structure, particularly in paleorefugial springs (e.g., chapter 6); springs often are subject to various forms of natural and anthropogenic disturbance that limit integrative biological interactions. Like some regulated rivers, the interruption of disturbance can disrupt springs ecosystem functionality (Kodric-Brown and Brown 2007).

2. Many springs researchers are struck by the individuality of springs: distinctive physical and biological differences among springs that lie in close proximity (e.g., Nebeker et al. 1977; Malanson 1980). Which of the many physical and biogeographic factors most strongly affect the individuality of species assemblage composition and structure?

3. Many springs, particularly those with "harsh" geochemistry and steady flows, support unique species. How does endemism arise at springs?

4. Springs are among the few terrestrial environments in which the analysis of the impacts of interactions between productivity and ecological disturbance on biodiversity can be distinguished. How do such interactions affect biodiversity?

These are just a few of the many basic and applied research topics awaiting exploration in springs ecosystem ecology.

Springs in the Anthropocene Epoch

Humanity is racing headlong and ever faster into a future with little ecological continence, the "Anthropocene Epoch" (Crutzen and Stoermer 2000), in which few natural ecosystems seem likely to persist. Ecosystems everywhere, large and small, are at risk, with destruction of biodiversity, loss of critical species, dysfunction of critical processes, interruption of the delivery of ecosystem goods and services, and rampant invasion of alien species. As mentioned above, one of the most pressing issues is the imperiled but largely undocumented condition of springs. Numerous isolated and unique organisms exist at springs, yet many springs ecosystems are threatened by groundwater extraction, modification of source areas, withdrawal of surface flow and alteration of disturbance regimes. The conservation of these ecosystems and the species they support requires a comprehensive, global conservation strategy, efficient assessment protocols, education at all levels of society, and commitment to implementation of effective conservation and restoration measures.

Springs ecosystems have a lengthy history of poor management, and many have been so thoroughly altered that ecological restoration must be based on regional modeling and informed guesswork. Nonetheless, springs are often manageably small, highly diverse, and productive ecosystems, and with care for the aquifers that feed them, a modicum of conservation attention and foresight is likely to result in substantial protection of regional biodiversity and ecological integrity.

With this book, we hope to encourage a paradigm shift in human valuation and management of springs ecosystems. There are many important tasks facing us in the twenty-first century, as human populations burgeon, as litigation and warfare over critical resources intensify, and as globalization drags us ever further from appreciation of, and reliance on, natural ecosystems and place-based societal grounding. Initiatives are needed to protect, conserve, and restore springs ecosystems at all spatial scales, from individual landscapes to national and international scales. To be successful, these initiatives must address gaps in the existing data and concepts, as well as provide education and incentives for the protection and restoration of springs and associated groundwaters across political divisions. Such initiatives also require the development of new, more sustainable approaches to springs protection and restoration. We suggest that these efforts are among the most reasonable and worthy conservation tasks. They may quickly result in tangible, long-term protection of our natural heritage. We hope that this volume will illuminate these issues and foster discussion and further research into springs ecosystem ecology and conservation. This book is an invitation for all of us to help guarantee the long-term protection and sustainability of these diverse, threatened, and remarkable ecosystems.

Flagstaff and Bloomington
June 2008

Chapter 2

The Demise of Desert Springs

PETER J. UNMACK AND W. L. MINCKLEY

Why should we have concern for desert springs and biotas? Some simply say "because they are there" (Rolston 1991). Because surface water is quite uncommon in desert lands, springs often attract attention mostly as anomalies. These isolated oases harboring aquatic organisms in the midst of seas of arid land are also often described as "aquatic archipelagos," similar to oceanic islands that allow dry-land plants and animals to live surrounded by seawater. Most islands are, however, populated by organisms dispersing from elsewhere, such as an adjacent mainland. Conversely, biotas of desert springs often are remnants, left behind as surface water disappears with the expansion of deserts (Hubbs and Miller 1948; Hubbs et al. 1974; Dumont 1982; Por et al. 1986; Lévêque 1990). Thus, spring biotas are often relicts of wetter times, providing clues to conditions in the distant past (Scates 1968; Por 1984; Roth 1987; Hershler et al. 1999).

In many cases, individual species are restricted to one or only a few springs. Because dry land is a greater barrier to obligate aquatic organisms than water is to creatures that fly, swim, or float, their purity of isolation presents unmatched opportunities for evolutionists to study processes of speciation (e.g., Miller 1948, 1950; J. H. Brown 1971; Nxomani et al. 1994; Duvernell and Turner 1999). Field research on other important questions, such as thermal or physicochemical preferences or tolerances of species, can be undertaken in the unique up- to downstream gradients of spring-runs (Brues 1928, 1932; Mason 1939; Brain and Koste 1991).

A spring is difficult to define more precisely than a place where water rises at an intersection of groundwater and land surface (but see chapter 4). Springs vary from tiny seeps to the outflows of underground rivers, with the former being far more common. But even if tiny or ephemeral, desert springs are invariably characterized by the presence of plants and animals that cannot survive without surface water.

Organisms in springs and other perennial desert waters are often considered remarkable in their tenacity (or luck) in surviving. However, it must be remembered that they are in fact aquatic. As long as sufficient water of suitable quality is present, aquatic plants and animals care little or not at all about their terrestrial surroundings. With the obvious exceptions of those with life stages designed to resist drying (e.g., drought-resistant eggs and seeds), one should think of aquatic organisms as living "in" rather than being "of" deserts (Deacon and Minckley 1974; Smith 1981). Relative to other aquatic habitats, springs may even ensure far greater predictability in possessing the constancy of flow, temperature, and chemistry that characterizes many groundwater basins.

Disappearances of desert springs and their biotas have accelerated in the recent past (Shepard 1993; Minckley and Unmack 2000). Major reasons for this trend, and threats to remaining springs, are reviewed here in two parts. The first documents the loss of springs that is attributable to human activities, which demands an urgent program to minimize future losses. The second part discusses the problems faced by remaining springs; some of these problems may yet be corrected. We describe the impacts of spring alterations, demonstrating the magnitude and universality of ecological catastrophes involved as background for conservation efforts. Each system is unique, so anecdotes presented here will serve to alert others of what some danger signals may be and where to watch for them. Coverage is not comprehensive but emphasizes our aridland experiences in Australia and western North America, with the addition of world literature and some organisms other than fishes.

Groundwater Extraction: The Killer of Springs

Desert basins and uplands alike can hold great volumes of groundwater that rise as base flow in streams and springs (Maxey 1968). Water stored in such aquifers may have originated in the distant past, with the aquifer's volume being maintained by slow, sometimes long-distance percolation from distant recharge zones (Winograd and Thordarson 1975; Winograd and Pearson 1976; Habermehl 1980). Most natural events influencing water levels of aquifers—thus, spring outflows—are slow, accompanying global trends in climate: for example, from wetter to drier. Natural increasing aridity since

Cretaceous times (Axelrod 1979; Burkle 1995) doomed innumerable springs and their inhabitants to extinction. Springs that are "turned off" suddenly or that appear miraculously after earthquakes (DuBois and Smith 1980) or tilting (e.g., the floor of Death Valley, California) that dries some and creates other springs over a few thousand years (Hunt and Mabey 1966) are exceptions that likely have little influence on this overall trend.

"Fossil" springs (fig. 2.1) are abundant (Waring 1965) and are marked by remnant, freshwater calcium deposits termed tufas or travertines and by beds of peat formed in marshes formerly fed by groundwater outflows (Meyer 1973; Roberts and Mitchell 1987; Boyd 1990a, 1990b, 1992, 1994). Each remnant also documents the former presence of species and communities now extinct, representing lost letters that may render sentences unintelligible in the volumes of our earth's history:

> [The naturalist] looks upon every species of animal and plant now living as the individual letters which go to make up one of the volumes of our earth's history; and, as a few lost letters may make a sentence unintelligible, so the extinction of the numerous forms of life which the progress of cultivation invariably entails will necessarily render obscure this invaluable record of the past. It is, therefore, an important object [to preserve them].
>
> If this is not done, future ages will certainly look back upon us as a people so immersed in the pursuit of wealth as to be blind to higher considerations. They will charge us with having culpably allowed the destruction of some of those records of Creation which we had it in our power to preserve; and, while professing to regard every living thing as the direct handiwork and best evidence of a Creator, yet, with a strange inconsistency, seeing many of them perish irrecoverably from the face of the earth, uncared for and unknown. (Alfred Russel Wallace 1863:264)

Subsurface water resources proved irresistible to the European-origin colonists of the western United States. Exploitation began with hand-dug wells or windmills: low-volume extraction with little effect on aquifer volumes except when uncontrolled flow from artesian bores led to depletion (Habermehl 1980; Brune 1981). Major impacts in the western United States closely followed the Rural Electrification Act of 1936, after which

a

b

Figure 2.1. Fossil spring deposits come in a range of forms. *a*, Scattered travertine deposits with an extinct spring mound in the background at Chimney Hot Springs, Nevada. *b*, The top of Coward Springs mound, South Australia, showing a round depression reminiscent of an old spring pool. A dying spring outflow is present in the foreground, while the main spring outflow comes out from lower down the mound. *c* and *d*, Spring deposits from Cuatro Ciénegas: *c* is a fossil spring outflow channel; *d* is a fossil spring source. (Photos by P. J. Unmack, 1994–95)

c

d

Figure 2.2. Comanche Springs was once at least the eighth-largest spring in Texas and has been a valued community resource since at least the 1500s by providing water for human utilization, habitat for aquatic organisms, and recreational opportunities. These were all lost when declining groundwater levels caused the spring to dry in 1961 (Brune 1981). The extinct source of the main Comanche Spring is present just in front of the building. (Photo by P. J. Unmack, 1996)

high-volume electric pumps could be deployed. Agriculture expanded exponentially, even in remote areas, and rates of groundwater extraction soon became unsustainable (Postel 1992; Pringle and Triska 2000); far more water was removed than could be replenished by natural recharge.

Later, use of land for housing became lucrative in the western United States, and urbanization began to supplant agriculture. The population of Arizona grew from 2.7 million to 4.7 million people between 1980 and 1997 (U.S. Census Bureau 1999), and in adjacent and comparably arid northern Mexico population grew from 5.5 million to 8.3 million from 1980 to 1990 (Contreras-B. and Lozano-V. 1994). Although urbanization may be less demanding on water supplies than is irrigated agriculture, groundwater use has continued to rise with expanding human populations (Swanson 1989).

In Texas, 63 of 281 historically significant springs had gone dry by the 1970s (fig. 2.2), and far more suffered reductions in volume (Brune

Figure 2.3. Pahrump Spring, Nevada, dried between 1955 and 1957. It was one of three habitats in which the various Pahrump poolfish *Empetrichthys latos* subspecies were found, as well as several likely unique invertebrates. All have lost their natural habitat because of excessive groundwater extraction (Soltz and Naiman 1978). (Photo by P. J. Unmack, 1997)

1975, 1981). Unknown numbers of fishes and other biotic groups must have been impacted (Edwards et al. 1989; Longley 1981, 1992; Bowles and Arsuffi 1993). In northern Mexico, a known minimum of 92 springs (of an unknown total) failed in the two decades ending in 1993 (Contreras-B. and Lozano-V. 1994). Examples from other parts of this region (figs. 2.3 and 2.4) are summarized in Miller 1961, Minckley and Deacon 1968, Williams et al. 1985, Meffe 1989, and Minckley and Deacon 1991.

Such human-induced catastrophes can occur swiftly or take longer, depending on the volume of water extracted. Springs at several Australian sites dried or nearly dried soon after water extraction began (Ponder 1986; Ponder and Clark 1990), or they suffered progressive flow reductions because of groundwater removal (Harris 1981, 1992). A well-documented example is Elizabeth Springs, Queensland, which has suffered a reduction of approximately 95 percent since the 1880s (Habermehl 1982). Four pupfishes (*Cyprinodon* spp.) along with uncounted, unstudied invertebrates disap-

Figure 2.4. Part of the Laguna Chiqueros system in Cuatro Ciénegas, which is the source of the Río Garabatal and part of the type locality of the Cuatro Ciénegas killifish (*Lucania interioris*). Lowered groundwater levels have reduced it to a tiny pool a few meters in area at the spring source. (Photo by P. J. Unmack, 1995)

peared when springs in two small Mexican basins were pumped dry in only a few years (Lozano-V. and Contreras-B. 1993; Contreras-B. and Lozano-V. 1996). In the Middle East, Al Kahem and Behnke (1983) caught a minnow (*Cyprinion* sp.) in a spring in Khaybar City, Saudi Arabia, that was dry four years later. Several other springs reportedly failed in that same area and during that same period of time. By contrast, some places in that region occupied since the Pleistocene by humans with primitive extraction capabilities have persisted much longer; for example, water and endemic snails persist at the Palmyra Oasis in Syria despite human use for thousands of years (Schutt 1987).

In theory, if recharge balances extraction, then a large aquifer with an intact watershed can yield significant volumes of water indefinitely. Yet where reliable records exist, most springs fed by even the most extensive aquifers are affected by exploitation, and spring flow reductions relate directly to quantities of groundwater removed (Dudley and Larson 1976). As already noted, the great Edwards Aquifer of Texas is declining (Brown et al. 1992), with potentially catastrophic results for the indigenous biota of its springs (Longley 1981; Edwards et al. 1989; Longley 1992) as well as for humans depending on the resource. Similar situations are evident in the vast Ogallala aquifer of the North American High Plains (Winter et al. 1998), in comparably large aquifers of Saudi Arabia (Krupp et al. 1990), and elsewhere. Further development in Australia threatens aquifers: a mining venture extracting 1.5×10^4 cubic meters per day of water from the Great Artesian Basin in the 1990s increased to 3.3×10^4 cubic meters per day after the mine went into full production (Harris 1981, 1992). This has come to pass and bodes ill for the region's springs.

Reduction in discharge of a spring or spring group warns of unsustainable extraction somewhere. Because spring flow emerging many kilometers away may not be affected by groundwater interception for years, centuries, or even longer (Winograd and Pearson 1976), reduced discharge may portend uncorrectable failure in the future. Overgrazing, deforestation, and other land abuses are intuitive but largely undocumented factors reducing infiltration capacity and other features of a watershed, including recharge, which will ultimately influence the sustainability of the groundwater it feeds (Pringle and Triska 2000). Unfortunately, recharge rates, underground porosity, aquifer volume, and even the geographic extent of many aquifers are unknown. Such data are usually compiled only after some surface effect is detected, such as discharge fluctuations coincident with or related to pumping schedule, reduced flow, or even drying of springs (Dudley and Larson 1976; Habermehl 1980, 1982; Deacon and Williams 1991).

Less-than-Fatal Alterations, Some Correctable

Conservation Efforts

Early efforts at native-fish conservation involving desert springs, the training field for many of us now dedicated to such endeavors, resulted from events in Owens Valley, California, and Ash Meadows, Nevada. Spring development, artificial impoundment, and construction of canals to supply water for Los Angeles, followed by the stocking of non-native species for sport fisheries and release of bait, endangered the native biota of the long-isolated Owens Valley, with its four endemic fishes. This stimulated formation of a native-fish sanctuary in 1967–68, a story in itself (Miller and Pister 1971; see also Perkins et al. 1984 and Minckley et al. 1991; for recovery perspectives, see U.S. Fish and Wildlife Service 1984, 1989).

In Ash Meadows, agriculture and urban development—especially groundwater pumping and clearing—threatened springs and their biotas and with them an evolutionary drama in progress for millennia (Pister 1991). A group of agency, academic, and private people deemed this unacceptable. They were able to rally support sufficient to stop the action and ultimately to result in the formation of Ash Meadows National Wildlife Refuge, but only after carrying their concerns to a favorable decision by the U.S. Supreme Court. These efforts and other details pertaining to saving and managing springs in arid lands, including litigation, are summarized in Dudley and Larson 1976, Deacon and Williams 1991, and Pister 1991.

As a generalization, changed volumes or patterns of flow of a spring or spring system have a "domino effect," involving numerous, diverse, and intertwined biotic and physicochemical shifts. Three major factors determine the severity of impact of reduction in flow or spring diversion: (1) proportion of flow lost; (2) reduction in downstream extent of the system as a result of less water or distance of interception from the source; and (3) new connections made by diversions between nearby spring outflows. Clearly, the smaller the proportion of water involved, the less the impact, and if flow reduction is small or diversion occurs far downstream, marshes can persist. When diversion is near the spring head, however, downflow marshes are drained and significant wetlands may be lost. This kind of damage has broader impacts than is often realized, because—in addition to containing indigenous, often-endemic resident biota—large, isolated desert marshes often are critical resting and feeding habitats for waterfowl and other migratory and resident birds (Minckley and Brown 1994).

Alterations of Head-Springs

Human alterations of head-springs, to concentrate or increase discharge, negatively impact spring systems and invariably result in loss of biota. Many springs in western Queensland, Australia, were excavated to increase water supply, with no success (Ponder and Clark 1990). Sometimes explosives have been used in an effort to increase discharge. Pumps inserted

Figure 2.5. Jackrabbit Spring was repeatedly pumped dry over about a year during 1969–70. Because of the continuing poor condition of the habitat caused by nearby pumping, the spring was not restocked with fishes until 1977 (Williams and Sada 1985). (Photo by E. P. Pister, 1969)

directly into springs, or wells sunk nearby, dry them and destroy their biota. Jackrabbit Spring in Ash Meadows, Nevada, was repeatedly pumped dry for irrigation (fig. 2.5) (Pister 1991), a situation rectified when much of its biota was reintroduced and became reestablished (Soltz and Naiman 1978; Williams and Sada 1985). The common practice of capping springs and piping water away from the source for domestic or livestock water also destroys fishes, although seepage may continue to support small arthropods and snails (e.g., Shepard and Threloff 1997; Hershler 1998). Raycraft Spring in Pahrump Valley, Nevada, was simply filled with dirt in the 1950s to get rid of mosquitoes, and unintentionally the world's only population of Raycraft poolfish (*Empetrichthys latos latos*) was also eliminated (Minckley and Deacon 1968; Soltz and Naiman 1978).

In some cases, spring biotas persist to recover after major disturbance. One unusual catastrophe related to blasting had such an unexpected result. A drill centered in a hot spring for geothermal exploration near El Doctor, Sonora, Mexico, apparently penetrated a high-pressure zone that caused an explosion resulting in human injury and loss of life, destruction of the rig, and ejection of water, substrate, and adjacent marshes into the surrounding desert. A year later, WLM found that native desert pupfish (*Cyprinodon macularius*), snails (*Tryonia* sp.), and a suite of introduced fishes had recolonized the spring—a remarkable feat!

Kings Pool and Point-of-Rocks springs and their outflows and marshes in Ash Meadows suffered repeated dredging, ditching, and diversion (fig. 2.6) (Soltz and Naiman 1978), some even after they had become part of Ash Meadows National Wildlife Refuge (Polhemus 1993), yet much of their original biotas persisted. Big Bend gambusia (*Gambusia gaigei*), adapted to thermal springs in Texas, essentially disappeared when its habitat was dammed in 1953 to make a fishing pond. The endemic fish survived sport-fish predation by living along weedy margins, but some were brought into captivity just before the species' elimination in nature when mosquitofish (*G. speciosa*) gained access (Hubbs and Broderick 1963). The pond leaked, spring conditions were re-created in 1972, and by 1983 the endemic gambusia had reestablished naturally from a nearby, but unknown, refuge (Hubbs et al. 1986).

It bears noting that alteration for recreation or domestic water supply can also, sometimes, result in inadvertent protection of springs and their biotas. Survival of endemic Mexican fishes, crustaceans, snails, and other groups was reported by Contreras-B. (1991) in four of five large springs in Chihuahua, Mexico, that were modified into reservoirs, swimming pools, or thermal baths. He further noted that three other springs (one each in the desert states of Coahuila, Durango, and Nuevo León) were also modified for recreation in ways that resulted in protection for endemic biota. Five other large Mexican springs, some modified and some not, suffered reduced flow because of pumping or lost their endemics after non-native fishes appeared. In cases when springs mangers zealously protect their local water supply, associated organisms may be maintained in a seminatural state (fig. 2.7) (Schütt 1987; Shepard 1993).

Alterations of Outflows

The effects of inserting a barrier across an outflow can vary, from system destruction if the spring head is inundated (fig. 2.8) to both up- and

Figure 2.6. Kings Pool was totally remorphed in 1969 using heavy earthmoving equipment. All of the spring's outflow was diverted into canals for use in irrigating alfalfa to feed cattle (Soltz and Naiman 1978). Today the spring pool has been restored by the U.S. Fish and Wildlife Service. (Photo by E. P. Pister, 1969)

Figure 2.7. In the 1930s, San Solomon Spring in Texas was converted into a swimming pool and its outflow diverted into a concrete channel, from which the water is taken for irrigation. Despite this, the native fauna has persisted. The area has been protected by Balmorhea State Park since 1968. More recently, a ciénega was created as a refugium for native pupfish with a public underwater viewing area (Garrett 1999). (Photo by P. J. Unmack, 1995)

downstream changes in chemistry, physics, and biology related mostly to barrier size and its distance from the source. Similar impacts occur when levels in a spring source decline below natural outflow elevation because of diversion or other reduction in flow.

Riparian vegetation has an early and evident response to reduction in flow. Plants characterizing spring outflows are typically stable for extended periods. These plants include sedges, grasses, and allies, adapted for life in waterlogged, reducing soil (Hendrickson and Minckley 1985; Boyd 1990a, 1990b, 1992, 1994). As the water level declines, soil dries and becomes aerated, so forbs, shrubs, and trees can invade. New water depths and soil conditions may also allow stands of cattails (*Typha* spp.) and reeds (native *Phragmites* spp. and African *Arundo donax*, widely introduced in North America and elsewhere) to expand, sometimes choking the habitat (see later discussion).

Figure 2.8. Poza de la Becerra was significantly reduced in size over a few days in 1964 upon completion of the canal shown from the upper right to the upper left of the remaining spring pool (Minckley 1969, 1992). The outline of the former spring pool/marsh can still be seen behind the present spring pool. (Photo by P. J. Unmack, 1995)

An example of the beheading of a large, complex system was photographically documented by WLM (Minckley 1969, 1992) for Poza de la Becerra in the Cuatro Ciénegas basin, Coahuila, Mexico, where a marshland covering approximately 10 square kilometers was reduced in a few days to less than 10–15 hectares by diversion for irrigation (fig. 2.9). No species was known to be lost, but habitats for populations of three turtles, seven fishes, and at least eight snails, all endemic to the basin, were eliminated, and numerous other indigenous creatures died. In a comparable event, Carson Slough in Ash Meadows was drained to facilitate the mining of a thick deposit of peat, resulting in significant loss of habitat for endemic speckled dace (*Rhinichthys osculus nevadensis*), Ash Meadows pupfish (*Cyprinodon nevadensis mionectes*), and other organisms (Soltz and Naiman 1978). An example in Israel involved draining a lake and associated marshes for flood control and soil reclamation, followed by diversion of the spring that resulted (among other things) in a loss of refuge from winter cold for warm-adapted fishes

Figure 2.9. Big Spring at Panaca, Nevada, was dammed and its outflow ditched sometime between 1939 and 1959. Damming flooded the original spring source, and outflow marshes were lost when the outflow was channelized, resulting in extinction of the cyprinid fish, *Lepidomeda mollispinis pratensis* (Miller and Hubbs 1960). Today the spring contains only non-native fishes, some of which were introduced prior to 1959. None of the three original native fishes persists here today. (Photo by P. J. Unmack, 2000)

(Goren and Ortal 1999). Only two of seventeen indigenous fishes survived. Another example at an opposite extreme, the flooding of a spring, followed the closure of Amistad Reservoir on the Rio Grande, the boundary for the states of Texas and Coahuila, that inundated Goodenough Spring and destroyed endemic Amistad gambusia (*Gambusia amistadensis*) (Peden 1973; Minckley et al. 1991). Other examples relating to spring outflow diversions are shown in figures 2.10 and 2.11.

Downstream connections of diverted water, like the flooding of spring heads and breaching of critical barriers, may also provide access for other species from distant places. In Death Valley, Tecopa pupfish (*Cyprinodon nevadensis calidae*) occupied a warm spring isolated by tiny but impassable waterfalls from the adjacent Amargosa River. When the falls were altered while building public baths, Amargosa pupfish (*C. n. amargosae*) gained ac-

Figure 2.10. Little Warm Spring in Railroad Valley, Nevada, drains via several artificial outflow ditches. Depending on where water is needed, an outflow may be cut off and dried, with the loss of any fauna that has colonized it. The head-spring has been left somewhat intact, allowing some fauna to persist and recolonize the outflows. (Photo by P. J. Unmack, 1994)

cess and hybridized Tecopa pupfish out of existence (Soltz and Naiman 1978). In Chihuahua, Conchos pupfish (*Cyprinodon eximius*) invaded thermal springs occupied by big-head pupfish (*C. pachycephalus*) through artificial channels, which also led to hybridization (Smith and Chernoff 1981; Minckley and Minckley 1986). Artificial canals in Cuatro Ciénegas allowed two other pupfishes (*C. atrorus*, *C. bifasciatus*) to intermingle and hybridize between their distinctive habitats (Miller 1968; Minckley 1984; Carson and Dowling 2006). Far more subtle are situations in which special conditions in a spring head, often not well understood (Hubbs 1995; Minckley and Unmack 2000), are abrogated to allow access to aliens. In Clear Creek, Texas, partial flooding of a spring head allowed invasion of the habitat of endemic Clear Creek gambusia (*Gambusia heterochir*) by western mosquitofish (*G. affinis*), again resulting in hybridization (Hubbs 1957, 1959, 1971).

    In another case, temperatures in excess of 38°C in some spring heads in Soldier Meadows, Nevada, exceed the thermal tolerances of en-

Figure 2.11. The outflow of Preston Big Spring at Preston, Nevada, is diverted into pipes for irrigation 1.7 kilometers from the spring source. This type of diversion is as problematic as that shown in figure 2.10 but is made far worse when diversion occurs close to the spring source. This has resulted in the extirpation of three of four native fishes in nearby Lund Town Spring at Lund, Nevada, when it was diverted 50 meters from the spring source. (Photo by P. J. Unmack, 2000)

demic desert dace (*Eremichthys acros*). The fish lived downstream from hot-water sources, and the water was cooled by flowing a considerable distance before entering the habitat of the dace (Ono et al. 1983). When water was diverted too near a source, cool water was unavailable downstream, extirpating many populations. Similarly, WLM visited a hot spring (the temperature was more than 50°C at the source) in Oman where resident minnows (*Garra barreimiae, Cyprinion micropthalmum*) occurred below, but not above, a vertical barrier measuring 1.5 meters high, formed by a road. A knowledgeable resident reported both that fishes invaded upstream of the barrier during overland runoff and that they retreated downstream or suffered thermal death when base flow resumed. No thermometer was available, but base-flow temperatures (approximately 40°C) seemed identical to the hand above and below the road crossing; yet fishes were absent upstream.

Subtle but rarely mentioned changes may also occur with altered flow of carbonate-rich waters. Underground water is often charged with carbon dioxide (CO_2) from decomposing organics in strata through which it moves. CO_2 combines with water to form weak carbonic acid, which dissolves calcium carbonate ($CaCO_3$) rocks to form the more soluble bicarbonate (HCO_3^-). Concentrations of groundwater gases vary with pressure and temperature, so when a spring emerges, CO_2 is driven off by reduced pressure, temperature changes, and channel irregularities that cause agitation. Biology also comes into play because photosynthetic microphytes use both gaseous CO_2 and that carried in HCO_3^-. Acid-base relations change through all of these processes, resulting in deposition of insoluble $CaCO_3$ as travertine (Minckley 1999a). Precipitation is greatest near areas of channel roughness that promote agitation and enhance microphytes, thus resulting in an even greater loss of CO_2. These deposits, in turn, create more turbulence and thus more deposition, and a carbonate "dam" may form, sometimes extensive enough to impound and isolate a spring or modify a channel by creating pools and other complexity (e.g., Roberts and Mitchell 1987; Drysdale and Gale 1997). The length of spring outflow over which carbonate precipitates varies with distance from a spring source as a function of ion concentration, temperature, and so forth. Thus, speed and amount of carbonate precipitation vary with changes in volume and patterns of water flow, channel alterations such as artificial dams reducing or creating turbulence, or diversion and channelization that concentrate volume and change speed of flow.

Biological Problems

Other major impacts leading to biotic damage can result from livestock (Ponder and Clark 1990; Krupp et al. 1990; Harris 1992; Hershler 1998). Large ungulates—domestic, feral, or otherwise—seek springs for food and water, both limiting in deserts. Unless specifically protected, the majority of springs that we have visited in deserts of North America, Australia, and the Middle East have clearly suffered from livestock damage. Palatable plants are eaten and trampled, local soils compacted, and banks collapsed. Increased input of organic wastes increases nutrient concentrations. Some nutrients (e.g., nitrogen compounds) can themselves be toxic, and their presence can result in increases in potentially damaging microbes (Taylor et al. 1989). Domestic animals can also be trapped in soft spring deposits, with disastrous results when they die and decompose, resulting in the decline or loss of fauna (Unmack 1995). Outflows with livestock bones scattered over boggy areas are mute testimony to such accidents.

Nonetheless, most springs were originally grazed by native herbivores, so their natural biota evolved in concert with grazing pressure. Changes in grazing regime are followed by vegetation change, especially in smaller systems. Typically, the removal of grazing results in invasion by cattail, reeds, or both. Examples of this phenomenon include Mexican Spring in Ash Meadows, formerly supporting the smallest-known self-sustaining vertebrate population in the world (between twenty and forty-seven warmsprings pupfish, *Cyprinodon nevadensis pectoralis* [J. H. Brown 1971]), which dried in 1973 by evapotranspiration when cattails expanded after fencing to exclude grazing (Soltz and Naiman 1978). Soon after being enclosed by a fence, a similarly tiny spring near Bylas, Arizona, was choked and desiccated by cattails, eliminating a population of the endangered Gila topminnow (*Poeciliopsis occidentalis*) (Marsh and Minckley 1990). Overgrowth by cattails required periodic clearing of Corn Creek Spring, Nevada, a refuge for Manse Spring poolfish (*Empetrichthys latos concavus*) (Minckley et al. 1991); the same problem existed in Australian springs that became densely vegetated with *Phragmites* soon after fencing (fig. 2.12) (Harris 1981; Wager and Unmack 2000). Alternatively, some springs have had surrounding vegetation physically removed to reduce water loss, but the concomitant reduction in shade cover resulted in increased temperature and evaporation from exposed water surfaces, as well as a decrease in organic inputs, all of which also may negatively influence the biota.

Thus, grazing and other manipulations of natural vegetation are double-edged swords, dangerous at extremes. But controlled grazing and other clearing methods can be used to manage amounts and kinds of vegetation, allowing manipulation of evapotranspiration to result in more or less water and moderation of organic inputs to maintain a spring nearer its natural state.

Finally, accompanying (even, in some cases, barring) changes in discharge and impacts other than desiccation, the single most destructive force influencing desert-spring biotas of western North America, and increasingly elsewhere, is the ongoing invasion by alien organisms. Floating water plants such as water hyacinth (*Eichornia crassipes*) can cover the surface, as in some

Figure 2.12. Outside Spring in South Australia was fenced in 1988 and exemplifies the complexities involved in spring conservation. The area inside the fence contained dense stands of reeds in 1994, with unknown effects on the flora and fauna, yet outside the fence the spring outflow was devastated by cattle grazing, with obvious effects. How to compromise between these two problems is not yet clear. (Photo by P. J. Unmack, 1994)

springs of Cuatro Ciénegas (Contreras-B. 1991). Large, non-native plants such as salt-cedar (*Tamarix* spp.) and giant African reed are ever-increasing problems in North America. Palm trees have become established in locations such as Dalhousie Springs, South Australia (*Phoenix dactylifera*) (Mollemans 1989), and warm springs on the Moapa River, Nevada (*Washingtonia filifera*). They develop dense stands, shade out native vegetation, reduce solar input, contribute far more organic material than natural vegetation does, increase evapotranspiration, change outflow channel morphology, and also provide fuel for devastating wildfire (White et al. 1995).

Alien aquatic animals naturalized to western North American springs include a number of large invertebrates, especially the snail (*Thiara* [*Melanoides*] *tuberculata*) and crayfishes (mostly *Procambarus clarkii*), that prey on, compete with, or otherwise displace native biota (Williams et al. 1985; Contreras-A. et al. 1995). Among vertebrates, bullfrogs (*Rana cates-*

beiana) have proven especially damaging. Adults feed voraciously on native frogs, snakes, and toads, often extirpating them, and eating fishes as well (Rosen et al. 1995). To our knowledge, the impacts of bullfrog tadpoles, which often swarm and may remain aquatic for up to two years, are as yet unassessed.

Introduced fishes have appeared in springs of arid zones throughout the world (e.g., Coad 1980; Ben-Tuvia 1981; Ross 1985; Moyle et al. 1986; Krupp and Schneider 1989; Courtenay and Moyle 1992; Coad and Abdoli 1993; Wager and Unmack 2000), with tropical species even appearing in warm springs in Canada (Nelson 1984). Introductions of small taxa often originate from the disposal of unwanted aquarium and bait fishes, but others are stocked for control of mosquitoes, as escapees from commercial facilities, or with the intent to harvest them later for profit. Larger species originate from stocking for sport or food or as escapees from aquaculture ventures (Courtenay and Stauffer 1994). The primary impacts of alien fishes are predation, competition for space or other resources, and hybridization.

In western North America, many native populations and species of fishes have disappeared in the face of large, efficient non-native predators such as largemouth bass (*Micropterus salmoides*), other centrarchids (e.g., green sunfish, *Lepomis cyanellus*), a number of African and Central American cichlids, and others (Minckley et al. 1991). Smaller but equally efficient predators, mosquitofish (*Gambusia affinis*, *G. holbrooki*), have been translocated worldwide for control of pestiferous insects. They have also been implicated, repeatedly, in the disappearance of Gila topminnow (Meffe 1985; Courtenay and Meffe 1989; Minckley 1999b) and numerous other taxa (Myers 1965). Impacts of the remarkable number of alien predatory fishes established in the western United States (Fuller et al. 1999) are yet to be quantified.

Various levels of competition must exist between native and non-native species (Schoenherr 1981; Douglas et al. 1994) but are difficult to prove; relationships are rarely clear-cut. By contrast, hybridization between related native and non-native species is readily documented by molecular and other methods and is clearly of concern. In addition to those discussed already, examples include Owens tui chub (*Siphateles bicolor snyderi*), now restricted to isolated springs in Owens Valley, as those in other habitats of that area become hybridized with non-native chubs brought as bait from elsewhere (U.S. Fish and Wildlife Service 1989). Leon Springs and Comanche pupfishes (*C. bovi-nus, C. elegans*) are similarly imperiled in Texas because of hybridization with an alien pupfish, the sheepshead minnow (*C. variegatus*) (Hubbs 1980; Echelle and Echelle 1994, 1997), also likely originating as discarded bait (Stevenson and Buchanan 1973; Garrett 1999). Pecos pupfish (*C. pecosensis*) is also being replaced by hybrids with sheepshead minnow in that same part of Texas and northward in New Mexico (Echelle and Conner 1989).

Transmission of alien diseases and parasites along with introduced non-native fishes and other organisms may also constitute a significant threat now and in the future (Hoffman 1970; Hoffman and Schubert 1984; Langdon 1988). Williams and colleagues (1985) suggested that anchor-worm (*Learnea* sp.), an ectoparasite appearing along with alien fishes, contributed to the decline of endemic springfish (*Crenichthys baileyi grandis*) in Hiko and Crystal Springs, Nevada. Asian tapeworm (*Bothriocephalus acheilognathi*) has become widespread in the western United States (Heckman et al. 1993; Clarkson et al. 1997) and has been recorded in Australia (Dove et al. 1997), originating along with non-native species and even with native fishes restocked from hatcheries for recovery purposes (WLM, personal observation). Amin and Minckley (1996) found an alien nematode (*Hysterothylacium* sp.) in an endangered Gila topminnow. Although not previously reported in Arizona, the nematode was common in non-native fishes of a Colorado River reservoir. Few studies have as yet been performed in these areas of concern, and the disciplines of fish disease and parasitology beg for additional work immediately.

The Big Picture

A relevant special issue of *BioScience* (Rosenberg et al. 2000) was dedicated to "global-scale environmental effects of hydrological alterations." Although emphasizing the damming of major rivers, many authors in that special issue focused on the intimate connections between surface and subsurface waters, as well as the broad linkages in physical, chemical, and biological relationships that are being dramatically modified as human development proceeds. Reflection of these changes in the biodiversity of whole river basins and subcontinental regions, even in well-watered zones such as the tropics, is of major concern (Pringle 2000; Pringle and Triska 2000). The loss of springs is but one symptom of the broader problem of our ignorance in using Earth's resources.

In the "big picture," the disappearance of springs may seem a minor event, but to depauperate deserts, springs are often essential for much of the biota and thus become a major focus of biodiversity. Because of their regional importance, uniqueness, and vulnerability, recognition of the problem of loss and decline of springs and their biotas came early. Their conservation lagged behind, but recent attention in the literature, an establishment of refuges and protected areas in North America and Australia (Zeidler and Ponder 1989; U.S. Fish and Wildlife Service 1990; Williams 1991; U.S. Fish and Wildlife Service 1995; Abell et al. 2000), and greater emphasis on such habitats in other world deserts (Scates 1968; Skelton 1990) suggest that springs are now receiving increasing interest.

In most instances, managing springs is an exercise in informed guesswork. Things that one should know in advance include definitions of aquifer and recharge zones, their geographic extent, recharge volumes, and subsurface transmission rates, all of which are critical in evaluating vulnerabilities and potentials for damage by water extraction. Such information is often unavailable or becomes so too late, after a change in spring flow is detected—or even worse, after the spring is but a memory. An aggressive international program is needed that defines springs and educates people regarding their uniqueness and value as part of our world heritage; this requires that we take action to set aside, restore, and perpetuate as many examples as possible, before they all are gone.

ACKNOWLEDGMENTS

We thank the many people who contributed unpublished information, literature, and thoughts, as well as the agencies, organizations, and individuals supporting our interest in desert springs, their environs, and biota. Prominent among individuals is Virginia M. Ullman of Phoenix, Arizona, who graciously arranged for WLM to travel in the Middle East. Edwin P. ("Phil") Pister of the California Department of Fish and Game (retired), provided the quotation from the pen of Alfred Russell Wallace. Thanks to Vicky Meretsky, Rachael Remington, and an anonymous reviewer for their constructive comments on the manuscript.

Chapter 3

The Hydrology of Desert Springs in North America

DAVID K. KREAMER AND ABRAHAM E. SPRINGER

The springs of the arid and semiarid regions of North America are a direct result of local, and in many instances regional, geologic and hydrologic conditions. Stratigraphic and structural geological features determine water's subsurface pathways, travel times, flow quantities, and water quality. Specifically, faults, folds, and the vertical sequence of aquifers (geologic water-bearing strata) and aquitards (low-permeability regions where water flow is retarded) are some of the factors that determine where springs emerge, the temporal variability of their flow, and their temperature and chemistry. In turn, the quantity and quality of spring water directly influence dependent ecosystems.

Human population growth and agricultural development in these desert regions of North America have resulted in increasing exploitation of groundwater, leading to concerns over the long-term sustainability of many springs. Many of the groundwater systems tapped by pumping wells also serve as the same hydrologic source for springs, and declining water tables can signal concurrent diminishment of spring water quantity and possibly quality. The repercussions on ecosystems that depend on these springs can be severe.

The Hydrology of Springs

Springs and seeps are regions where groundwater discharges onto the surface of the earth. From a geological point of view, there are many different types of springs (Fetter 1994). One type is a depression spring, which is formed when a high water table reaches a low point on the ground surface. Because high water tables are uncommon in arid regions of the world, depression springs are unusual in deserts. Another kind of spring more common to deserts occurs when low-permeability geologic strata (aquitards or aquicludes) underlie higher-permeability strata (aquifers); water can be forced along the contact between the two beds and can emerge at an outcrop as a contact spring. Faulting can create a spring by placing an impermeable barrier rock adjacent to an aquifer, forcing regional groundwater flow upward to form a fault spring. A fault itself can be a conduit to flow through open fractures or can be a barrier to flow if the fracture plane is sealed with fine, low-permeability material. Yet another sort of spring is an artesian spring, which can be formed when water under pressure in confined aquifers is released to the ground surface through fractures, joints, or collapse features associated with sinkholes.

Groundwater systems that supply springs are composed of aquifers and confining beds beneath the surface of the land. There are two general types of aquifers: unconfined, or water table, aquifers; and confined, or artesian, aquifers. Water enters groundwater systems from recharge areas where precipitation and surface water infiltrate into the earth and migrate underground to discharge areas such as springs, seeps, rivers, lakes, and the ocean. Groundwater systems both store water and transmit it, and travel times for water through the subsurface are dependent on the configuration of geologic strata and the hydraulic conductivities of those underground formations. Subsurface travel times can range from a few days, where a recharge area is close to a discharge area, to millennia, where the distances are long and the hydraulic conductivity is low (Heath 1989). Most recharge areas are laterally extensive, whereas discharge of groundwater typically occurs over smaller areas with associated locally high hydraulic conductivity. In the deserts of North America, the characteristic timing of recharge versus discharge is also quite different, with recharge occurring during intermittent precipitation events and discharge releases being more continuous unless hydraulic heads drop below the elevation at which discharge occurs. Spring flow can be diminished or even stop when water tables or piezometric (pressure) surfaces decline, but this process typically manifests itself more slowly than the great variability of rainfall typical in desert environments might suggest.

There are many examples of stratigraphic and structural geologic control of springs flow, and these can be exemplified by the springs of many national parks in the southwestern United States. When downward-percolating recharging water in this subsurface environment encounters the low permeability of an aquitard, it is deflected horizontally and often emerges on a canyon wall.

Perhaps nowhere is stratigraphic control more evident than in Grand Canyon National Park. Here, a series of gently dipping sedimentary strata common to many locations in the Colorado Plateau (Heath 1984) are exposed in the canyon wall, and those strata have different hydraulic conductivities. Figure 3.1 is a photo of a travertine deposit in the Grand Canyon that represents precipitated minerals from a spring or seep, which emerged from the contact between the permeable Coconino Sandstone and the underlying low-permeability Hermit Shale. Lower down in the Grand Canyon, many springs likewise emerge from the Muav, Temple Butte, and Redwall Limestones, underlain by the fissile Bright Angel Shale aquitard.

Springs in the Grand Canyon and in the deserts of North America are also associated with structural geologic features. In addition to exerting stratigraphic control, folds and faults act as ways through which subsurface water is intercepted, gathered, and in the case of the Grand Canyon, sent toward the canyon wall. O'Neill Spring in the Grand Canyon is located below the Grandview monocline, which is a fold in the otherwise relatively horizontal strata. The dark-colored rock of the Muav Limestone is not only tilted where the spring emerges but also presents itself many meters above the spring, above the monocline. The monocline acts as a tilting bed that directs groundwater down-dip. In Bryce Canyon National Park, another example of the effect of geologic folding on spring flow is evident. Here an anticline, which is an upfold in a wave of crustal folding, has an axis that parallels the length of the park in a north–south trend (Gregory 1951; Marine 1963; Bowers 1990). Groundwater discharges to the east and west of Paunsaugunt Plateau, migrating along the shallow dipping limbs of the anticline. Again water follows outward-sloping bedding planes down-dip and away from the higher central axis of the fold.

Faults can also serve as conduits or barriers to groundwater flow and can be of central importance in the emergence of springs. In the Basin and Range provinces of the Mojave and Amargosa deserts, many of the springs of Ash Meadows National Wildlife Refuge and neighboring Death Valley National Park emerge from a deep carbonate aquifer of Paleozoic age, which is faulted and folded (Winograd and Thordarson 1975; Kreamer et al. 1996). Underlying this aquifer is a confining basement unit of Cambrian and Precambrian siltstones and quartzites. Parts of the regional carbonate aquifer are overlain by confining layers of welded and unwelded tuff of Tertiary age,

Figure 3.1. A travertine deposit in the Grand Canyon, showing the residual precipitated minerals from a spring or seep located on the contact between the permeable Coconino Sandstone and the underlying low-permeability Hermit Shale.

and by alluvial valley sediments. Water flows horizontally beneath much of southern Nevada until it reaches Ash Meadows in the Amargosa Desert near the California-Nevada border, where a normal fault with at least 160 meters of displacement forms a hydraulic barrier, which forces groundwater to the surface at the outcrop of the carbonate aquifer. This example from the southern Great Basin is illustrative of many fault-controlled springs in the deserts of North America.

Understanding and Predicting Spring Hydrology

The arid regions of North America are subject to great scientific scrutiny because of limited water resources, and groundwater and springs are key because of the scarcity of surface water. Tremendous influences on spring flow can occur with groundwater pumping or climatic changes such as prolonged drought. Mapping recharge and discharge areas is vital to an understanding of both a groundwater system and a spring's long-term sustainability. To better predict the impacts of increasing groundwater use and climatic change on spring flow, many approaches are utilized to define a groundwater flow system. Indicators of groundwater flow patterns include (1) topography, (2) piezometric patterns, (3) environmental tracers including hydrochemical trends and environmental isotopes, and (4) soil and land features (Freeze and Cherry 1979). Although groundwater flow does not always follow land surface slope, recharge areas typically are at higher elevations and discharge at relatively lower elevations. Another important hydrologic tool is the measurement of changing hydraulic head. Subsurface hydraulic head can be measured in wells and piezometers and indicates the direction and quantity of subsurface flow. Aquifer tests performed on pumping wells can also provide key information on the hydraulic conductivities of rock and sediment strata, which then can be used to predict quantity and speed of groundwater flow. However, in desert environments, where the water tables are often quite deep, measurement of piezometric patterns can require expensive deep drilling.

Much research is being conducted to define and model geologic controls on individual springs and the area of groundwater recharge basins. Hydrologists are developing many techniques for quantifying and modeling spring flow and predicting the occurrence and movement of associated groundwater. Hydrological researchers have also sought to determine water's subsurface pathways and travel times by analyzing new chemical and isotopic parameters in spring water. Many physical properties and chemical constituents can serve as environmental tracers that enable age-dating of spring water relative to when it fell as rain or snow. This is done to identify the amount of time that spring water has spent in the aquifer system.

One of the simplest indicators of groundwater pathway is the physical parameter of temperature. If groundwater subsurface residence time is long and flow is deep, very often (but not always) it can assume the elevated temperature of deep adjacent rocks. Higher total dissolved solids in groundwater can also indicate long travel times, as dissolution of minerals into water occurs. There is a long history of analyzing dissolved aqueous chemical parameters in groundwater and spring water as indicators of movement and source. Historically, major anions and cations in water have been measured, because the chemical analysis of these compounds is straightforward and they exist in the aqueous phase in large concentrations. With advancements in analytical chemistry, more attention is being turned toward trace elements, such as the rare earth elements, in spring waters (Johannesson et al. 1997). For example, springs in Death Valley have shown dramatic differences in trace element chemistry depending on whether the source of the springs is in Paleozoic limestones, Tertiary volcanics, or Quaternary sediments (Kreamer et al. 1996). Because geochemical interpretation requires that a large number of representative wells, piezometers, and springs be sampled in an area, and because the newer analytical chemistry techniques are more sensitive and measure concentrations of many parameters, statistical techniques are increasingly utilized to interpret data.

Figure 3.2 shows graphically how the statistical technique of principal component analysis groups six Death Valley springs according to trace element geochemistry. These groupings also indicate the differences in the types of rocks through which the spring water has moved (Kreamer et al. 1996).

Environmental isotopes have demonstrated utility in tracking groundwater (Montgomery and Perry 1982). Stable isotopes of oxygen and hydrogen have long been recognized as fractionating through the evaporation or precipitation of water, producing a measurable isotopic signature to

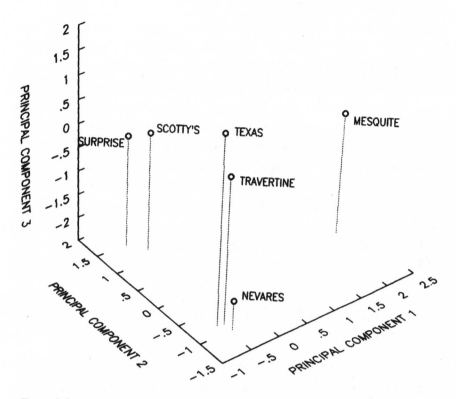

Figure 3.2. The results of principal component analysis of six Death Valley springs according to trace element geochemistry (after Kreamer et al. 1996). These groupings indicate the differences in the types of rocks the spring water has moved through and emerges from. Texas, Travertine, and Nevares springs emerge from limestones and pass through thin Quaternary sediments before surfacing. These springs are grouped together in the first two principal components. Nevares Spring emerges closer to its limestone source than do Texas and Travertine and therefore is slightly different in its third principal component. Surprise and Scotty's springs issue from Tertiary volcanic rock and are grouped together. Mesquite flows through a long sequence of Quaternary alluvium and is grouped separately.

infiltrating groundwater, which is often unique. Mixing ratios of these non-radioactive isotopes from different sources can help identify the origin of groundwater and spring water. Radioactive isotopes in water (for example, tritium, carbon-14, chlorine-36, and uranium isotope disequilibrium) can also be extremely important in identifying the age and pathway of groundwater.

In the arid regions of North America, springs are primary indicators of regional groundwater discharge. In these deserts, the water table is deep, and most streams quickly lose any collected precipitation to the subsurface. Therefore, any upwardly discharged groundwater to the surface of the land is a key characteristic of the local or regional groundwater system. These contact, fault, or artesian springs are often associated with geologic outcrops or lineaments and can be strikingly apparent in aerial photography. In addition to geologic surface expressions of discharge, the phreatophytes associated with these springs are usually dramatically different from the surrounding desert vegetation. However, there can be many environmental and human impacts on regional groundwater systems. The quantity and quality of spring discharge can reflect climatic variability as well as the pattern and degree of groundwater withdrawals.

Hydrological Investigations in Grand Canyon National Park

Grand Canyon National Park in northern Arizona is a good example of how increasing groundwater withdrawal can be understood in a management context for protection of unique springs within the canyon. The Grand Canyon has been declared a World Heritage Site and is one of the most popular tourist destinations in the United States, with nearly five million visitors each year. The majority of visitors come to Grand Canyon Village on the park's South Rim through the town of Tusayan, approximately sixteen kilometers south of the rim. The park is surrounded by the Kaibab National Forest and has no reliable water supply nearby. Water for the South Rim comes from springs on the north side of the canyon via a cross-canyon pipeline. Water in Tusayan comes from a combination of deep wells drilled into the Redwall–Muav Aquifer over seven hundred meters below ground surface, water delivered from Grand Canyon National Park, and water trucked in from surrounding areas. The aquifer was first tapped in 1989, and groundwater use has increased steadily.

Because the wells tap the same geologic strata that feed springs that emerge in the Grand Canyon, there is concern over the potential for diminished flow in South Rim springs and seeps. Reduction of spring flow would directly affect riparian habitat in the canyon and could result in the loss of recreation opportunities, wilderness benefits, and Native American cultural and religious values (U.S. Department of Agriculture 1999). Total aquifer discharge to the South Rim springs of the Grand Canyon is estimated at approximately 61,675,000 cubic meters per year (50,000 acre-feet per year) and involves about thirty perennial springs and other ephemeral seeps. Springs and their perennial streams support a variety of vegetation and wildlife, with many of the species uncounted. Seeps with hanging gardens are known to support at least forty-seven native species of plants and two introduced plant species (U.S. Department of Agriculture 1999).

Studies are being conducted using hydrochemical parameters, environmental isotopes, and a mathematical model to establish and quantify the hydrologic connection of Grand Canyon springs with the regional Redwall-Muav Aquifer beneath the Coconino Plateau. For example, tritium isotope dating, combined with uranium isotope disequilibrium studies, has indicated that water from springs below the South Rim of the Grand Canyon ranges in age from decades to more than 3,500 years old (Fitzgerald 1996; Monroe et al. 2005). Figure 3.3 shows the activity ratio of uranium isotopes 234 versus 238, plotted against the tritium ratio for those springs. Elevated activity ratios and low tritium ratios indicate old groundwater, that is, groundwater that began as infiltrating precipitation and surface water a minimum of several decades ago. This lends credence to the idea that the springs are hydraulically connected to the old waters of the regional aquifer below the Coconino Plateau; thus, any diminishment of groundwater feeding the springs may take a long time to be replenished. The springs represent the terminal point of longer flow paths, and any disruption of those flow paths will eventually manifest itself at the springs—the effects of disruption on spring flow may be rapid or may take some time. Groundwater pumping is a primary example of this disruptive action. However, once disruption reduces spring flow, a long time will have to pass before the South Rim springs recover and return to a normal state, when and if there is a cessation of pumping.

When physical and chemical spring analysis methods are combined, one can gain insight into how these arid-region aquifers respond to human use. Numerical groundwater flow models can be used to delineate the areas of the aquifers that contribute water to the springs (capture zones). To characterize the impacts of groundwater withdrawals from this regional aquifer, a geologic framework model and a numerical groundwater flow model were

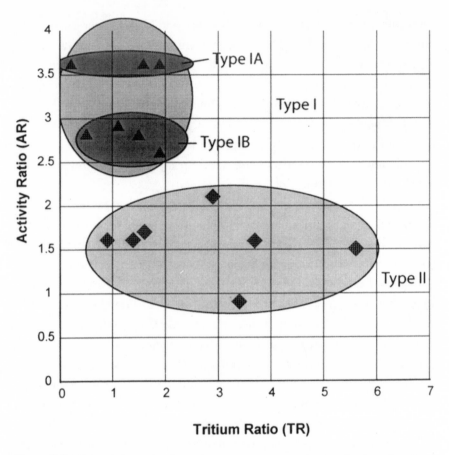

Figure 3.3. The activity ratio of uranium-234 versus uranium-238, plotted against the tritium ratio for selected Grand Canyon springs (after Fitzgerald 1996). Elevated activity ratios and low tritium ratios indicate old groundwater that began as infiltrating precipitation and surface water a minimum of several decades ago.

constructed (Wilson 2000). Capture zones of the major springs were delineated to determine which portions of the aquifer could be influenced by which wells.

The purpose of building a framework model is to create a three-dimensional representation of geologic features to be used for visualization and as the basis for a groundwater flow model. If digitized, the framework model can be an extremely useful foundation for future modeling efforts for a particular geologic region, as it provides a detailed characterization of the hydrologic system. The objective of building a framework model for the

South Rim of the Grand Canyon and the entire Coconino Plateau ground-water subbasin was to provide the database for a new generation of more-accurate groundwater flow models of the area.

Groundwater flow through the Coconino Plateau groundwater sub-basin and Redwall-Muav Aquifer on the South Rim of the Grand Canyon was simulated with a three-dimensional numerical, groundwater flow model, MODFLOW (Harbaugh and McDonald 1996). MODFLOW was selected because it is relatively well suited to simulate flow in heterogeneous and anisotropic systems such as the Redwall-Muav Aquifer and because it is well documented and widely used in the public domain. The heterogeneity in the aquifer arises from many different depositional environments in the Paleozoic strata of the canyon, and anisotropy (the unequal hydraulic con-ductivity in horizontal and vertical directions) arises as a result of the man-ner in which sediments have been deposited, lithified, and weathered. Al-though the model was originally developed to simulate flow through porous media, it has been applied successfully to similar regional fractured aquifer systems.

The purpose of the groundwater flow model was to characterize the impacts of groundwater pumping on the three major springs draining this aquifer in the Grand Canyon. The flow model had two main objectives. The first was to calibrate a steady-state groundwater flow model to prepumping (pre-1989) hydrogeologic data. This included eight wells, three springs, and limited measurements of recharge and specific capacity. The second was to delineate capture zones for the three major springs in the region. A nu-merical postprocessor to the numerical flow model (Pollack 1994) was used to track water particles and delineate capture zones for each of the major springs. This provided insight into which regions are most likely to impact specific springs if more wells are drilled. The approach used for building the numerical groundwater flow model included building a conceptual model, selecting a numerical model, constructing the model, calibrating to steady-state conditions, and conducting a sensitivity analysis.

The simulated area of the Redwall-Muav Aquifer is approximately 9,500 square kilometers. A two-layer, three-dimensional grid was con-structed to simulate flow and delineate capture zones. The top layer repre-sented the Supai Formation, and the bottom layer represented the Redwall and Muav Limestones. The grid was oriented 60 degrees west of north to

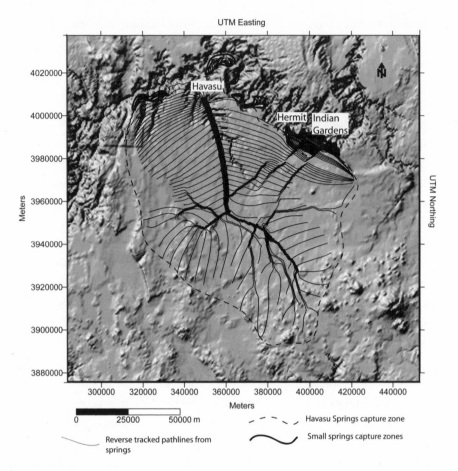

Figure 3.4. Steady-state, pre-development capture zones for Havasu, Hermit, and Indian Garden springs on a shaded relief map of the Coconino Plateau and the Grand Canyon.

parallel the principal direction of groundwater flow toward Havasu Springs (fig. 3.4). Data from the digital geologic framework model were imported for the top and bottom elevations of both layers.

A total recharge of 160,079 cubic meters per day (m³/d) was applied to the model. Simulated (and target values in parentheses) steady-state, pre-development spring discharges for the Redwall-Muav Aquifer were 157,090 (157,000) m³/d for Havasu Springs, 1,644 (1,640) m³/d for Hermit Spring, and 1,640 (1,640) m³/d for Indian Garden Spring. The root mean squared error for hydraulic head values from the steady-state simulation was 41.2

meters, or 16 percent of the total head change across the region. This was considered an acceptable error for this regional aquifer.

A capture zone analysis with hydraulic heads from the steady-state predevelopment numerical model indicated that capture zones for Hermit and Indian Garden Springs do not extend far from the South Rim (fig. 3.4). The model indicates that water is actually drawn to those springs from a distance of about only five kilometers south of the South Rim. Havasu Springs captures most of the groundwater in the aquifer and has the most extensive capture area. There are two existing wells within the Indian Garden Springs capture area at Tusayan, and as a result, pumping from these wells will likely influence this spring. All of the other existing pumping wells in the region are within the capture area for Havasu Springs.

The creation of a three-dimensional geologic framework model can prove very useful for deep, large regional aquifers in the Sonoran Desert. A single hydrogeologic database can be visually revealing and interpretively significant and can aid in developing the conceptual model of the flow system. Additionally, a digital framework can easily be updated and modified as new data are collected. In many desert aquifers, the only hydrologic information about the aquifer exists at the spring. As more information on the aquifer system becomes available, it can be added to the database and model. This is especially true for large regional aquifers in areas such as the Coconino Plateau, where there currently is a paucity of data. Growth and development will increase the knowledge base through new wells, new site investigations, and more monitoring of groundwater conditions.

Tools for Understanding Flow

Human population growth and agricultural development in desert regions of North America have resulted in increasing exploitation of groundwater, with many pumping wells tapping the same hydrologic source for springs. This can result in the diminishment of spring water quantity and quality, with repercussions to ecosystems that depend on these springs.

The hydrology of desert springs can be better understood and predicted by mapping recharge and discharge areas and defining groundwater flow. Indicators of groundwater flow patterns include topography, piezometric patterns, hydrochemical trends, environmental isotopes, and soil and land features. Specific water-quality parameters and environmental isotopes can be key indicators in tracking groundwater flow patterns and travel times. Another important approach involves numerical groundwater flow models, which—when coupled with conceptual models—are useful tools for improving our knowledge of flow in the aquifers feeding springs. Some of the insights to be gained include distribution of hydraulic conductivity, distribution of recharge, and connectivities of the fracture systems. Particle tracking on water levels derived from the numerical groundwater flow models can allow capture zones for springs to be defined. The capture zones can be useful for land and water managers to guide management decisions for aquifers discharging to desert springs.

ACKNOWLEDGMENTS

The authors wish to acknowledge James Fitzgerald of North State Resources, and Klaus Stetzenbach of the Harry Reid Center for Environmental Studies, University of Nevada, Las Vegas for their support of this work.

Chapter 4

A Comprehensive Springs Classification System

Integrating Geomorphic, Hydrogeochemical,
and Ecological Criteria

ABRAHAM E. SPRINGER, LAWRENCE E. STEVENS,
DIANA E. ANDERSON, RODERIC A. PARNELL,
DAVID K. KREAMER, LISA A. LEVIN, AND
STEPHEN P. FLORA

Integration of the hydrological, geological, and ecological characteristics of springs ecosystems provides a much-needed means of classifying the types and distribution of these important landscape features. Efforts since the early 1900s have produced general classifications of the physical, chemical, and thermal properties of springs, but these classification systems have varied with the intent of classification needs and the quantity and quality of information collected about the springs. Although classification systems have focused on water up to, and at, the point of discharge, more recent needs have arisen to classify springs beyond the point of discharge for characterizing the geomorphology of the spring channel, the biogeographic context of the spring, associated biota, and cultural values, users, and management. The lack of a comprehensive springs classification system has resulted in insufficient and inadequate inventories, ecological assessments, and conservation measures for these ecosystems. As a preliminary step in overcoming these challenges, we have developed an improved classification system incorporating geomorphic, hydrological, geochemical, ecological, and management criteria of springs ecosystems. A central database, ideally using all of these criteria, is needed to facilitate springs ecosystem inventory, classification, and conservation.

An Introduction to Springs Ecosystems

A Definition of Springs

Springs are places where groundwater is exposed at the earth's surface, often flowing naturally from bedrock or soil onto the surface of the land or into á body of surface water (Wilson and Moore 1998). Springs emerge in most of the ecosystems on Earth, including a wide array of subaerial terrestrial settings and as subaqueous discharge from the floors of freshwater and marine bodies of water. Springs are important ecological and cultural resources, but scientific study and conservation have been hampered by the lack of a comprehensive classification system through which to quantify types of springs (Alfaro and Wallace 1994). Such a classification system would greatly improve our understanding of the distribution of different types of springs and would lay the groundwork for regional, national, and international conservation efforts. The failure to develop such a classification system has undoubtedly played a role in the global demise of these extraordinary ecosystems: recent texts fail to describe the fundamental geomorphic and ecological differences between springs and other kinds of riparian ecosystems (Malanson 1993; Stevens et al. 2005), and recent reviews of the services provided by natural ecosystems largely fail to mention springs as important ecosystems (Postel and Carpenter 1997). These oversights are attributable not only to the absence of a synthetic classification system but also to the limited understanding of springs ecology, poor communication between the sciences of hydrology and ecology, the limited attention paid to groundwater extraction—a form of resource exploitation that we cannot directly observe (Glennon 2002)—as well as to the ambiguous legal status of springs in groundwater and surface-water laws (see chapter 16). Clearly, the need for (and value of) such a classification system is great.

The Distribution of Springs in the Western United States

Stevens and Meretsky (see chapter 1) describe the distribution of named springs in the western United States, with states in the Great Plains having the lowest springs densities and states with large plateaus dissected by deep canyons (e.g., Oregon and Arizona) having the highest densities.

However, more than 50 percent of springs have not been currently recognized or mapped, particularly in topographically complex terrain, and thus are not located on USGS topographic maps. Also, numerous springs either have dried up or were mistakenly assumed to be perennial at the time of mapping. Whereas the distribution and biota of marine geothermal vents have received much recent attention (e.g., Van Dover et al. 2002), less information is available on nonthermal subaqueous springs emerging in marine or freshwater settings.

Springs Classification Systems

Geologists have traditionally classified the physical parameters of springs up to their point of discharge (e.g., Bryan 1919; Meinzer 1923), but little attention has been paid to springs after the point of discharge. Other geochemical, geomorphic, biological, and cultural classification systems have been developed for surface waters and riparian systems downstream from the point of discharge (e.g., Hynes 1970; Rosgen 1996); however, geomorphically based classification systems have largely ignored differences between spring–fed channels and surface runoff–dominated systems, and biologically oriented analyses have focused either on individual springs or on individual taxa. An integrated springs classification system should include the major physical, biological, and sociocultural variables. Such a classification system will permit assessment of the distribution of different kinds of springs within ecosystems, thereby improving resource inventory and development of conservation and restoration strategies (e.g., Sada et al. 2001; see chapter 11, this volume).

Alfaro and Wallace (1994; Wallace and Alfaro 2001) updated and reviewed the historical spring classification schemes of Fuller (1904), Keilhack (1912), Bryan (1919), Meinzer (1923), Clarke (1924), Stiny (1933), and others. Of the previously proposed systems, Meinzer's (1923) classification system has been the most persistently recognized. He included eleven characteristics of springs based on various physical and chemical variables. Although Meinzer's (1923) scheme has been widely used, it is not comprehensive. Clarke (1924) considered three criteria to be most important for springs classification: geologic origin, physical properties, and geochemistry. However, none of the classifications proposed thus far include ecologically relevant variables, such as considerations of spatial and temporal degree of isolation, microhabitat distribution, biota, and surrounding ecosystem context. Thus, no comprehensive classification system has yet been developed or accepted (Wallace and Alfaro 2001).

In this chapter, we build on previous classification efforts and present an integrated springs classification system (see appendix 4.1), with the understanding that testing and refinement of this classification system will require much future work. We update nine of Meinzer's (1923) classes, integrate Alfaro and Wallace's (1994) recommendations, and then propose additional ecological elements. We propose an organizational structure that integrates springs data, and we reiterate Alfaro and Wallace's (1994) recommendation to develop a global database on springs using this comprehensive classification system. This classification system should permit better management and conservation of springs ecosystems, and the proposed structure should serve as the basis for development of the comprehensive springs database.

A Proposed Springs Classification System

Geomorphic Considerations

THE HYDROSTRATIGRAPHIC UNIT. Meinzer's (1923) characterization of the aquifer lithology and geologic horizon can be reduced to the rock type(s) of the hydrostratigraphic unit (igneous, metamorphic, or sedimentary). Sedimentary units can be consolidated rock or unconsolidated sediments. Seaber (1988) defines a hydrostratigraphic unit as "a body of rock distinguished and characterized by its porosity and permeability." This classification requires that the nature and boundaries of the stratigraphic unit be mappable. As with some other spring classifications, such information may not be available without a detailed investigation of the aquifer. For instance, the spring may issue from a bedrock aquifer but may travel through one or more other units (e.g., alluvium) before discharging to the surface. Such information may be even more difficult to obtain in deep marine and other subaqueous settings.

THE EMERGENCE ENVIRONMENT. The environment in which the spring orifice exists varies widely, from the special case of in-cave springs whose flow subsequently may or may not reach the surface, to subaerial

Figure 4.1. A rheocrene spring: Hermit Creek Spring, Grand Canyon National Park, Arizona.

emergence in a wide array of geomorphic settings, to springs that emerge below glaciers, in subaqueous freshwater lentic and lotic settings, on the floor of estuaries, and in a wide variety of subaqueous marine settings.

ORIFICE GEOMORPHOLOGY. Spring orifices occur in several specific geomorphic environments (Meinzer 1923). Groundwater may be exposed in, or flow from, filtration settings (poorly consolidated, permeable materials), bedrock fracture joints, or tubular solution passages. We modify the fracture spring list to include springs that exist as groundwater exposed at the surface although they may not flow above the ground surface (e.g., Devils Hole in Ash Meadows, Nevada; Wilson and Blinn 2007). We also include stratigraphic contact environments in which springs, such as hanging gardens, emerge along geologic stratigraphic boundaries.

THE "SPHERE OF DISCHARGE." The "sphere" into which the aquifer is discharged as described by Meinzer (1923) was greatly simplified by Hynes (1970) into three different classes. We re-expand these historical schemes to include twelve classes of springs, some of which are shown in figures 4.1–4.4 (see also fig. 6.1). These classes include (1) springs that emerge in caves, (2) limnocrene surficial lentic pools (see fig. 5.1b), (3) rheocrene lotic channel floors (fig. 4.1), (4) mineralized mounds (figs. 5.1c, 6.1b), (5) helocrene wet meadows, (6) gushets (fig. 4.3), (7) hillslope springs (fig. 4.4), (8)

Figure 4.2. A hanging garden spring: Cliff Spring, Grand Canyon National Park, Arizona.

Figure 4.3. A gushet spring: Vaseys Paradise, Grand Canyon National Park, Arizona.

Figure 4.4. A hillslope spring: Lower Butte Fault Spring, Nankoweap Canyon, Grand Canyon National Park, Arizona.

contact hanging gardens (fig. 4.2), (9) geysers, (10) artesian fountains, and (11) hypocrene buried springs, and (12) exposure springs. In addition, we recognize paleosprings ("fossil springs"), which flowed in prehistoric times but no longer flow (see also chapters 2 and 5). Both Meinzer's (1923) original and Hynes's (1970) limited classification schemes become complicated when multiple spheres of discharge are present or if the spring has a highly variable discharge rate and creates multiple spheres over time. Therefore, all major spheres of discharge should be noted during each site visit, and the importance of each should be described.

SPRING CHANNEL DYNAMICS. In the special case of subaerial springs that create or flow into channels, such discharge may support distinct geomorphic characteristics within the channel. If a subaerial flowing spring feeds the stream headwaters and there is little to no runoff contributing to the stream flow, the stream is classified as a spring-dominated stream (Whiting and Stamm 1995). If the spring discharge is relatively constant and permanent, then the morphology of the channel will be distinctive. These types of channels often flow at bank-full stage, or slightly above, 20 percent of the time (Whiting and Stamm 1995). If the spring discharges to a channel that has significant components of runoff, it is classified as a runoff-dominated stream (Whiting and Stamm 1995). Such systems may be classified using stream-channel geomorphology terminology (e.g., Rosgen 1996). Some spring systems have components of both spring and runoff domination, such as some spring rills in Ash Meadows, Nevada.

Forces Bringing Water to the Surface

The classes for the forces that bring water to the surface may not be evident on a single visit or without information on subsurface water from surrounding wells. Meinzer (1923) categorized springs on the basis of the pressure exposing or forcing water out (i.e., gravity, thermal) and other pressures. Gravity-fed spring systems are fed by down-gradient groundwater flow within the aquifer. Artesian springs discharge water under pressure or may issue from an aquifer that has an upper confining layer, subjecting the flow to fluid pressures in excess of the pressure attributable to gravity at the point of discharge. Thermal springs emerge when groundwater comes in contact with magma or geothermally warmed crust and is forced to the surface, sometimes explosively as in geysers. Water is forced to the surface by explosive release of CO_2 in the geyserlike "Coke-bottle" springs of Utah (Meinzer 1923; Shipton et al. 2004). Fluid discharge in submarine springs associated with methane seepage is often forced out by diurnal tidal variation in the pressure of overlying water (Tryon et al. 2001). Some springs do not flow and therefore are not subject to pressurized discharge, while other springs may have multiple forcing mechanisms. Anthropogenic factors, such as groundwater loading around large reservoirs, also may create forces that affect the emergence of springs.

Flow

PERSISTENCE. Springs may function as refugia across ecological and evolutionary time scales. We follow Nekola 1999 in distinguishing between springs that have recently developed or been exposed to the atmosphere

(Holocene neorefugia) versus those that have existed since the Pleistocene or longer (paleorefugia; see chapter 6). Nekola (1999) predicted that paleorefugia were likely to exhibit high levels of endemism, unique species, and well-sorted assemblages. In contrast, neorefugia were predicted to support more weedy species, with low levels of unique taxa. His studies of land snails at springs affected or not by Pleistocene ice sheets supported these predictions, and Blinn's (chapter 6) study of Montezuma Well in central Arizona provides additional support for the paleorefuge concept. Paleosprings that do not presently flow may contain important paleoclimatical, paleontological, or archaeological remains (see chapter 5).

FLOW CONSISTENCY. Meinzer (1923) defined two classes of perenniality exhibited by springs. Springs are considered to be perennial if they discharge continuously, or intermittent if their discharge is naturally interrupted or sporadic. Intermittent springs may flow regularly — whether hourly or daily (e.g., some geysers), seasonally, annually, or interannually — or only on an erratic basis. Human impacts, such as groundwater extraction and well drilling, may affect discharge consistency. As with flow variability (below), multiple observations of a spring are required to determine the permanence of discharge.

FLOW RATE. Meinzer (1923) developed eight discharge classes, defined by the magnitude of discharge from a spring at the time of measurement; however, his numeric scheme is reversed from the intuitive scale (low discharge should have a low value). We propose reversing this numeric system in a scale that accommodates the full range of springs discharges known, from seeps with near-zero flow (a score of 1) to springs with a flow of greater than 10 cubic meters per second (a score of 8; e.g., Ra El Ain Spring in Syria, with a discharge of 36.3 cubic meters per second [Alfaro and Wallace 1994]). Because the discharge of many springs varies temporally, the flow rate class will change depending on the time of measurement. Fluid flow rates are measured in different units by marine hydrogeologists. For marine cold seeps, fluid flow rates have been estimated at 10 liters per square meter per day on the Alaska margin at a depth of 5,000 meters (Suess et al. 1998), and up to greater than 1,700 liters per square meter per day on the Oregon margin (Linke et al. 1994), with intermediate values off Peru (440 $L/m^2/d$

[Linke et al. 1994]; 1,100 L/m²/d [Olu et al. 1996]). Hydrothermal vents may have flow rates that are orders of magnitude larger than those of these marine cold seeps.

FLOW VARIABILITY. Springs discharge may be variable at different temporal scales. Short-term variability may be related to loading effects (such as the siphon effect, in which the filling of groundwater solution channels creates periodic surging of a spring's discharge). Short-term hydrologic alterations may include individual storms or droughts, while longer-term flow variation may result from interannual climate variation or Pleistocene-Holocene climatic and hydrologic changes. Variability in springs discharge may affect the distribution of associated microhabitats. For these reasons, the classification of discharge variability should be based on repeated discharge measurements, sometimes over long time periods.

Meinzer (1923) considered three levels of springs discharge variability: constant (steady), subvariable, and variable. This classification requires multiple measurements to characterize diurnal, seasonal, annual, interannual, and long-term variability. Netopil (1971) and Alfaro and Wallace (1994) used flow duration statistics to calculate a discharge variability ratio (DVR): DVR = $Q_{10\%}/Q_{90\%}$. In this ratio, $Q_{10\%}$ is the high flow exceeded 10 percent of the time and $Q_{90\%}$ is the low flow exceeded 90 percent of the time. Of course, calculation of these flow rates requires monitoring over a multiannual period. Steady discharge results in a DVR of 1 (extraordinarily balanced), while wildly varying flows may produce a DVR of greater than 10 (extraordinarily unbalanced). Intermittent springs have an infinite DVR.

Water Quality

Classification of spring water quality is often specific to an individual study, but several comprehensive approaches have been suggested. Most traditional classifications are based on water temperature and/or the dominance (concentration) of ions.

WATER TEMPERATURE. Five classes for water temperature in springs have been recognized, based on a comparison of spring water temperature with the mean annual air temperature (modified from Alfaro and Wallace

1994): cold, normal, warm, hot, and superthermal springs. Cold-water springs are, by convention, greater than 12.2°C cooler than the mean annual ambient temperature. Spring waters within 12.2°C of the mean ambient temperature may be (but are not necessarily) responding to ambient atmospheric temperatures. This is to be expected in springs that emerge from shallow aquifers; these may have temperatures that vary seasonally with air temperature. Springs with warm (>12.2°C above the mean ambient air temperature but <37.8°C) and thermal water (>37.8°C) are connected either to very large aquifers with long flow paths or to geothermal sources of heat. Superheated geothermal springs are commonly reported in tectonically active areas, such as geyser fields or marine sea floor settings. The upper temperature limit presently known for life is 121–130°C for bacteria-like extremophile Archaea in Pacific Ocean vents (Anonymous 2003). Variability in spring water temperature may also be important but can be assessed only from multiple visits or by using recording thermistors. Deep submarine springs are typically characterized as cold seeps or hot vents based on their relationship to ambient ocean water temperatures. Because the temperature of seawater in the deep sea is relatively invariant (Gage and Tyler 1991), even small changes above ambient in venting waters represent a significant warming effect with major biological consequences.

GEOCHEMISTRY. Numerous schemes have been developed to classify water geochemistry, primarily through the surface-water pollution literature, but few studies attempt a comprehensive classification of spring-water geochemistry. Clarke (1924) classified the waters of mineral springs based on the dominance of seven ion groups—calcium, sodium, potassium, magnesium, chloride, sulfate, carbonate (and combinations of these constituents)—silica dioxide (SiO_2), borate (B_4O_7), nitrate, and phosphate, as well as acidity. Furtak and Langguth (1967, cited in Alfaro and Wallace 1994) classified Greek springs as belonging to (1) normal earth alkaline (hydrogen-carbonatic) waters; (2) normal earth alkaline, hydrogen-carbonatic sulfatic waters; or (3) enriched alkali earth alkaline (primarily hydrogen-carbonatic) waters. Dinius (1987) used an expert-based decision process to develop an index of surface-water quality to compare levels of pollution in bodies of fresh water. The twelve variables derived from that analysis include specific conductance (micromhos/cm at 25°C), pH, alkalinity (concentration of equivalent calcium carbonate), water color (platinum units), and the concentrations of chloride and nitrate (NO_3^-), which may be relevant to the water-quality classification of springs, as well as several variables that may not be relevant, including dissolved oxygen concentration, biological oxygen demand, and bacterial concentration. Davies-Colly and Smith (2001) and others consider turbidity to be important to surface water quality, and we include it in our list. More recent classifications have emphasized more-comprehensive geochemical analyses, rare-earth element analyses (Kreamer et al. 1996), and isotopic analyses, all of which may be informative for distinguishing among springs. Also, more-recent studies have emphasized more-elaborate statistical analyses (e.g., principal component analysis or cluster analysis; Kreamer et al. 1996) to integrate major and minor element relationships, and such approaches will be fruitful when a large database on spring-water geochemistry has been developed.

A comprehensive analysis of spring-water geochemistry awaits development of the springs database recommended below. Meanwhile, we base our selection of water-quality variables on the above springs studies and on relevant surface-water quality studies. We recommend that eight groups of geochemical variables be measured during springs inventories: (1) major anions and cations (chloride, sulfate, carbonate, calcium, sodium, magnesium); (2) minor constituents (iron, borate, silica dioxide, carbonate/chloride, triple waters [i.e., water with a high concentration of carbonate, chloride, and sulfate]); (3) pollution indicators (selenium, fecal coliform); (4) useful tracers (stable isotopes, radioactive isotopes, rare-earth elements), (5) alkalinity; (6) the concentration of total dissolved solids (or specific conductance); (7) pH; and (8) nutrient concentrations (nitrate and phosphate).

Fluids associated with marine hydrothermal vents and cold seeps contain variable concentrations of dissolved or gaseous methane, sulfides, and hydrogen that exert a large influence on chemosynthetic biological processes. These are typically measured in submarine springs to characterize the emerging fluids.

Habitats

SYNOPTIC CLIMATE. Synoptic climate strongly affects the ecosystem development, processes, and biodiversity of springs. Climate variables that are often available or can be regionally modeled include air temperature (seasonality and mean annual), precipitation (seasonality and mean annual), growing season length, and relative humidity. The seasonality index for temperature is the ratio of the mean temperature of the hottest month to the mean temperature of the coldest month in degrees Celsius. The precipitation seasonality index is the ratio of the average total precipitation for the three wettest consecutive months to the average total precipitation for the three driest consecutive months (Bull 1991). While there is no climate relevant to submarine springs, seasonal variation may occur in the ambient water temperature at depths below 200 meters, in current strength, seafloor storms, or in inputs of photosynthesis-based production falling to the seabed.

SURROUNDING ECOSYSTEMS. The kinds of ecosystems and vegetation types that surround springs often influence habitat conditions, biogeographic processes, wildlife and human uses, and other characteristics. In general, steep ecological gradients of environmental stability (disturbance intensity), geochemistry, moisture availability (in arid regions), productivity, and other factors most strongly affect the biodiversity, endemism, and habitat use of springs (Malanson 1993; Huston 1994; Alexander et al. 1997).

In many springs, the dominant disturbance regime from the surrounding landscape may strongly affect spring microhabitats. Plenet and colleagues (1992) examined a spring on the Lône des Iles Nouvelles backwater of the Rhône River floodplain. Hypogean (hyporheic) and epigean benthic macroinvertebrates were sampled; these responded to flooding by entering the hypogean zone. Epigean density decreased during high flows, whereas the density of hypogean organisms remained much more consistent and did not respond strongly to surface flows. Similar levels of influence exist in forest springs, where the regional fire regime may alter spring habitat dynamics (see chapter 15). Disturbance regimes in submarine springs may include an increase or decrease in extrusion of new magma at hot vents, slumping, mass wasting or turbidity flows in sedimented margins, and storm-driven resuspension and deposition of sediments. The dominant disturbance regime(s) occurring in the surrounding ecosystem(s) should be noted.

BIOGEOGRAPHIC ISOLATION. Island biogeographic theory provides a convenient framework for understanding species distribution at springs

(MacArthur and Wilson 1963; Lomolino et al. 2006). Colonization is likely to be rare and extirpation is common in small springs or those that are far from other springs or wetland source areas. Conversely, colonization is common and extirpation is less likely in large springs or those that are near other springs or wetland source areas. The configuration of springs, such as hanging gardens or travertine springs, along geologic contacts may result in archipelagoes, a distribution that facilitates colonization and gene flow and reduces the probability of extirpation. From a biogeographic perspective, the extreme isolation of springs in North America's southern Great Plains indicates that these paleorefugia are likely to support relatively high levels of endemism and may be excellent sites for biological research and conservation. Island biogeographic theory has yet to be rigorously tested on spring biota at ecoregional or continental scales. Hydrothermal vent sites tend to occur like strings of pearls along midocean ridges where seafloor spreading occurs. Researchers believe that planktonic larvae of vent species move between vent sites along the ridge axis, moving in plume-driven superhighways (Mullineaux and France 1995).

HABITAT SIZE. The area of aquatic, wetland, and riparian spring habitats is important to our understanding of biogeographic impacts on biotic assemblages. We propose that habitat area should be measured and categorized within eight size classes for each of these three important habitats. These habitat size classes range from extremely small (less than 2 square meters) to extremely large (more than 100 hectares).

MICROHABITAT DIVERSITY. In addition to the aquatic, riparian, or terrestrial habitats that springs may support, their associated spheres of discharge are capable of creating unique microhabitats. Microhabitats may be created by specific physical or chemical characteristics, such as temperature, depth of water, dissolved ion or oxygen composition, disturbance regime, or a suite of physical variables. Emphasis on microhabitats is warranted, as some spring microhabitats support high levels of rare or endemic species (e.g., Erman 1992; Sada et al. 2001; Hershler and Sada 2002; Polhemus and Polhemus 2002).

Spring microhabitats that we consider to be important include cave environments, wet walls, madicolous (fast-flowing water) habitats, hyporheic (saturated subfloor) habitats, open-water pools, spring streams (including those partially or more completely dominated by surface flow), wet meadows, riparian habitats, waterfall spray zones, and barren rock habitats adjacent to springs. Microhabitat diversity may be calculated using the standard Shannon (1948) diversity index (H'; Magurran 2004), using the proportional area (p_i) of each microhabitat: $H' = -\Sigma (p_i * \log_{10} p_i)$. Any logarithmic base may be used for this calculation, as long as the use is consistent, but for ease of interpretation, we recommend the use of base 10.

Biota

SPECIES COMPOSITION. We recommend that all species of plants, invertebrates, vertebrates, and other biota observed at the spring during each site visit be recorded, as such information will generally contribute to the inventory. Threatened, endangered, and endemic species are most likely to be of immediate management concern, and these should be carefully documented. Research conducted in Grand Canyon (Grand Canyon Wildlands Council 2004) indicates that a relatively thorough baseline inventory of plants can be conducted in two to three site visits, whereas a 95 percent inventory of aquatic macroinvertebrates may require five or more site visits over two or more years. Detection of vertebrates is likely to require more site visits and significant survey intensity.

VEGETATION. The observer should identify and measure the area of each distinctive patch of vegetation at the spring and visually estimate the percentage of cover of each species in each of four strata: ground cover (annual, <1 m in height), shrub cover (perennial, 1–4 m in height), woodland/midcanopy (perennial, 4–10 m in height), and tall canopy (>10 m in height). The site should be photographed, and aerial photographs should be used if available. A standard Shannon-Weiner diversity index can be used to calculate percent cover within strata at each springs.

FAUNAL DIVERSITY. Invertebrates have been widely used and tested as indicators of water quality. Excellent metrics have been developed for assessing the health of streams using aquatic invertebrate sampling, such as the Wisconsin Index, the Index of Biological Integrity (Karr 1991), and

AusRivAS (Smith et al. 1999); however, no such criteria exist for springs. The wide variation in flow, water quality, microhabitat diversity, and biogeographic issues will render such a metric difficult to develop. Development of such criteria for commonly occurring spring microhabitats (e.g., limnocrene pools) requires compilation and analysis of the large database that this chapter recommends be collected. Springs that represent extreme environments often support large mats of cyanobacteria, sulfide- or iron-oxidizing bacteria, or consortia of bacteria and archaea. Very often, these define the nature of the venting fluids and form the visually dominant life-form. We suggest that dominant microbial forms be documented. There is also a growing interest in characterizing microbial diversity through gene sequencing, particularly in extreme-environment springs where conditions may resemble those of the early Earth or other planets.

Other Habitat and Biological Criteria

Other habitat and biological criteria may eventually be proven useful in the classification of springs; however, because sufficient research has not yet been conducted on most of these topics, their inclusion here is not recommended. Contemporary spring soils have received limited attention in relation to spring biota and habitats; however, litter accumulation and soil development may occur on low-gradient slopes and may play important roles in some species distributions. Ecological processes, such as productivity and decomposition, are likely to be closely related to biodiversity, particularly in relation to the steepness of the slope gradient across the springs-to-uplands interface (see chapter 11). Also, we presently regard springs ecosystems as being dominated by physical processes and characterized primarily by "bottom-up" ecological processes; however, biological interactions may have as yet unrecognized importance in some types of springs and create important "top-down" trophic cascades. More research is warranted on these and many other springs ecosystem ecology topics. While this may be addressed only in later stages of research, the degree to which spring biota support life-forms of the surrounding environments (by exporting production via vagrants) may be important. In the deep sea, this is often measured with stable isotopic signatures (Carney 1994; MacAvoy et al. 2002).

Springs Management and Use

The cultural significance, land management history, ownership, and other human factors may strongly affect geomorphology, flow, geochemistry, and the ecological condition of springs and surrounding ecosystem landscapes, making management and use critical elements of a comprehensive springs classification system (White 1979; Alfaro and Wallace 1994). The authority to manage springs may fall to private, public governmental, or tribal managers and may be subject to common or legislative law. For example, groundwater use is often governed by common law or precedents established by legal decisions. Groundwater law typically varies from state to state in the United States, with appropriative water law governing water rights in the western United States and riparian rights governing water rights in the eastern U.S. and California. In the United States, threatened and endangered species may be protected under the Endangered Species Act, and traditional cultural uses of springs may be governed by federal laws, including the Native American Religious Freedom Act or the Antiquities Act. Federal actions on lands containing federally owned springs take place under the National Environmental Policy Act if contests among multiple stakeholders arise over resource use. Individual states have authority over stream channels and water quality within their borders, and the U.S. Army Corps of Engineers has authority over the surface-water quality and navigability of interstate waters. Jurisdiction over deep-sea springs outside the exclusive economic zone of countries is a more difficult issue. Some guidelines have been developed to establish these as fragile environments protected from damage by energy exploitation, and scientists have begun to establish no-research conservation areas to protect certain sites from damage by underwater vehicles.

The wide array of human values of springs includes numerous utilitarian purposes, such as culinary water supplies, livestock watering, municipal or industrial use, recreation, wildlife, conservation, scientific research, and other purposes. Some springs may have had prehistoric or early historic use or modification that should be documented. The impacts of anthropogenic alterations vary by degree and may be classified as undisturbed, partially diverted/disturbed, or fully diverted/disturbed. Full diversion at the point of discharge may not provide any potential for support of dependent ecosystems. "Beneficial uses" may involve partial or complete extraction or abstraction before or after emergence. Springs are also commonly used or regarded in a religious or ceremonial fashion as traditional cultural properties, and such values commonly conflict with extractive utilitarian uses.

Springs Information Management and Database Development

Springs research and classification requires precise measurement of physical, biological, and cultural information, followed by compilation of those data. The spring should be appropriately georeferenced and photographed, and various physical characteristics should be measured during each visit. Elevation and the aspect of the spring should be measured, as these characteristics are likely to be of biological relevance in springs, particularly at higher latitudes. If possible, Solar Pathfinder (1994) should be used to determine the solar energy budget of the site, because aspect influences important physical properties of the study sites, such as temperature, the amount of light available for photosynthesis by wetland vegetation, the duration of freezing in winter, and evaporation and relative humidity in the summer months. The slope (dip angle) of the site similarly should be documented. Some of the proposed classes can be determined through visual observation of a spring during a single visit, but other variables (e.g., flow and geochemical variability) require multiple observations or information in addition to that gathered during a single site visit. Other variables, such as aquifer dynamics, may require additional research and synthesis of numerous studies. Rigorous quality-control standards should be applied to the samples and data collected at springs, and these data should be placed in an integrated information management system.

Integrated information is needed from reconnaissance, classification, and ecosystem health assessment efforts at springs. Considerable time, resources, and expertise are needed for detailed classifications of springs, and the impact of climate changes on groundwater dynamics that supply water to springs is but one of many active areas of research. The classification system proposed here should serve as a template for development of a comprehensive global database on springs. Analyses of integrated spatial and temporal data are likely to reveal hitherto unrecognized patterns in springs hydrogeology, distribution, ecology, and conservation, and are likely to result in clarification and modification of data collection protocols.

Improving Springs Management and Conservation

Springs are greatly threatened. by human impacts, but rarely are these productive, biologically diverse ecosystems managed for long-term, ecological sustainability. Development of the unified classification system and lexicon proposed here may help clarify the distribution, condition, and conservation of springs ecosystems. Existing classifications of springs have thus far been concerned with water only to the point of emergence, and not the dependent downstream ecosystem. The physical classifications of Bryan (1919) and Meinzer (1923) require modification because of logical inconsistency and changes in geologic and hydrologic theory. The biological classification criteria of springs proposed above are an important addition to these classifications and emphasize the biogeographic and refugial status, ecoregional setting, and steepness of ecological gradients with respect to surrounding upland environments. Management authority and cultural uses of springs are important variables for status and conservation analyses, although additional understanding of traditional cultural knowledge, history, and uses of springs is often needed. Development of a well-managed information database will require rigorous quality-control protocols. We recommend the use of the above classification system to develop this database globally. Analyses of such a database will provide fruitful future research into springs ecology, making the database essential for the conservation and sustainability of these ecosystems.

ACKNOWLEDGMENTS

The authors thank the following agencies and organizations for project support: the U.S. Forest Service; the U.S. National Park Service; the U.S. Bureau of Land Management, Grand Staircase–Escalante National Monument; the U.S. Geological Survey; the Salt River Project; and the Museum of Northern Arizona, Flagstaff.

Appendix 4.1

PROPOSED SPRINGS CLASSIFICATION CRITERIA AND CHECKLIST

Spring Name		Latitude	Elevation
Sampling Date	Observers	Longitude	Aspect
Map No.	Photo ID No.	GPS Accuracy	Slope (Dip Angle)

Comments:

Class Variable	Type / Criterion	Value / Descriptor	Ref.	Score
Geomorphic Considerations				
Hydrostratigraphic unit	Parent rock of aquifer(s)	Sedimentary (bedrock or uncon- solidated sediments)	M23	
		Igneous	M23	
		Metamorphic	M23	
		Mixed (combination of above)	—	
Emergence environment	Cave	Special case, not usually considered as a spring because it may not be directly exposed to the atmosphere	—	
	Subaerial	Aboveground emergence—note geomorphic setting (e.g., floodplain, prairie, piedmont, canyon floor or wall, mountainside)	—	
	Subglacial	Aboveground emergence beneath a glacier	—	
	Subaqueous— lentic freshwater	Aquatic emergence into a pond or lake—note substratum (organic ooze, silt, sand, rock)	—	
	Subaqueous— lotic freshwater	Aquatic emergence into a stream or river—note substratum (organic ooze, silt, sand, rock)	—	
	Subaqueous— estuarine	Aquatic emergence in an estuary— note substratum (organic ooze, silt, sand, rock)	—	
	Subaqueous— marine	Aquatic emergence in a marine setting—note substrate (e.g., silt, sand, coral)	—	
Orifice geomorphology	Seepage or filtration spring	Groundwater exposed or dis- charged from numerous small openings in permeable material	M23	
	Fracture spring	Groundwater exposed or dis- charged from joints or fractures	M23	

	Tubular spring	Groundwater discharged from or exposed in openings of channels, such as solution passages or tunnels	M23	
	Contact spring	Flow discharged along a stratigraphic contact (e.g., a hanging garden)	—	
"Sphere of discharge"	Cave	Emergence in a cave	—	
	Limnocrene — emerges from lentic pool(s)	Emergence in pool(s)	M23* H70*	
	Rheocrene — lotic channel floor	Flowing spring, emerges directly into one or more stream channels	M23* H70*	
	(Carbonate) mound-form	Emerges from a mineralized mound	—	
	Helocrene (marsh) or cienega (wet meadow)	Emerges from low-gradient wetlands; often indistinct or multiple sources	M23* H70*	
	Hillslope spring	Emerges from a hillslope (30-60° slope); often indistinct or multiple sources	—	
	Gushet	Discrete source flow gushes from a wall	—	
	Hanging garden	Dripping flow emerges usually horizontally along a geologic contact	—	
	Geyser	Explosive flow	—	
	Fountain	Artesian fountain form	—	
	Hypocrene	A buried spring where flow does not reach the surface	—	
	Exposure	Groundwater is exposed at the land surface but does not flow	—	
Spring channel dynamics (if any)	Spring-dominated stream	Little external flow impact	WS95	
	Intermediate stream	Spring and runoff channel morphologies	—	
	Runoff-dominated stream	Dominated by external flow impacts	WS95	
Flow-forcing mechanisms	Gravity-driven springs	Depression, contact, fracture, or tubular springs	M23	

	Increased pressure due to gravity-driven head pressure differential	Artesian springs	M23	
	Geothermal springs	Springs associated with volcanism	M23	
	Springs due to pressure produced by other forces	"Coke-bottle" springs and springs associated with gas release in deep-seated fractures	M23 So4	
	Springs due to pressure produced by anthropogenic forces	Anthropogenic artesian or geyser systems (e.g., hot springs associated with Hoover Dam, Arizona-Nevada)	—	
Flow Characteristics				
Persistence	Neorefugium	Holocene (<12,000 yr old), zero endemic species	N99	
	Paleorefugium	Pleistocene or older (≥12,000 yr old), 1 or more endemic species	N99	
	Paleospring	Pleistocene but not apparent recent flow, travertine or other paleo-flow indicators	—	
Flow consistency	Perennial	Continuous flow	M23	
	Intermittent—regular	Regular—flow occurs regularly on hourly or daily basis (e.g., some geysers), seasonally, annually, or interannually	M23	
	Intermittent—erratic	Flow occurs only on an erratic basis, can be noted with vegetative indicators	—	
	Intermittent—dry	No flow at all times of measurement	—	
Flow rate (mean)	Unmeasurable	No discernable flow to measure	—	
	First	<0.12 gpm (<10 ml/s)	M23*	
	Second	0.12–1.0 gpm (10–100 ml/s)	M23*	
	Third	1.0–10 gpm (0.10–1.0 L/s)	M23*	
	Fourth	10–100 gpm (1.0–10 L/s)	M23*	
	Fifth	100–448.8 gpm (10–100 L/s)	M23*	
	Sixth	448.8–4,488 gpm (0.10–1.0 m³/s)	M23*	
	Seventh	4,488–44,880 gpm (1.0–10 m³/s)	M23*	
	Eighth	>44,880 gpm (>10 m³/s)	M23*	

Flow variability $(DVR = Q_{10\%}/Q_{90\%})$	Steady (extraordinarily balanced)	1.0–2.5	M23 N71 AW94
	Moderately (well) balanced	2.6–5.0	M23 N71 AW94
	Balanced	5.1–7.5	M23 N71 AW94
	Moderately unbalanced	7.6–10.0	M23 N71 AW94
	Highly unsteady (extraordinarily unbalanced)	>10.0	M23 N71 AW94
	Ephemeral	Infinite	—
Water Quality — Temperature			
	Cold	Below mean annual ambient temperature	AW94
	Normal	Within 12.2°C of the mean ambient temperature	AW94
	Geothermal — warm	>12.2°C warmer than mean annual ambient temperature but <37.8°C	AW94
	Hot	Significantly warmer than mean annual ambient temperature, 37.8°C–100°C	AW94
	Superheated (usually pressurized)	>100°C	—
	Ambient temperature	Taken at time of spring water temperature measurement; temperature at time of measurement varies with daily mean temperature	—
Water Quality — Geochemistry			
Dominant cation type	Magnesium type	Magnesium (Mg^{2+}) is dominant cation	B66*
	Calcium type	Calcium (Ca^{2+}) is dominant cation	B66*
	Sodium type	Sodium (Na^+) is dominant cation	B66*
	No dominant type	No dominant cation	B66*
Dominant anion type	Sulfate type	Sulfate (SO_4^{2-}) is dominant anion	B66*
	Bicarbonate type	Carbonate (CO_3^-) and bicarbonate (HCO_3^-) are dominant anions	B66*

	Chloride type	Chloride (Cl⁻) is dominant anion	B66*	
	No dominant type	No dominant anion	B66*	
Minor constituents	List of minor constituents	E.g., borate, iron	M23 C24	
Pollution indicators	Polluted— mineral	E.g., selenium, fecal coliform	—	
	Polluted— biological	E.g., fecal coliform	—	
	Polluted—human	E.g., debris	—	
	Polluted— multiple	Combination of above indicators	—	
Tracers	List of tracers	E.g., stable isotopes, radioactive isotopes, rare-earth elements	K96	
Alkalinity	List value	Concentration of equivalent $CaCO_3$	C24 FL86	
Total dissolved solids (or specific conductance)	Hyperfresh	0–100 mg/L TDS	—	
	Fresh	100–1,000 mg/L TDS	F94	
	Brackish	1,000–10,000 mg/L TDS	F94	
	Saline	10,000–100,000 mg/L TDS	F94	
	Brine	>100,000 mg/L TDS	F94	
pH	Acidic	pH < 5.0	—	
	Moderately acidic	5.0 < pH < 6.0	—	
	Neutral	6.0 < pH < 8.0	—	
	Moderately basic	8.0 < pH < 10.0	—	
	Basic	pH > 10.0	—	
Nutrient concentra-tions	Low nitrate	NO_3-N < 0.3 mg/L	—	
	Moderate nitrate	$0.3 < NO_3$-N < 5 mg/L	—	
	High nitrate	NO_3-N > 5.0 mg/L	—	
	Low phosphate	PO_4^{2-} < 50 μg/L	—	
	Moderate phosphate	50 μg/L < PO_4^{2-} < 500 μg/L	—	
	High phosphate	PO_4^{2-} > 500 μg/L	—	
Habitats—Climate Variables				
Mean annual air temperature	Pergelic	<0°C	B91	
	Frigid	0°C–8°C	B91	
	Mesic	8°C–15°C	B91	
	Thermic	15°C–22°C	B91	
	Hyperthermic	>22°C	B91	
Air temperature seasonality	Nonseasonal	Seasonality index <2	B91	
	Weakly seasonal	Seasonality index 2–5	B91	
	Moderately seasonal	Seasonality index 5–15	B91	
	Strongly seasonal	Seasonality index >15	B91	

Mean annual precipitation	Extremely arid	<50 mm	B91	
	Arid	50–250 mm	B91	
	Semiarid	250–500 mm	B91	
	Semihumid	500–1,000 mm	B91	
	Humid	1,000–2,000 mm	B91	
	Extremely humid	>2,000 mm	B91	
Precipitation seasonality	Nonseasonal	Seasonality index 1–1.6	B91	
	Weakly seasonal	Seasonality index 1.6–2.5	B91	
	Moderately seasonal	Seasonality index 2.5–10	B91	
	Strongly seasonal	Seasonality index >10	B91	
Growing season length	Short	Number of growing degree days <2000	—	
	Medium	number of degree days 2000–6000	—	
	Long	Number of degree days >6000	—	
Habitats—Biological Variables				
Surrounding ecosystem(s)	Terrestrial	Barren lands	—	
		Grasslands	—	
		Herb lands	—	
		Shrublands	—	
		Woodlands	—	
		Forest	—	
	Freshwater	Lentic	—	
		Lotic	—	
	Marine	Euphotic (shallow)—note substrate (e.g., silt, sand, coral)	—	
		Aphotic (deep)	—	
	Urban/suburban/rural	Note human development	—	
Biogeographic isolation	Nearest spring very near	<1–100 m between springs	—	
	Nearest spring nearby	100 m–1 km between springs	—	
	Moderately isolated	1–10 km between springs	—	
	Isolated	10–100 km between springs	—	
	Highly isolated	>100 km between springs	—	
Habitat size (associated aquatic, wetland, and riparian areas)	Extremely small	<2 m²	—	
	Very small	2–10 m²	—	
	Small	10–100 m²	—	
	Medium-small	100–1,000 m²	—	
	Medium-large	0.1–1.0 ha	—	
	Large	1–10 ha	—	
	Very large	10–100 ha	—	
	Extremely large	>100 ha	—	

Microhabitats	Cave	Permanently dark zone	—	
		Twilight zone	—	
		Entrance	—	
	Wet wall	Wet, seeping, or dry wall(s)	—	
	Madicolous	Falling or fast-flowing stream water	—	
	Hyporheic	Habitat beneath the floor of the stream	—	
	Open-water pool	Mud-, ooze-, sand-, gravel-, boulder-, or bedrock-floored pond	—	
	Spring stream	Fine-grained (sand or silt) floor	—	
		Gravel floor (note embeddedness of gravels)	—	
		Cobble-boulder floor	—	
		Bedrock floor	—	
	Wet meadow	Ciénega—low-slope wetlands	—	
		High-slope wetlands	—	
	Riparian	Note area by vegetation cover type	—	
	Spray zone	Areas watered by spray from waterfalls	—	
	Barren rock	Cliffs, slopes, or relatively flat	—	
	Microhabitat diversity	Shannon diversity index calculated using proportion of each micro-habitat area: $H' = -\sum (p_i * \log p_i)$	S48	
Biota	Plant species richness	Species observed, with percent cover of each species in each major vegetation patch by stratum—ground cover, shrub, mid-canopy/woodland, and tall canopy; note any sensitive species	—	
	Vegetation diversity	Shannon diversity index, using percent cover for each stratum in each patch: $H' = -\sum (p_i * \log p_i)$	S48	
	Invertebrate species richness	Number and species observed; note any sensitive species	—	
	Invertebrate diversity	Shannon diversity index, using quantitative measures of diversity for aquatic and terrestrial taxa separately: $H' = -\sum (p_i * \log p_i)$	S48	
	Animal species richness	Number and species observed; note any sensitive species	—	
	Animal diversity	Shannon diversity index, using quantitative measures of diversity for aquatic and terrestrial taxa separately: $H' = -\sum (p_i * \log p_i)$	S48	

Management Aspects				
Land use and man-agement	Landownership	Federal, state, local, private	—	
	Legal authorities	Applicable laws, including water rights and environmental protec-tion laws	—	
	Land-use history	Land use history should be refer-enced or compiled	W79	
	Prehistoric/early historic use/ modification	Document any use of spring by prehistoric or early historic cultures	—	
	Primary use	Culinary, livestock watering, recre-ation, religious, wildlife, conserva-tion, research, other	—	
	Secondary use	Culinary, livestock watering, recre-ation, religious, wildlife, conserva-tion, research, other	—	
	Other uses	Culinary, livestock watering, recre-ation, religious, wildlife, conserva-tion, research, other	—	
	Groundwater modification	Extraction, augmentation, pollution	—	
	Emergent flow regulation	Dewatering, abstraction, diversion, pollution	—	
	Microhabitat modification	Piping, fencing, tanks, ponds, spring house, etc.	—	
	Surrounding eco-system health	Condition of surrounding ecosys-tem	—	
Information management	Metadata criteria	Data management criteria and authorities (federal, state, local, private)	—	
		Location and forms of original data	—	
		Data quality control protocols	—	

Sources: To make the checklist more usable in the field, the references have been abbreviated as follows: AW 94 = Alfaro and Wallace 1994; B66 = Back 1966; B91 = Bull 1991; C24 = Clarke 1924; F94 = Fetter 1994; FL86 = Furtak and Langguth 1986; H70 = Hynes 1970; K96 = Kreamer et al. 1996; M23 = Mein-zer 1923; N71 = Netopil 1971; N99 = Nekola 1999; S48 = Shannon 1948; So4 = Shipton et al. 2004; W79 = White 1979; WS95 = Whiting and Stamm 1995. An asterisk following an abbreviated reference code indicates that the original data have been modified. A dash indicates that the data derive from the current study.

Chapter 5

Quaternary Cauldron Springs as Paleoecological Archives

C. VANCE HAYNES JR.

Cauldron springs are major attractants for wildlife and people seeking prey as well as water and provide remarkable archives of paleoecological and archaeological information reflecting regional climate change. Fossil-bearing springs have played a significant role in paleontological and Paleo-indian investigations in North America since the excavations by Charles Wilson Peale in New York in 1801. In 1840, Albert C. Koch reported the discovery of artifacts with mastodon remains in a spring in the Missouri Ozarks. Investigations in the 1970s proved the association invalid but demonstrated an age of about 30,000 years for a composite mastodon skeleton that Koch misassembled and called the "Missouri Leviathan." Megafauna bones in fossil springs ten miles north of Las Vegas, Nevada, once thought to be 28,000 years old and associated with artifacts, turned out to be more than 40,000 years old and without any evidence of human presence. However, the Tule Springs project provided a wealth of paleoecological information on the late Quaternary of the Las Vegas Valley. The indisputable association of humans with fossil springs began with the discovery in a gravel pit in 1935 of two mammoth skeletons and Clovis points in spring deposits near Clovis, New Mexico. Since then, other spring sites with Paleoindian evidence have been found. The concentration of mammoths and other megafaunal elements at springs during dry periods indicates that drought may have been a factor in Pleistocene extinctions, in addition to predation by Clovis hunters and rapid climate change.

Description of Cauldron Springs

The cauldron (or limnocrene) springs discussed in this section were selected for the faunal and floral paleoecological data they have provided in geochronological contexts. The Spanish term *ojo*, or eye (fig. 5.1a), is an apt description of the surface expression of springs where groundwater rises to the surface through a narrow neck to feed a funnel-shaped bowl or cauldron (fig. 5.2a) filled with crystal-clear water (the pupil) (fig 5.1b) and surrounded by a raised ring of green vegetation (the iris) through which water seeps away from the mound in all directions. At the throat, where the narrow conduit or feeder joins the bowl, there is a significant reduction in velocity as water enters the bowl. Here, in many examples, roiling sand can be seen at the bottom of active springs. With reduction in spring discharge, usually climate induced, silt and clay may settle out, cover the sand-filled throat, and eventually fill the bowl (fig. 5.2b) as a smaller vent directs the reduced discharge to the surface. With further reduction in artesian pressure, the vegetation mat may completely cover the cauldron and create a "bouncy" or "quaking" bog (fig. 5.2c). Additional desiccation causes compaction and may leave little more than a shallow depression that holds water only after a rain. Eventually the extinct spring may be eroded and buried to become a fossil spring (fig. 5.2d).

Bryan's (1919) classification of springs is based mainly upon hydrologic source and geologic structure and does not include spring morphology. Therefore, cauldron springs do not occur in his classification but could derive from several of his categories, such as fissure springs or fracture springs.

Active cauldron springs can occur in two forms: (1) as open pools in relatively flat terrain (also referred to as limnocrene springs) or (2) as mounds or knolls (fig. 5.1c) with a pool at the top (figs. 5.1a, 5.1b, and 5.1d) that varies in size (also called mound-forming springs) (Meinzer 1911). Mound springs develop from pools as encompassing vegetation traps wind-borne sediment to form an aggrading substrate for more vegetation growth and, in turn, eolian sand entrapment (fig. 5.3). In this way, the upward growth of the mound is limited by the piezometric surface. Some mounds display leakage at the base. If the artesian pressure declines, the mound can dry out and die (fig. 5.4a). Eventually it may be reduced by deflation (Meinzer and Hare 1915). In some areas, crusts of tufa (Weed 1889) have formed around springs (fig. 5.4a). Also, wet meadows are commonly spring fed (fig 5.4b). Many of these forms of springs in southern Nevada, observed during one of the Wheeler expeditions (1869–79), were described and illustrated by Lieu-

a

b

Figure 5.1. Selected views of active springs. *a*, At Corn Creek Springs in Nevada, the author is showing a small cauldron spring to R. E. Taylor during the 1962–63 Tule Springs interdisciplinary expedition (photo by William Belknap, 1963). *b*, Crystal Spring in Ash Meadows in Nevada is a cauldron approximately 12 meters in diameter that is filled with clear blue water and is host to desert pupfish (*Cyprinodon* sp.). From the throat of the feeder, seen right of center, water emerges with enough velocity to keep it clear of sand and form a slight bulge on the water surface. Satellite feeders contain well-sorted roiling sand (photo by Jay

c

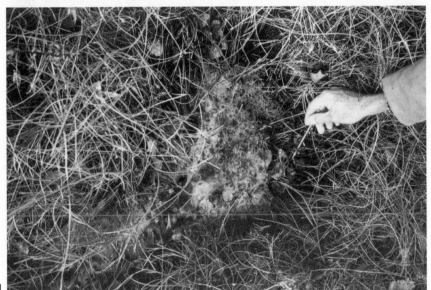

d

Quade, 2000). *c*, At Tule Springs State Park (now called Floyd Lamb State Park) in Nevada in 1962, springs of the Gilcrease alignment, probably along a fault, displayed several stages of activity, from active to barely active to dead to eroded. This mound is essentially inactive but retains enough moisture to support a large cottonwood tree (photo by the author, 1962). *d*, At the top of the barely active spring mound at Corn Creek, Nevada, a small pool is choked with algae and surrounded by grass (photo by William Belknap, 1963).

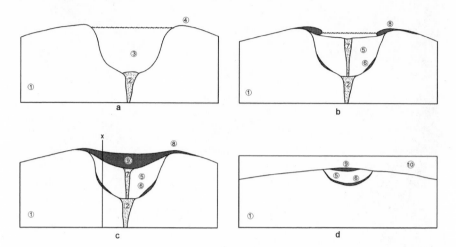

Figure 5.2. Stratigraphic cross sections of four stages of cauldron spring activity, decline, and death. *a*, High artesian pressure forces water through a weakness in the host rock or mudstone (1). Ascending water erodes a conduit or feeder (2) and initially winnows out clay and silt until water filling the cauldron or bowl is clear (3), as in figure 5.1b. *b*, As artesian pressure declines, clay and silt settle out in the bowl (5) and bury waterlogged organic matter (6). With compaction, a smaller subconduit (7) forms to feed a shallow pond surrounded by an organic mat (8) that may thicken by entrapment of wind-borne sediment. The spring of figure 5.1a is at this stage. *c*, With further decline in artesian pressure, spring discharge ceases. A thickened vegetation mat (9) is essentially a peat deposit preserving fossil pollen, insects, and plant parts. Sediments filling the bowl commonly preserve an abundance of shells of a molluscan fauna. All of the above have paleoecological value. *d*, A dead spring may eventually be eroded and buried by younger sediment (10). If scientific excavation produces a cross section at X of figure 5.2c, the black oxidized lenses may be mistaken for lenses of charcoal of an artificial fire hearth (see figure 5.7).

tenant David A. Lyle (1878). The zoology of these and others were the basis for the classical paleoecological study by Hubbs and Miller (1948).

Through time, cauldron springs have been major attractions for wildlife and people seeking prey as well as water. They also became death traps for animals. Clayey deposits commonly preserve fossil pollen as well as invertebrate shells, and the vegetation mat or peat deposit preserves plant macrofossils and beetle fauna as well as pollen. The plant remains, especially seeds, provide excellent opportunities for radiocarbon dating indi-

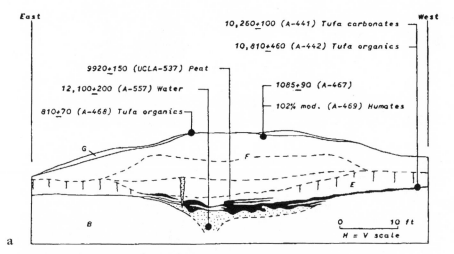

East   West

10,260±100 (A-441) Tufa carbonates ─

10,810±460 (A-442) Tufa organics ─

9920±150 (UCLA-537) Peat ─

12,100±200 (A-557) Water ─   ─ 1085±90 (A-467)

810±70 (A-468) Tufa organics ─   ─ 102% mod. (A-469) Humates

G

F ─ ─ ─ ─ ─

E

B   0   10 ft
                                                 H = V scale

a

b

Figure 5.3. Gilcrease Spring mound no. 4 is one of several aligned along a
probable fault through Tule Springs State Park north of Las Vegas, Nevada.
a, The stratigraphic profile of a trench wall through the mound revealed an
organic black mat that radiocarbon ages indicate formed during the Younger
Dryas cold period (from Haynes et al. 1966). *b,* The mound grew by accretion of
eolian sand and phreatogenic calcium carbonate as a small subfeeder that supplied
adequate moisture to support vegetation. The spring became inactive in 1958 after
deep wells in the Las Vegas Valley drew down the water table (Haynes et al. 1966;
Haynes 1967). A vertical series of tags spaced every ten centimeters marks the
pollen sampling of Mehringer (1967). The late Philip Jenny stands in for scale.
(Photo by the author)

a

b

Figure 5.4. Features of dead springs and artists' renditions from paleoecological findings. *a*, Tufa surrounds a small vent atop a dead spring mound in the southeast quadrant of the intersection of Smoke Ranch Road and the Tonopah highway in northwestern Las Vegas, Nevada (photo by the author, 1968). *b*, Excavators removing fossil bones and teeth of extinct animals from the sand-filled conduit of a buried cauldron spring at the Murray Springs Clovis site in southeastern Arizona. This subaqueous spring fed a late glacial-age lake when it was most active, 22,000 BP to 17,000 BP. It had all but dried up by the time Clovis folk arrived 13,000 years ago, when they killed at least one mammoth and twelve bison (*Bison antiquus*) at a nearby water hole. Megafaunal extinction was complete before Younger Dryas cooling caused groundwater to rise and form the black

c

d

wet-meadow soil ("black mat") from 12,900 to 11,200 years ago (photo by the author, 1971). *c*, An Illinois State Museum painting by R. G. Larsen depicts the Breshears Valley in the Missouri Ozarks as it may have appeared 16,000 years ago. The vegetation and fauna are based on scientific investigations referenced in the text (courtesy of the Illinois State Museum). *d*, A painting by National Geographic artist Jay Matternes depicts Tule Springs in Nevada as it may have appeared about 15,000 years ago. The 1962–63 Tule Springs expedition failed to confirm human presence there before 13,000 years ago. However, this depiction is based on the wealth of paleoecological data recovered by the scientific team. (Courtesy of the National Geographic Society) vidual species. The result is that fossil springs provide remarkable archives of paleoecological and archaeological information reflecting regional climate change. Provided that there has been no significant change in catchment, spring discharge can be a quantitative, as well as qualitative, gauge of local rainfall and groundwater recharge (Holmes et al. 1981). Recent studies of organic deposits, "black mats," associated with springs in southern Nevada have provided stable isotope data and radiocarbon ages indicating that most formed during the Younger Dryas cold paleoclimate period of the Pleistocene-Holocene transition (Quade et al. 1998).

Fossil-bearing springs have played a significant role in paleontological investigations in North America since the middle eighteenth century (Silverberg 1970) and the mastodon excavations of Charles Wilson Peale in New York in 1801. President Thomas Jefferson was so intrigued by these finds, as well as earlier ones from Big Bone Lick, Kentucky, that he advised Lewis and Clark to check on reports of living examples in the western territory that they were about to explore in 1803 (Schultz et al. 1963). Major contributions to vertebrate paleontology have been made by excavations of bones from Tertiary and Quaternary fossil springs in Meade County, Kansas (Hibbard and Riggs 1949).

In this overview, specific spring localities of the continental United States have been selected as examples on the basis of their documented paleoecological contributions, preserved geomorphology and stratigraphy, associated archaeology, and my experiences in working out the geochronology of some of those in the presentation that follows. (For coverage of some Old World spring deposits, see Ashley 2001.)

Specific Examples

Koch Spring in Missouri

In 1840 Albert C. Koch reported on the discovery of artifacts with mastodon remains in a spring by the Pomme de Terre River in the Ozarks of Missouri (McMillan 1976). Koch, obsessed with collecting large animal skeletons, particularly mastodon, went to great lengths to follow up on reports of large bones that he might acquire for his small museum in St. Louis, Missouri. In 1838 recent immigrants to the Ozark Plateaus in southwestern

Missouri had reported finding large bones in a spring pond by the Pomme de Terre River. Two years later, with local help, Koch dug out many bones of what we know today as the American mastodon (*Mammut americanum*). However, using bones of at least three mastodons, Koch assembled a monster skeleton that he named the "Missouri Leviathan." He took it on tour to raise more money for his search for more exotic bones. In Philadelphia, scientists at the Academy of Sciences criticized his mounting and declined to purchase the bones. Koch, while on tour in Europe, eventually sold the Leviathan to the British Museum, where the skeleton was reassembled as the best-preserved skeleton of *M. americanum*. It remains there on display.

In 1965 R. B. McMillan, director of the Illinois State Museum, rediscovered the location of Koch's spring while conducting archaeological investigations in the Breshears Valley (Wood and McMillan 1976). In 1971 excavations around the spring revealed numerous stone artifacts of the Archaic period, including projectile points similar to the one Koch had found under mastodon bones (McMillan 1976). Excavation of the spring revealed a platform made of walnut and oak fence rails, from which the Koch crew in 1840 had groped for bones in the mud. In addition to finding fragments of bone and tusk, the 1971 crew was able to reconstruct the stratigraphy and radiocarbon-date the peat deposit as being 32,000 years old (Haynes et al. 1983; Haynes 1985). Clearly, the artifacts that Koch found in 1840 had entered the spring through the activity of Archaic people some 28,000 years after the death of the mastodon.

Four other mastodon- and mammoth-bearing springs in the area produced a remarkably complete geologic record of the late Pleistocene of the Ozark Highlands (Brakenridge 1981; Haynes 1985; Saunders 1988). Jones Spring contained bones of both mammoths and mastodons. It revealed two episodes of discharge: one covered by a peat of interglacial Sangamon age and one of peat of mid-Wisconsin glacial age (fig. 5.5). Nearby Trolinger Spring provided a wealth of paleoecological data for the middle of the late glacial Wisconsin period, and Boney Spring (fig. 5.6) about two miles northwest completed the late Wisconsin and Holocene records (Saunders 1988). The geochronological investigations indicated that periods of maximum spring discharge correlated with glacial stades and periods of reduced discharge and peat formation correlated with interstades (Haynes 1985). The paleoecological reconstruction shown in a painting of the Breshears Valley of

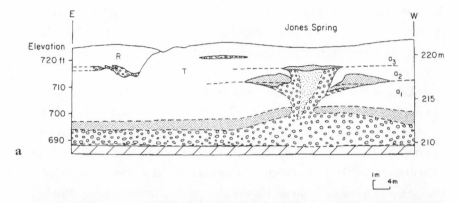

Figure 5.5. Jones Spring, Pomme de Terre Valley, Missouri Ozarks. *a*,
Stratigraphic cross section shows a last interglacial (Sangamon) spring-laid peat
lens (a_2) penetrated by an early Wisconsinan spring conduit topped by a mid-
Wisconsinan peat lens (a_3). Gravel swept up from a buried channel indicates an
intense initial discharge followed by sand during reduced discharge (from Haynes
1985; copyright Geological Society of America). *b*, Excavation of the conduit filled
with well-sorted medium sand revealed concentric rings of the organic matter.
A mastodon tusk occurs in the wall behind the bucket in the left foreground.
(Photo by the author, 1975)

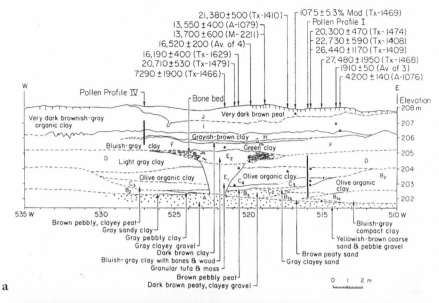

21,380±500 (Tx-1410)
13,550±400 (A-1079)
13,700±600 (M-2211)
16,520±200 (Av. of 4)
16,190±400 (Tx-1629)
20,710±530 (Tx-1479)
7290±1900 (Tx-1466)

107.5±5.3% Mod. (Tx-1469)
Pollen Profile I
20,300±470 (Tx-1474)
22,730±590 (Tx-1408)
26,440±1170 (Tx-1409)
27,480±1950 (Tx-1468)
1910±50 (Av of 3)
4200±140 (A-1076)

W ———— E

Pollen Profile IV ┐   ┌ Bone bed

Elevation
208 m

Very dark brownish-gray
organic clay   Very dark brown peat   207

J
I   206
Grayish-brown clay
Bluish-gray clay   F   G   H   Green clay   F   205
D   E₂
Light gray clay   204
E₁   C₄   B₂
Olive organic clay   Olive organic clay   C₃   203
B₃   Olive organic
C₃   C₂   B₁   clay   B₁ₐ
B₂   A   B₁ₐ   202

535 W   530   525   520   515   510 W

Brown pebbly, clayey peat
Gray sandy clay
Gray pebbly clay
Gray clayey gravel
Dark brown clay
Bluish-gray clay with bones & wood
Granular tufa & moss
Brown pebbly peat
Dark brown peaty, clayey gravel

Bluish-gray
compact clay
Yellowish-brown coarse
sand & pebble gravel
Brown peaty sand
Gray clayey sand

0   1   2 m

a

b

Figure 5.6. Boney Spring, Pomme de Terre Valley, Missouri Ozarks. *a*, A
stratigraphic cross section showing a Farmdalian-age peat (C₃) penetrated by a
late Wisconsinan-age sand feeder (E₁) surrounded by a ring of mastodon bones
(in stratum E₂) overlain by Holocene clay (F) (from Haynes 1985; copyright
Geological Society of America). *b*, The water table was drawn down by round-the-
clock pumping to allow excavators to map and recover bones of mastodons ranging
in age from mature adults to babies. (Photo by the author, 1971)

16,000 years ago (fig. 5.4c) was based on the palynological investigations of King (1973) and the vertebrate faunal studies of Saunders (1977).

Tule Springs in Nevada

In 1962 a radiocarbon date of approximately 28,000 years ago for human occupation of the Las Vegas Valley, Nevada, provided the impetus for what is probably the most ambitious earth-moving archaeological project ever conducted in North America (Shutler 1967). At the Tule Springs site, what had been interpreted as a fire hearth of human construction turned out to be a hearth-shaped lens of naturally oxidized carbonized plant remains in a fossil spring deposit more than 40,000 years old (Haynes 1967). The initial off-center position of the excavation wall (fig. 5.2c, see cross section at X) is what caused the hearth-shaped appearance (fig. 5.2d). In subsequent excavations, the wall went through a central sand-filled feeder (fig. 5.7b). What were thought to be bone tools were more likely naturally fragmented bones rounded and polished by spring action, not human use. And the one stone artifact found in situ was actually in a colluvial slope-wash deposit of Holocene age.

The National Geographic Society felt the project unworthy of coverage in their magazine once the earliest human presence could not be demonstrated. However, they overlooked the fact that the Tule Springs expedition provided a quantum leap in knowledge of the paleoclimate of southern Nevada. At that time, it was the most significant paleoecological information ever gleaned from the stratigraphic record of the late Quaternary of western North America (Wormington and Ellis 1967). It became the subject of three Ph.D. dissertations. The vegetation changes of the past 30,000 years were reconstructed from fossil pollen (Mehringer 1967); the mammalian fauna for this period was ascertained from bone concentrations in spring conduits (Brooks 1967; Mawby 1967), as was the molluscan fauna (Taylor 1967); and the accurate reconstruction and radiocarbon dating of the stratigraphic framework put it all into perspective (Haynes 1967). The largest bulldozer in the world at that time made a geologic trench that extended across 3,700 feet (1,130 meters) of the Las Vegas Valley and provided a remarkable record of the late Quaternary. Extinction of the Pleistocene megafauna occurred about 13,000 years ago (based on modern calibration of radiocarbon ages), and the

a

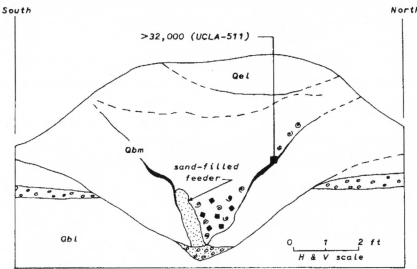

South   North

\>32,000 (UCLA-511)

Qel

Qbm

sand-filled
feeder

Qbl

0   1   2 ft

H & V scale

b

Figure 5.7. Buried springs at the Tule Springs archaeological site north of Las Vegas, Nevada. *a*, Locality 2 is an extinct cauldron spring, in excess of 40,000 years old, that was eroded, buried, and re-exposed by erosion before excavations exposed the internal stratigraphy shown in this cross section. Carbonized wood fragments (black squares) and black organic lenses were originally interpreted as artificial fire hearths before chemical analysis showed them to be natural features. Snail shells (spirals) are unburned, further supporting a natural origin. From Haynes 1967 (photo by the author, 1971). *b*, The stratigraphic profile of figure 5.7a is the back wall of this excavation. Another conduit in the center of the photo is filled with a jumble of bones of large animals that fell into the spring over an unknown period of time. As each animal attempted to extract itself, it crushed bones of previous victims, producing splintered bone fragments abraded by roiling sand. Some fragments were originally misinterpreted as artifacts.

first human appeared at Tule Springs about two centuries later. National Geographic did permit the free use of the painting that artist Jay Matternes had done for the ill-fated magazine article (fig. 5.4d). Although it incorrectly portrays Paleoindians attacking a camel at one of the cauldron springs about 15,000 years ago, it accurately shows the vegetation and fauna of the terminal Pleistocene of the Las Vegas Valley based on the project's findings.

The Mammoth Site of Hot Springs in South Dakota

The largest and most spectacular (fig. 5.8a) discovery of mammoth skeletons (perhaps a hundred or more) in fossil spring deposits is that of the Mammoth site of Hot Springs, South Dakota (Agenbroad 1984). Whereas the giant cauldron is a sinkhole caused by solution, collapse, and brecciation of the Permian Minnelusa Formation, the floor of the collapse became a spring-fed oval pond, measuring forty-five meters in the long axis, that attracted mammoths and other Pleistocene mammals at least 26,000 years ago (Laury 1980). The sand-filled conduit (fig. 5.8b) located at the northeastern edge of the bowl fed a warm-water lake in which the skeletal remains sank and were preserved by the laminated lakebeds until water table lowering caused by stream entrenchment brought an end to spring discharge, and eventually the lake dried up (Laury 1990). Study of ostracodes, mollusks, fish bones, and aquatic pollen are consistent with a water temperature of approximately 35°C (Mead et al. 1990). Today the site is a major ongoing research facility with a remarkable display of mammoth skeletons that attracts thousands of visitors each year (fig. 5.8a).

The Clovis Type Site in Blackwater Draw in New Mexico

The indisputable association of humans with fossil springs began in a gravel pit in 1936 with the discovery of two mammoth skeletons, Clovis projectile points, and long, slender, pointed bone rods in spring-laid sand near Clovis, New Mexico (Cotter 1937, 1938). In 1962 the skeletons of five more mammoths with Clovis points and other stone artifacts were discovered (Warnica 1966) near 13,000-year-old spring vents (Haynes and Agogino 1966; Hester 1972; Green 1992; Haynes et al. 1992) at the north wall of the gravel pit located along Blackwater Draw, an extinct tributary of the Brazos

a

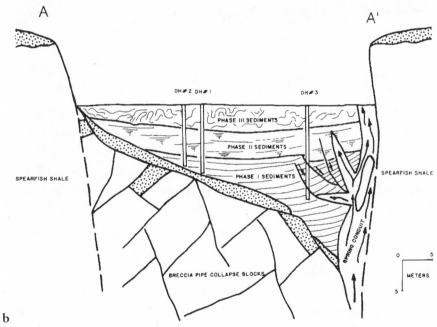

b

Figure 5.8. The Mammoth Site of Hot Springs, South Dakota. *a,* The spectacular occurrence of perhaps one hundred or more mammoth skeletons in spring-fed lake deposits provides an on-site laboratory for ongoing scientific investigations as well as an important attraction for public education (photo provided by the Mammoth Site of Hot Springs, South Dakota). *b,* This geologic cross section of the sinkhole at the Mammoth Site shows the stratigraphic relationship of the spring conduit to the mammoth-bearing lakebeds (from Agenbroad 1984; copyright University of Arizona Press).

River. Many highly polished stone artifacts were found within the spring vents, raising the question of accidental entry or deliberate deposition, perhaps ceremoniously (Haynes and Agogino 1966). This, the type site for the Clovis projectile point (Sellards 1952), is the most important Paleoindian site ever found because of a relatively complete stratigraphic framework (Haynes 1995) containing several post-Clovis Paleoindian levels as well as early Archaic deposits (Boldurian and Cotter 1999). The site became a focal point for the paleoecological studies of the late Quaternary of the southern High Plains (Wendorf and Hester 1975). Unfortunately, the most important parts of the Clovis site were destroyed by more than a decade of gravel mining, but much remains to be seen and explored (Stanford et al. 1990), and today an on-site museum vividly and accurately displays the archaeological horizons and artifacts as well as bones of the megafauna that became extinct 13,000 years ago.

The Murray Springs Clovis Site in Arizona

Arizona's upper San Pedro Valley contains six stratified Clovis sites. From south to north, these are Naco, Leikem, Navarrete, Lehner, Escapule, and Murray Springs. The Murray Springs Clovis site in Curry Draw (Haynes 1987), a tributary of the San Pedro River of Arizona, revealed Clovis points and other tools among the bones of two mammoths and a dozen bison (Hemmings 1970) along a spring-fed fossil stream that had dried up during a brief period of drought (Haynes 1991). Mammoths had dug a water hole in the dry streambed at the time of their discovery by Clovis folk who attacked and killed one of them 13,000 years ago. The most spectacular artifact is a broken shaft wrench made of mammoth bone (Haynes and Hemmings 1968). It was found among mammoth tracks near the water hole, where it probably had been stepped on by a mammoth.

The killing of at least twelve bison probably occurred during another hunt soon before or after the mammoth hunt. In any case, the processing of the many kills required a temporary encampment, the discovery of which yielded many Clovis stone scrapers and knives as well as broken Clovis point bases (Haynes and Huckell 2007).

Murray Springs, along with the five other Clovis mammoth-kill sites, were all preserved relatively intact by the deposition 12,900 years ago of a black organic clay that preserved the bones and accumulated along valley floors as the water table rose to form spring-fed ponds and wet meadows between 12,900 and 11,200 years ago. The base of the "black mat" marks the abrupt termination of the Pleistocene megafauna in the San Pedro Valley (fig. 5.4b). This black mat covers the skeletons of at least nine other mammoths in the valley, mostly young adults that showed no evidence of human involvement in their death. The indicated rise of the water table appears to be more a result of reduced evaporation attributable to cold temperatures than to increased precipitation, as low-intensity rains are more effective than high-intensity rains in recharging water tables. The black mat of the San Pedro Valley is the same age as a dark gray clay in northern Europe that represents a very sudden return to glacial climate during the Younger Dryas chronozone that followed the Allerød warm period (Mangerud et al. 1974).

Another important find at Murray Springs was the discovery of four pre-Clovis extinct spring vents filled with hundreds of bones and teeth of extinct Pleistocene fauna. What had been pool springs before desiccation between 15,000 and 13,000 years ago were covered by the Younger Dryas–age black mat before differential compaction converted the conduit sand to a mound, inverting the topography (fig. 5.4b). From about 25,000 to 22,000 years ago, these springs supported an abundant vertebrate fauna (Rancholabrean land-mammal age) around ponds and marshy ground long before people had arrived in the San Pedro Valley. After 22,000 years ago, the springs became submerged as the lake they fed (represented by Coro marl in Haynes 1987), expanded up the valley. Erosion followed desiccation of the lake about 15,600 years ago. When Clovis folk entered the valley 13,000 years ago, they found soggy ground but no water where these springs had formerly been active. What appear to be stepping-stones had been placed along the north side of one of the spring conduits where people may have obtained water from seeps in the wet meadow. Future excavations in this area should include a search for Clovis wells similar to one found at the Clovis type site (Haynes et al. 1999).

The Hiscock Site in New York

In eastern North America, the Hiscock site near Batavia, New York, is another spring-fed pond deposit containing Paleoindian artifacts (Laub

a

Figure 5.9. The Hiscock Site near Batavia, New York. *a*, A view northward showing the spring-fed pool before draining and excavations (photo by Richard S. Laub). *b*, A schematic cross section showing the stratigraphic sequence of spring deposition. Bones of mastodon as well as other animals, including a Pleistocene condor (*Gymnogyps californianus*), occurred at the base of the peat deposit along with several reworked Clovis points. (From Muller and Calkin 1988; reproduced with permission of the Buffalo Museum of Science, Buffalo, N.Y.)

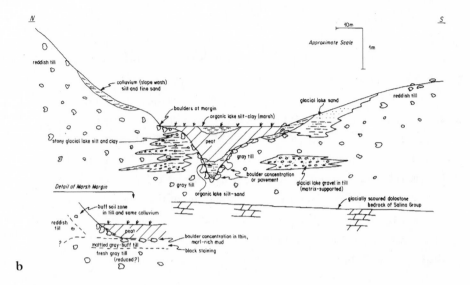

et al. 1988). A depression eroded into the latest Pleistocene till (fig. 5.9) contains a heterogeneous gravelly silt overlying glacial till and lake deposits, which (where they reach shallower depths) are capped by a cobble-to-boulder lag gravel (Muller and Calkin 1988). The gravelly silt contains bones of mastodon and other late Pleistocene animals (Laub et al. 1988; Steadman 1988), dating to about 13,000 years ago. Five Clovis points appear to have been reworked into tools (Gramly 1988; Laub 1994), and expedient tools of megafaunal bone were apparently used for butchering and hide processing (Laub 2000). Evidence of a depressed water table indicates that drought may have been a factor in the demise of mastodons at Hiscock (Laub and Haynes 1998) just as at Murray Springs and Blackwater Draw (see Holliday 2000 for a different paleoclimate interpretation).

Pleistocene Extinction

The spread of Clovis culture throughout North America occurred during a period of rapidly changing global climate at the end of the Pleistocene glaciation. At most Clovis sites, the climate changed from dry to wet to dry in less than five hundred years during the Bølling-Allerød warm periods. In post-Clovis Folsom time, the climate turned glacially cold for about a thousand years during the Younger Dryas period, when water tables rose to produce ponds and wet meadows at many Folsom sites located in topographic low areas (Haynes 1998). The Pleistocene megafauna became extinct at the end of the Allerød warm period and before the beginning of the Younger Dryas cold period.

The sudden extinction of many genera of Pleistocene fauna exactly at the onset of the Younger Dryas deep freeze invokes climate change as a significant factor. However, extinction also occurred during a time of depleted watering places and at the same time (Stafford 1990) as predation by Clovis folk (Martin 1967). Therefore, what may have depleted the fauna may not have been a single factor but a combination of three factors: (1) drought brought to an end by the sudden and dramatic change to glacial climate combined with (2) hunting pressure on weakened animals concentrated at the few remaining watering places that (3) may have become deeply frozen at the onset of the Younger Dryas.

Conclusions

More than a century of scientific investigation of prehistoric cauldron spring deposits in the United States has demonstrated that springs have been major attractants for wildlife seeking water and for people in search of sustenance. The organic sediments have preserved a variety of normally perishable indicators of past climates and ecologies. Such indicators include plant parts, pollen, molluscan fauna, and bones of mammalian fauna ranging in size from shrews to mammoths. Further, the organic preservation provides ample material for accurate radiocarbon age dating. Thus, spring deposits are natural archives of data that provide windows into the past.

In deposits 13,000 years old or younger, there are often artifacts indicating human interactions with fauna. Therefore, spring deposits are potentially important paleoecological archives that need to be scientifically tested and studied before their destruction by the advance of civilization. The Tule Springs locality in Nevada, for example, still contains a wealth of scientific data in late Quaternary deposits but is being encroached upon by development as the Bureau of Land Management makes the public lands north of Las Vegas available to private industry and housing development.

As indicated in this study, the geochronology and paleoecology from late Pleistocene cauldron spring deposits have made substantial contributions to our understanding of the transition from the Pleistocene to the Holocene. These investigations indicate that organic deposits can form under two climatic regimes. Peat deposits formed over the previously mentioned Missouri springs during interstadials as spring discharge waned or stopped altogether. By contrast, black organic mats or wet meadow soils in many regions formed during the Younger Dryas cold period as water tables rose and spring discharge increased or returned after a period of desiccation. The indicated drought may be one of three factors responsible for extinction of the Pleistocene megafauna at the end of the Allerød warm period — the other two being the appearance, a few centuries earlier, of Paleoindian hunters and the sudden onset of the Younger Dryas, whose deposits are devoid of megafaunal remains other than bison. Instead, Folsom bison hunting dominates the archaeological record of the Great Plains. Whereas no convincing evidence of pre-Clovis peoples in North America has been found in cauldron spring deposits, such deposits offer promising prospects for pre-Clovis geoarchaeology.

ACKNOWLEDGMENTS

This research was supported by various grants from the National Science Foundation, the National Geographic Society, and the University of Arizona Regents Professor Research Fund. Helpful comments on improving the manuscript were provided by Larry D. Agenbroad, Richard S. Laub, Robert L. Laury, R. Bruce McMillan, Vicky Meretsky, Jeffery J. Saunders, and Lawrence E. Stevens. Expert word processing was provided by Barbara Fregoso and technical editing by Carol Gifford, both of the Department of Anthropology, University of Arizona.

Chapter 6

The Extreme Environment, Trophic Structure, and Ecosystem Dynamics of a Large, Fishless Desert Spring

Montezuma Well, Arizona

DEAN W. BLINN

Springs provide a variety of aquatic habitats that frequently display relatively constant physical-chemical conditions. Some springs have extreme environmental conditions, including waters with low pH (<6), high water temperature (>35°C), and high concentrations of dissolved minerals (>3 g/L) and gases (>50 mg/L; Waring 1965; Van Everdingen 1991; O'Brien and Blinn 1999). As a result of such conditions, these springs may harbor organisms that have highly restricted distributions (i.e., endemics), especially those isolated in deserts that have fauna with weak dispersal abilities (Duffy 1993; Shepard 1993; Thomas et al. 1997, 1998; Govedich et al. 1999). Some endemic fauna change their behaviors to adjust to the extreme conditions in springs.

A robust literature exists on a diverse number of taxa that have undergone endemic adaptations in desert springs since the late Pleistocene. For example, the Devils Hole pupfish, *Cyprinodon diabolis* Wales, is native only to the small, cavernous spring system of Devils Hole, Nevada (Soltz and Naiman 1978; Deacon and Williams 1991). The isolated pupfish has a distinctive morphology that has been attributed to its separation from related pupfishes in the Ash Meadows/Death Valley region for nearly 20,000 years (Miller 1981). The Devils Hole pupfish is the smallest of all pupfish and lacks pelvic fins. Other endemic pupfish also occur in isolated desert springs (Soltz and Naiman 1978). In addition, many invertebrates native to isolated desert springs show endemic adaptations; these invertebrates include spring snails (Hydrobiidae; Hershler 1994), leeches (Govedich et al. 1998), elmid beetles (Shepard 1992), amphipods (Baldinger et al. 2000), and heteropterans (Polhemus and Polhemus 2002). Many endemic taxa in desert springs have become extinct or are in jeopardy of extinction because of the ever-increasing demands for water in desert regions (Soltz and Naiman 1978; Minckley and Deacon 1991; Shepard 1993; Wilson and Blinn 2007).

Despite their ubiquitous and abundant distribution throughout the world, springs have received little attention compared to other aquatic habitats (Williams and Danks 1991; Shepard 1993). Our present knowledge on the biodiversity of springs is fragmentary and incomplete at best, with information on energy flow in spring food webs even more scarce. Studies on the energetics of food webs in spring ecosystems include the classic works by Odum (1957), Teal (1957), and Tilly (1968), with a recent study on this topic as it pertains to Montezuma Well, Arizona (Runck and Blinn 1994). In 1982, the Scientific Committee of the Biological Survey of Canada initiated an attempt to study the biodiversity of springs across the Canadian provinces. This effort resulted in a special publication titled *Arthropods of Springs, with Particular Reference to Canada* (Williams and Danks 1991). Before the present volume, no similar initiative had been undertaken in the United States. Of even greater concern is the continued loss of these understudied habitats owing to increased municipal and agricultural development and changing climate caused by global warming.

The large, collapsed, travertine spring mound of Montezuma Well has received considerable attention on a regular basis over the past three decades. In fact, it may be the most continuously studied desert spring in the world, with an information base on archaeological, biological, geological, hydrological, and paleoecological endeavors. Information from Montezuma Well may provide a baseline for studies on other spring ecosystems in the southwestern United States and perhaps in other arid regions of the world. The intent of this chapter is to (1) synthesize our current knowledge on the physical-chemical and biological conditions of Montezuma Well in the hopes that this will inspire further research on the Well and on other spring ecosystems, and (2) review the importance of spring ecosystems in furthering our general knowledge on aquatic ecosystems.

A Description of Montezuma Well's Formation, History, and General Morphometry

The large circular mound within which Montezuma Well lies was formed by the slow accumulation of limestone precipitated from spring water discharged over several million years sometime between the Miocene and the Pleistocene in the Verde Formation of the Verde Valley, Arizona (Nations et al. 1981). In essence, the underground limestone rock was dissolved by water, transported aboveground in solution, and reprecipitated as a travertine (limestone formation) spring mound when dissolved carbon dioxide gas was lost to the atmosphere. The last limestone was deposited on the mound about 110,000 to 116,000 years ago. The present height of the mound is twenty to twenty-four meters. The collapsed circular nature of Montezuma Well is clear in figure 6.1b.

The dissolution and upward transport of limestone aboveground formed large underground, water-filled caverns. After years of dissolution, lateral fractures drained water from the voids and left partially empty underground caverns. Eventually the weight of the mound caused a series of collapses to form the sunken, circular spring pool we see today. The final collapse is thought to have occurred about 12,000 to 15,000 years ago. Minor collapses have occurred over the past fifty years. The most recent, in 1986, left the water a milky color for several months. Montezuma Well was once thought to have been formed by a meteorite (Hitchcock 1927). (For additional information on the formation of Montezuma Well, see Lange 1957; Hevly 1974; Nations et al. 1981; Davis and Shafer 1992; and Hevly et al. 1992.)

The Human History of Montezuma Well

The inner limestone walls that surround the Well were once occupied by Native Americans. Nearly 150–200 Sinagua lived in cliff dwellings between AD 50 and 1400 (Schroeder and Hastings 1958; P. Pilles, Coconino National Forest, pers. comm.). A few of the dwellings still remain today on the inner face of the collapsed dome. My colleagues and I (Blinn et al. 1994) reported the appearance of diatom microfossils in a dated sediment core from the Well. Several of the microfossils present during the Sinagua occu-

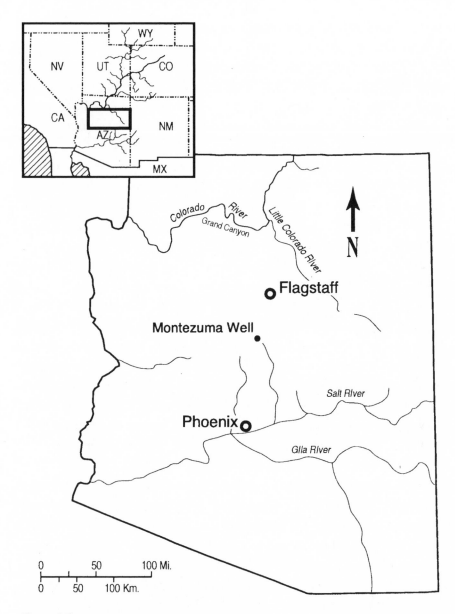

Figure 6.1. The location of Montezuma Well in Arizona and an aerial photograph showing the collapsed circular nature of Montezuma Well (photo by J. J. Landye). The swallet outlet and inner cave system are located at the point where the travertine mound is narrowest and are adjacent to Wet Beaver Creek.

pation suggested the organic enrichment of water; however, the microfossils disappeared from the core when the Native Americans left the area.

Montezuma Well was privately owned by several families. In 1883, Sam Shull took squatters' rights for two horses and built a shack on the present property (Beckman 1991). After several years, he sold the property to Abraham Lincoln Smith for forty dollars, a pair of chaps, and a horse. In 1887, William Back purchased the property for two horses. On April 4, 1947, Back sold the property to the U.S. government, and Montezuma Well became a satellite of Montezuma Castle National Monument with Al Schroeder as the first park ranger. Today nearly 200,000 people visit Montezuma Well annually (G. Henderson, Montezuma Castle/Tuzigoot National Monument, pers. comm.).

The Location and General Description of Montezuma Well

The Montezuma Well aquatic complex is located in the Upper Sonoran Desert grassland in Yavapai County, Arizona (34°39′ N, 111°45′ W), at an elevation of 1,122 meters (fig. 6.1a). The complex consists of a large spring pool and a cave system from which the water exits about 70 meters downstream into an irrigation ditch, originally of prehistoric Native American origin. The pool is a near-perfect circle, with a maximum water depth of 17 meters and a diameter of 112 meters.

Patches of emergent aquatic plants, including *Berula erecta* (water parsnip), *Hydrocotyle verticillata* (water-pennywort), *Eleocharis rostellata* (spikerush), and *Typha latifolia* (cattail), are found along the shoreline. The toxic roots of *B. erecta* are eaten by an amphipod (scud), *Hyalella azteca*, which lives within the root matrix. Studies have shown that the amphipod may use these toxic compounds (called coumarins) as an antipredator strategy with no harm to itself (Rowell and Blinn 2003).

The bottom contour of the littoral zone (region with rooted aquatic plants) consists of a gently sloping shelf (0.3–3.5 meters in depth) that extends 8–12 meters outward from the shoreline. Bottom substrates on the shelf consist of fine sediments, except where pieces of limestone have broken from the surrounding walls and fallen into the water. The rooted aquatic plants on the shelf consist primarily of pondweed (*Potamogeton illinoensis*), which encircles the entire well, mixed with common poolmat (*Zannichellia palustris*) and the macroalga *Chara vulgaris* in shallower waters. Pondweed is rooted in sediments up to 8 meters beneath the water surface at the sloping outer edge of the shelf and forms a sheer vertical wall of vegetation between the littoral and limnetic (open-water region without rooted plants) zones. (This may be the tallest reported aquatic macrophyte!) Stems of *P. illinoensis* are highly etiolated and produce leaves only on the upper 0.5 meters of the plant. The filamentous green alga, *Spirogyra parula*, forms dense floating mats over the pondweed beds in the spring and fall seasons. Beyond the shelf, the bottom drops precipitously to depths of 14–17 meters, where it is fairly uniform throughout the rest of the limnetic zone. The estimated water volumes for the littoral and limnetic zones are 8,842 cubic meters and 42,000 cubic meters, respectively. These and other morphometric features of Montezuma Well are given in table 6.1.

Water enters the Well primarily through four large vents at the bottom of the limnetic zone at a volume of 4,150–7,300 cubic meters per day (nearly 1.6 million gallons per day) and a near-constant temperature of about 24°C (±2°C). The vents are about 24 meters from the northeast shoreline at a depth of 15–16 meters. Although the water level in Montezuma Well is more constant than in most aquatic habitats in the Southwest, seasonal changes in water depth of up to 16 centimeters may occur, with the highest water levels in the winter and lowest in summer (Avery and Helmke 1993). The amplitude of change in water depth appears to be related to the amount of snowpack on the Mogollon Rim and water use in the region.

Substrates below the limnetic zone are composed of highly polished, fine, "squeaky" sands (especially near where water enters through the vents) and organic sediment. Scuba divers have observed churning sand at the bottom, which is caused by spring water moving up through sands and keeping them in suspension in a quicksandlike fashion (R. Carruth, U.S. Geological Survey, pers. comm.). In late January 2006, scuba divers observed a dense, fine, suspended sand layer at about 15–17 meters deep. Sand boiled up from bottom water vents and then cascaded downslope like a turbulent river of sand (D. Lanihan and D. L. Conlin, Submerged Resources Center, National Park Service, pers. comm.).

Water exits the Well at a shallow area (≤0.3 meters deep) in the southeast corner known as the inner outlet or swallet (fig. 6.1b). The con-

TABLE 6.1. Summary of morphometric and
chemical features of Montezuma Well, Arizona

Category	Measurement
Morphometric Features	
Area	0.76 ha
Maximum depth	17.0 m
Mean depth	6.7 m
Maximum length	112.0 m
Shoreline	345 m
Total volume	50,842 m³
Relative depth	17%
Shoreline development	1.04
Chemical Features	
Ca^{++}	117.0 ± 4.5
Mg^{++}	37.4 ± 2.2
Na^+	50.6 ± 0.5
K^+	5.1 ± 0.2
SO_4^{--}	10.8 ± 0.8
Cl^-	44.6 ± 3.4
HCO_3^-	595.0 ± 5.0
$NO_3\text{-}N^-$	0.02 ± 0.005
NH_4^+	0.007 ± 0.09
$O\text{-}PO_4^{-3}$	0.15 ± 0.02
SiO_2	21.9 ± 3.0
pH	6.5 ± 0.02
Specific conductance	985 μS ± 4.0

Note: Morphometric data were taken from Cole and Barry 1973.
Physical-chemical data (given in mg/L with standard error
noted) represent at least ten measurements for each season over
the past decade. Area is the total area of pool. Maximum length
is the maximum diameter of the pool. Shoreline is the measured
distance of shoreline. Shoreline development is a calculated
index of the shape of the pool (a circle would be 1.0).

tinual influx of water results in an approximate turnover rate of 10 percent in total water volume each day (Cole and Barry 1973). Dense patches of aquatic moss, *Fissidens grandifrons*, are attached to limestone rocks in the swallet the entire year, while the freshwater red alga, *Batrachospermum moniliforme*, occurs only between mid-December and June. Water passes through the swallet and enters a limestone cave system, reemerging to the surface about 46 meters downstream and 5 meters lower in elevation. It takes water about seven minutes to pass through the cave, suggesting a complex passageway of subterranean pools and riffles. Water emerging from the cave enters an irrigation ditch and flows parallel to Wet Beaver Creek for 1.6 kilometers. Patches of *B. erecta* and *B. moniliforme*, along with ferns (*Adiantum* sp.), are located along the upper 150 meters of the ditch. A description of morphometric features of the Well can be found in Cole and Barry 1973.

The Physical–Chemical Environment of Montezuma Well

The Open-Water Column

Because of the high turnover rate of water in Montezuma Well, the physical-chemical environment in the open-water column is relatively constant. Temperature throughout the 17-meter vertical column varies less than 3°C during any time of the year, with the greatest rate of change in the top 0.5 meters from solar input. The average annual temperature is 21°C (±4°C), ranging from a near-isothermal water column of 19–20°C in the winter to temperatures approaching 25°C at the surface in midsummer. Therefore, daily and seasonal thermal cues used by organisms for behavioral and reproductive strategies are minimal in the limnetic zone of the Well compared to other aquatic habitats. Instead, temporal and spatial changes in light are the main biological cues in the water column (see later section on diel interactions between organisms).

Water transparency (clarity) varies with season depending on the density of the plankton, particularly phytoplankton (floating microscopic algae). In the summer, dense plankton blooms reduce mean Secchi depth transparency to 2.2 (±0.3) meters, compared to 3.5 (±0.2) meters in the winter, when plankton blooms are low. The average compensation depth (where photosynthesis equals respiration) is 8.2 (±0.5) meters in the summer, compared to 10.1 (±0.2) meters in the winter.

Daytime dissolved oxygen concentrations in the summer range from 9 milligrams per liter at the surface to 6 milligrams per liter near the bottom. During summer nights, dissolved oxygen ranges from 8 milligrams per liter at the surface to 3.5 milligrams per liter at the bottom, owing to the lack of photosynthesis and the high respiration of plankton. Bottom waters are never anoxic, which suggests that the artesian water comes in contact with the atmosphere via cracks before entering the Well. This may occur at nearby Soda Springs, which is located about 0.5 miles northeast of the Well. Surface-to-bottom dissolved oxygen values during the winter commonly range from 8 to 6 milligrams per liter over a twenty-four-hour period during both day and night.

Soluble nutrients are relatively constant throughout the year, with average nitrate (NO_3-N) and phosphorus (PO_4^{-3}) concentrations at 0.02 (±0.005) milligrams per liter and 0.15 (±0.02) milligrams per liter, respectively. Some minor elements such as arsenic exceed 100 micrograms per liter in Montezuma Well, a concentration that is five times greater than standards established by the Environmental Protection Agency (Compton-O'Brien et al. 2003). Most nutrients enter the Well from bottom vents, since the area of the watershed surrounding the Well is less than one hectare. Average precipitation for the region is 29.6 centimeters, with approximately 40 percent of the moisture falling during the summer monsoonal rains (July–September). The pH (6.5 ± 0.02) of the water in the Well is very stable because of the high alkalinity (595 ± 15 mg/L $CaCO_3$) that buffers the diel metabolic activity of the aquatic plants and animals. These and other physical-chemical features of Montezuma Well are given in table 6.1.

The Littoral Vegetation Zone

Temperatures vary more in the weed bed than in the open-water column. Summer temperatures in the top ten centimeters exceed 30°C on clear sunny days because of the absorption of infrared light by plants but are 25°C at twenty centimeters below the surface and throughout the entire water column at night. The vertical and diel (over a twenty-four-hour period) changes in water temperature are less dramatic in the winter, with diel temperatures in the vegetation similar to those in the open water. Also, the dense vegetation causes dramatic diel changes in dissolved oxygen in the summer ranging from supersaturation, more than twelve milligrams per liter (>150 percent), at midday to less than four milligrams per liter (<50 percent) at night. These dramatic vertical and diel physical-chemical gradients in the littoral vegetation influence the metabolism of organisms in the Well (Oberlin and Blinn 1997).

Dissolved Carbon Dioxide Concentrations

One of the most influential parameters that structures the biotic communities in Montezuma Well is the elevated level of dissolved carbon dioxide. In this regard, Montezuma Well differs from many other spring pools. In the Well, water percolates through limestone and enters from deep bottom vents supersaturated with dissolved calcium carbonate ($CaCO_3$) and carbon dioxide (CO_2). The relatively low surface area to volume ratio of the Well reduces immediate degassing at the surface and allows high concentrations of CO_2 to remain in solution. Most limestone spring pools are more shallow and degas quickly, as has been reported for those found in the Great Artesian Basin of southern Australia (Zeidler and Ponder 1989).

Dissolved CO_2 enters the Well from the bottom vents at concentrations that seasonally exceed 500 milligrams per liter (fig. 6.2) — over one hundred times higher than the amount of dissolved CO_2 found in most aquatic habitats. Some dissolved CO_2 is lost in surface degassing and as water passes through submerged aquatic plants, especially in the summer, during periods of high photosynthesis (fig. 6.2). Water further degasses as it passes through the shallow swallet into the cave and along the irrigation ditch. As CO_2 is lost to the atmosphere, a shift in the carbonate equilibrium results in two dynamic changes: pH increases from 6.6 to 7.6 along the irrigation ditch and limestone is redeposited. Cole and Batchelder (1969) estimated that nearly 2,000 kilograms (4,410 pounds) of dissolved carbonate flow from the Well daily, of which nearly 89 kilograms (40 pounds) precipitate as limestone along the 1.6-kilometer irrigation ditch. The concentration of dissolved CO_2 is reduced to less than 20 milligrams per liter by the time water reaches the end of the irrigation ditch (fig. 6.2). The changes in physical-chemical conditions provide a myriad of opportunities to examine changing biotic communities along sharp abiotic gradients. (For additional physical-chemical information on Montezuma Well, see Cole and Barry 1973; Boucher et al. 1984; Blinn et al. 1994; Konieczki and Leake 1997; O'Brien and Blinn 1999.)

The Biotic Communities of Montezuma Well

The structure and dynamics of the invertebrate community in the Well have largely been shaped by the high concentrations of dissolved cal-

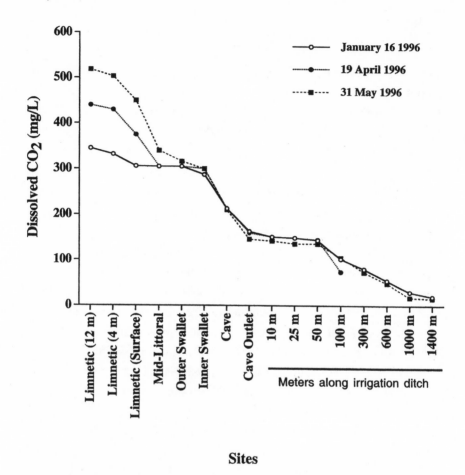

Figure 6.2. Concentrations of dissolved carbon dioxide along transects in the Montezuma Well. The depth collections in the limnetic zone over the bottom vents were taken 50 meters from the outer swallet, and the littoral measurements were taken 5 meters from the outer swallet. Water in the irrigation canal was diverted below 100 meters on 16 January 1996 (some data from O'Brien and Blinn 1999).

cium carbonate and carbon dioxide, the relatively low pH, and the lack of predatory fish. The absence of fish is attributable to the high concentrations of dissolved carbon dioxide (Cole and Barry 1973).

The extreme chemical conditions and lack of fish predators contribute to the occurrence of endemic organisms in the Well. These endemic organisms probably appeared in the last 12,000 years, because dissolved CO_2

concentrations were likely lower prior to the major collapse in the Well at that time. To date, one endemic diatom and four invertebrate species have been reported only from Montezuma Well. The diatom is *Gomphonema montezumense* (Czarnecki and Blinn 1979); it first appeared around 8000 BP, according to analysis of sediment cores (Blinn et al. 1994). The endemic invertebrates include a spring snail (*Pyrgulopsis montezumensis* [Hershler and Landye 1988]) in the swallet and other associated limestone and moss habitats along the shorelines; a water scorpion (*Ranatra montezuma* [Polhemus 1976]) in the littoral vegetation; and an amphipod (*Hyalella montezuma* [Cole and Watkins 1977]) and a leech (*Motobdella montezuma* [Govedich et al. 1998], formerly described as *Erpobdella montezuma* [Davies et al. 1985]) in the water column of the littoral and limnetic zones. The amphipod *H. azteca* in the shoreline vegetation is likely to be a new endemic species. All endemic species occur at exceptionally high densities and have unique behavioral, physiological, and/or morphological features.

For example, *Motobdella* is a newly described leech genus that feeds almost exclusively on the endemic *H. montezuma* by sensing water waves created by the swimming amphipod (Blinn et al. 1988a; Blinn and Davies 1989a). The three other leech species (*Helobdella* spp.) listed in appendix 6.1 have a more typical feeding behavior and forage on a variety of small invertebrates (Davies and Govedich 2001). *Motobdella montezuma* places its cocoons at depths below five meters on the long etiolated stems of pondweed to avoid heavy predation by insects at the surface (Davies et al. 1987).

Both *P. montezumensis* and *H. montezuma* have adjusted to the extreme chemical conditions in the Well. Neither species can live in ambient concentrations of dissolved CO_2: for survival, they require at least 50 milligrams per liter of this gas, or over ten times the concentration in most standing waters (O'Brien and Blinn 1999). These unique organisms can live in CO_2 concentrations exceeding 600 milligrams per liter for a week. (It is noteworthy that CO_2 gas is used as an anesthetic for invertebrates.) The endemic *H. montezuma* shows distinct genetic, morphological, physiological, and behavioral differences from other closely related allopatric species of *Hyalella* (Blinn and Johnson 1982; Wagner and Blinn 1987a; Thomas et al. 1994, 1997; Duan et al. 2000). For example, the concentration of carotenoid pigments in the carapace of the endemic *H. montezuma* is nearly three times higher than that of *H. azteca* located in the shoreline vegetation, owing to the greater exposure of *H. montezuma* to ultraviolet light in the open-water column.

Finally, the long, narrow body of the endemic *R. montezuma* mimics pondweed stems. It selectively buries its eggs in soft, partially decayed plant tissue to hide them from predators (Blinn and Runck 1989). During this process, *Ranatra* crawls onto the surface of the soft pondweed stems and is exposed to variable air temperatures. Therefore, because of freezing nighttime air temperatures during the winter, *Ranatra* oviposits eggs only between March and November, even though water temperatures in the Well remain constantly warm (Blinn and Runck 1989). During December through February, adults hibernate in the cracks of the limestone around the shoreline of the Well.

The near-constant conditions in Montezuma Well maintain a uniform phytoplankton assemblage for most of the year. The unicellular, microscopic (1–2 μm) species of *Coccomyxa minor* and *Nannochloris bacillaris* make up 60–65 percent of the phytoplankton community for most of the year in Montezuma Well. Other chlorococcalean species include *Ankistrodesmus*, *Chlorella*, *Monoraphidium*, *Nephrocytium*, and *Scenedesmus*, which comprise about 12 percent of the assemblage. In contrast, phytoplankton communities in seasonally variable aquatic habitats undergo a predictable succession of algal species. Assorted cyanobacteria (*Chroococcus*, *Gloeocapsa*, *Merismopedia*, and *Oscillatoria*) make up another 12 percent of the assemblage. Even though silicon dioxide (SiO_2) concentrations are high (22 mg/L), diatoms with siliceous walls make up less than 10 percent of the assemblage, with *Achnanthidium*, *Cocconeis*, and *Gomphonema* most common. The reason for the reduced abundance of planktonic diatoms is likely the constant warm water and the relatively low pH. The thermal, carbonate spring of Devils Hole has a similar algal assemblage (Shepard et al. 2000).

The zooplankton community is dominated by *H. montezuma*, which attains a length of 5 millimeters, and a tiny copepod, *Tropocyclops prasinus mexicanus*, which is less than 0.5 millimeters in length. *Hyalella montezuma* is the sole planktonic filter-feeder within the North American amphipod fauna. Both crustaceans forage on planktonic algae, and perhaps other zooplankton, and swim freely in the water column. Typically, freshwater amphipods occur in vegetation and on bottom substrates, and they commonly feed on dead organic matter and associated microbes; however, the unique

H. montezuma swims and feeds on phytoplankton at rates of 15.6 milliliters per day. These filtering rates are equal to or greater than rates measured for filter-feeding cladocerans (Blinn and Johnson 1982). The average annual densities of *T. prasinus mexicanus* are equal to or greater than 200 animals per liter, and each animal can filter up to 3 milliliters of water per day. Based on the combined densities and filtering rates of both zooplankters, nearly 60 percent of the entire water column in the Well may be filtered daily in the summer. Females of both species carry eggs the entire year because of the constantly warm water. A larger and predaceous copepod, *Macrocyclops albidus*, occurs at low annual densities of slightly more than two animals per liter. (For further information on planktonic communities in Montezuma Well, see Kidd and Wade 1963; Czarnecki 1979; Blinn et al. 1982, 1986, 1988a, 1988b; Boucher et al. 1984; Blinn and Davies 1990; Dehdashti and Blinn 1991; Oberlin and Blinn 1997; Ellsworth and Blinn 2003.)

Insects constitute fifty-seven taxa in sixteen families and display the greatest diversity of animals in the Well (Blinn and Sanderson 1989). All of the insects spend most of their time closely associated with the weed bed, except for the nightly migrations into the water column by the water scorpion, *Ranatra montezuma*, and the giant water bug, *Belostoma bakeri*, to feed on amphipods (Blinn et al. 1982).

Montezuma Well is an isolated desert aquatic habitat where several species have disjunct distributions. For example, three aquatic beetles in the family Hydrophilidae (*Anacaena signaticollis*, *Crenitulus* nr. *debilis*, *Laccobius ellipticus*) are new records for Arizona, and *Enochrus sharpia* is a new record for the United States (Blinn and Sanderson 1989). The water bug, *Microvelia rasilis* (Veliidae), feeds at the water surface; this is also a new record for the United States (Polhemus and Sanderson 1987). Finally, the fishing spider, *Dolomedes triton*, is a large, relatively rare southwestern species.

Aquatic dipterans make up 60 percent of the insect taxa, half of which are biting midges in the family Ceratopogonidae. Within this midge group, several undescribed adults of the genera *Brachypogon*, *Ceratoculicoides*, and *Dasyhelea* have been collected in light traps near the shore.

Four taxa constitute over 80 percent of the insect biomass in Montezuma Well. These insects include the unique *R. montezuma*; the giant water bug *Belostoma bakeri*; the damselfly *Telebasis salva*; and a dytiscid beetle, *Cybister ellipticus* (figs. 6.3 and 6.4). (For the ecology and behavior of these

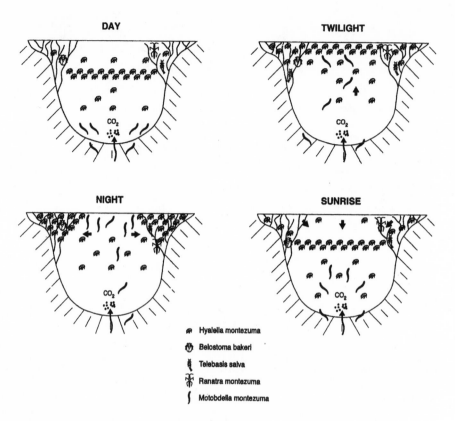

Figure 6.3. Diel migrations of invertebrate predators and prey in Montezuma Well. Amphipod prey move from the weed bed to the open-water column during the day to avoid visual insect predators and back to the weed bed at night to avoid the blind leech that feeds in the water column at night.

insects, see Blinn et al. 1982; Blinn and Runck 1989; Blinn et al. 1993; Runck and Blinn 1990, 1992, 1993, 1994, 1995.)

The aquatic vertebrates in the Well include the Sonoran mud turtle (*Kinosternon sonoriense*) and the introduced red-eared slider turtle (*Trachemys scripta*), both of which bask in the sun on floating logs and vegetation. Estimates suggest that more than two hundred turtles live in the spring pool (P. Rosen, University of Arizona, pers. comm.). Preliminary stable isotope analysis suggests that turtles feed primarily on invertebrates, especially amphipods and aquatic insects, and on vegetation. Also, the muskrat, *Ondatra zibethicus*, dens along the shoreline, particularly in the swallet, and is active during the night. Montezuma Well is the type locality for the Colorado River

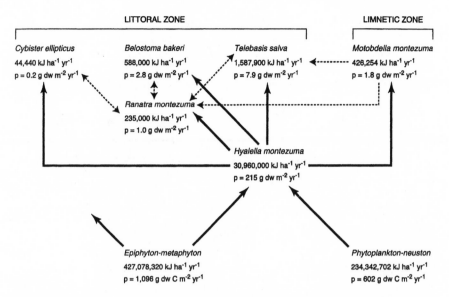

LITTORAL ZONE   LIMNETIC ZONE

Cybister ellipticus   Belostoma bakeri   Telebasis salva   Motobdella montezuma
44,440 kJ ha⁻¹ yr⁻¹   588,000 kJ ha⁻¹ yr⁻¹   1,587,900 kJ ha⁻¹ yr⁻¹   426,254 kJ ha⁻¹ yr⁻¹
p = 0.2 g dw m⁻² yr⁻¹   p = 2.8 g dw m⁻² yr⁻¹   p = 7.9 g dw m⁻² yr⁻¹   p = 1.8 g dw m⁻² yr⁻¹

Ranatra montezuma
235,000 kJ ha⁻¹ yr⁻¹
p = 1.0 g dw m⁻² yr⁻¹

Hyalella montezuma
30,960,000 kJ ha⁻¹ yr⁻¹
p = 215 g dw m⁻² yr⁻¹

Epiphyton-metaphyton   Phytoplankton-neuston
427,078,320 kJ ha⁻¹ yr⁻¹   234,342,702 kJ ha⁻¹ yr⁻¹
p = 1,096 g dw C m⁻² yr⁻¹   p = 602 g dw C m⁻² yr⁻¹

Figure 6.4. A model of the trophic structure, production, and energy flow in the littoral and limnetic zones in Montezuma Well. Values of p are annual production. Energy values are the energy equivalent of annual production based on microbomb oxygen calorimetry values. Arrows indicate trophic linkage, with solid arrows indicating strong energy flow and dotted arrows, weak energy flow. Note that 50 percent of the invertebrates displayed are endemic to Montezuma Well. (Sources for production values: phytoplankton and epiphyton, Runck and Blinn 1994; *H. montezuma*, Dehdashti and Blinn 1991; *R. montezuma*, Runck and Blinn 1990; *T. salva*, Runck and Blinn 1993; *B. bakeri*, Runck and Blinn 1994; and *C. ellipticus* and *M. montezuma*, Runck 1993)

otter, which is now feared to be extinct. The reason for the absence of otters and other reptiles or amphibians is unknown.

Waterfowl, primarily the American widgeon (*Anas americana*), number forty to fifty from mid-October until mid-March. Summer abundance of widgeons is less than five (J. Coleman, National Park Service, pers. comm.). Other waterfowl—including the pied-billed grebe (*Podilymbus podiceps*), the lesser scaup (*Aythya affinis*), and ring-necked duck (*Aythya collaris*)—are each seen in numbers of fewer than five between October and May.

The inner cave system between the spring mound and irrigation ditch has a poorly developed fauna compared to the spring pool. The most abundant invertebrates in the cave are the freshwater bryozoan *Plumatella repens* and a bivalve mollusc, *Pisidium* sp., both of which filter-feed on particulate carbon. The predaceous elmid beetle, *Microcylloepus similus*, is also common in the cave system. The cave has low numbers of *H. azteca*, tubificid worms (*Limnodrilus hoffmeister* and *Quistadrilus multisetosus*; Blinn and Oberlin 1996; Wetzel et al. 1999), and the water mite *Hydrozetes* sp. (V. Behan-Pelletier, Agriculture and Agri-Food Canada, pers. comm.). All of these organisms are also found in the weed bed of the spring pool. Distributions of organisms along the irrigation ditch are given in Blinn and Oberlin 1996.

Temporal Variation in Trophic Structure

The Seasonal Dynamics and Behavior of Organisms

One of the attributes of stable spring ecosystems is that they tend to have fewer environmental variables that change with the season. Because solar energy is the only abiotic parameter that undergoes major seasonal change in the Well, we can better evaluate the role of light on each trophic level without the confounding influences of other variables. For example, mean summer densities for phytoplankton exceed 160,000 cells per milliliter, compared to fewer than 30,000 cells per milliliter in the winter, when light intensities are low. The effect of low phytoplankton densities is transferred up the food web without changing the composition of the biotic assemblages. In turn, the planktonic, grazing amphipod *H. montezuma* attains average densities of more than sixteen animals per liter in the summer, but densities fall below two animals per liter in the winter, when light energy and algal food are low. The brief period of reduced light during the summer monsoonal season also reduces the number of organisms supported in each trophic level of the food web (Boucher et al. 1984).

The one time of the year when the phytoplankton assemblage changes is during the usually brief three to five weeks of summer monsoonal rains. At this time, unicellular, flagellated species of *Cryptomonas* and *Chroomonas* become abundant and concentrations of organic nutrients increase (Boucher et al. 1984). These algal species require organic compounds for reproduction and growth. Monsoonal storms deliver organic nutrients to the Well from precipitation and runoff and briefly interrupt nutrient ratios, which initiates a change in algal composition. After the monsoon season, the composition of the phytoplankton returns to the typical constant assemblage of species. This suggests that light is influential in determining cell density while nutrients modify the composition within the assemblage. This hypothesis needs to be tested with both field and laboratory experiments. The brief change in nutrients does not influence the composition of upper trophic levels in the Well.

Both *Hyalella montezuma* and *H. azteca* are infected by a digenic trematode that turns the amphipod body orange based on a recent discovery by C. O'Brien (personal communication). The infected amphipods swim into the surface waters earlier in the crepuscular vertical migration sequence (see discussion below) than uninfected amphipods, perhaps because of the protection from ultraviolet light provided by the orange pigment. Over 95 percent of the amphipods first seen at the onset of the upward migration are orange.

Waterfowl selectively feed on the more conspicuous orange morphs, and the parasite completes its life cycle in the intestinal wall of the definitive host. Preliminary observations show considerable reductions in egg number in infected amphipods, as well as a more lethargic behavior. Because *H. azteca* is abundant in the moss in the swallet, the area where water and parasitic eggs exit the Well, the incidence of parasitism is several times greater in this species.

Diel Interactions between Organisms

Fish often play strong top-down trophic roles as predators, initiating trophic cascades and affecting aquatic invertebrate abundance and assemblage structure. Therefore, one of the unique features of Montezuma Well as a natural laboratory is the absence of these important predators.

Because of the absence of fish predators and the exceptionally high densities of invertebrate predators throughout the Well, the amphipod *H. montezuma* undergoes a highly predictable migration between the littoral vegetation and the open-water column each day and night (fig. 6.3). Without these timely migrations, the endemic amphipod would likely become extinct.

Hyalella montezuma is preyed upon heavily in the aquatic vegetation during the day by insects with keen eyesight and by the nonvisual leech *M. montezuma* in the open-water column at night. Average combined densities of the insect predators *T. salva*, *C. elliptica*, *B. bakeri*, and *R. montezuma* in the vegetation exceed 4,400 animals per square meter of the water column, with summer numbers near 12,000 animals per square meter. The density of leeches in the surface waters at night exceeds 5 animals per cubic meter.

All four predatory insects rest on submerged leaves in the dense weed bed and capture amphipods with their forelegs as the amphipods swim by. *Belostoma bakeri* and *R. montezuma* use a sharp beak to pierce and suck internal body fluids from adult prey, while *C. elliptica* uses chewing mouthparts and *T. salva* engulfs its entire prey, especially juvenile prey (<3 millimeters). Therefore, because of the variety of feeding strategies employed, all size classes of amphipods are susceptible to insects in the weed bed.

Juvenile amphipods are most vulnerable because of the exceptionally high numbers of *T. salva*. These predators feed primarily on smaller amphipods, owing to the limitations of how wide they can open their mouths to engulf prey (Runck 1993). Numbers of damselfly nymphs average over 4,200 animals per square meter of the water column during the summer and fall. Runck (1993) showed that all four predatory insect species are most effective at capturing amphipods during the day when they can see their prey; success at night is reduced several-fold. Therefore, the best time for *H. montezuma* to be in the vegetation is during the night, when they cannot be seen by their predators. Considering the high numbers of insects combined with their average daily consumption rates, Runck (1993) estimated that about 13 percent of the amphipod population would be consumed daily by *T. salva* if they remained in the weed bed during the day. Field experiments have shown that each *T. salva* consumes an average of 4–6 amphipods a day. At this rate, the amphipod population would be driven to extinction within a relatively short period of time. The alternative solution is for amphipods to swim out of the weed bed into the open-water column, where predation is minimal during the day. Animals may also leave the vegetation to avoid the high summer daytime temperatures.

During the day, the upper portion of the open-water column is nearly predator-free, because the leech is sensitive to light and remains in the dark depths of the Well (fig. 6.3). Scuba divers have observed high densities of leeches (5–10 animals per square meter) swimming near the bottom sediments or partially embedded in the soft sediments during the day. *Hyalella montezuma* is also light sensitive and therefore occupies a subsurface level in the water column to avoid leeches in the deeper portion (fig. 6.3). The two meters of the water column nearest the surface are nearly free of amphipods, but densities between two and four meters below the surface exceed 100 animals per liter during the summer and early fall. Sensitivity to light is considered to be the primary cause of their subsurface position, since the amphipod layer moves closer to the surface on overcast days. Escape from diving waterfowl is an alternative hypothesis but needs further testing.

At night, conditions change dramatically. Approximately one hour prior to sunset (twilight), amphipods migrate to the surface waters and feed on the dense surface film of organisms (fig. 6.3; Blinn et al. 1988b). One to two hours after sunset, the endemic leech begins to ascend to feed on the surface accumulations of amphipods (Blinn and Davies 1990). Unlike their insect counterparts in the weed bed, leeches do not need light to locate prey. Instead, they use sensitive hairlike structures, concentrated around their mouths, to detect swimming amphipods through water vibrations (Blinn et al. 1986, 1988a; Blinn and Davies 1989a). Leeches swim freely in the surface waters and engulf the smaller juvenile amphipods (fig. 6.3). My colleagues and I (Blinn et al. 1988a) have found that juveniles and adult amphipods produce distinct swimming vibrations; leeches selectively choose the signals produced by juveniles. Also, larger amphipods have sharp dorsal spines that limit predation by leeches as they attempt to engulf their prey. Over 90 percent of the leech's diet is made up of this amphipod (Blinn et al. 1987). Its feeding behavior and several morphological features make this leech unique (Govedich et al. 1998). There is evidence that leeches search for dense patches of amphipods because higher densities of prey yield greater feeding success (Blinn et al. 1990). Each leech typically consumes as many as twelve to sixteen amphipods each night.

Because of the presence of leeches in surface waters at night, a major portion of the amphipod population retreats to the outer fringe of the weed bed, where predation by visual insects is reduced by darkness (fig. 6.3). The vegetation provides several advantages to the amphipod population. First, it reduces the number of swimming amphipods in the water column and therefore reduces capture success by leeches. Second, the swimming signals are lost while resting on plants and muffled when amphipods swim within the dense weed bed. Third, energy is conserved (by resting on vegetation) for when they return to the open water the following day. Davies and colleagues (1992) estimated that nearly 10 microliters of oxygen per milligram of dry weight of amphipod tissue are used for each hour of swimming, compared to only 4 microliters while resting in the weed bed. This equates to a savings of 0.065 joules per milligram of dry weight per hour, or a 25 percent savings in energy.

The onset of light each morning causes the amphipods to return to the water column to avoid diurnally foraging insects (fig. 6.3). The column is safe because leeches have retreated to the Well's depths to avoid light and to process their night's bout of feeding. Amphipod carcasses are seen floating to the surface each morning; leeches use enzymes to digest the internal fluids of amphipods and excrete the empty carcasses.

This migration behavior occurs every day of the year in Montezuma Well but is less dramatic during the winter months because of the reduced numbers of prey and predators. During the day, the surface waters are calm and devoid of organisms, with most found in deeper waters (fig. 6.3). However, at night a feeding frenzy ensues, with dynamic interaction between predator and prey. These diel migrations may occur in other fishless aquatic ecosystems, but the limited number of nocturnal studies precludes any knowledge of such patterns.

Owing to its paramount role in the food web, *H. montezuma* is considered a "keystone species" in the Montezuma Well ecosystem. In other words, the overall structure and dynamics of the Well would greatly differ in the absence of this amphipod because nearly all other organisms feed on it.

The strategy of the congener amphipod *H. azteca* in the submerged shoreline vegetation of the Well is the exact opposite of the migration tactic of *H. montezuma*. *H. azteca* remains sedentary within the roots and foliage of the shoreline plants and consumes toxic coumarin compounds in the roots of *Berula* to reduce predation by the dense aquatic insect populations (Rowell and Blinn 2003).

Energy Flow and Food Web Dynamics

The Montezuma Well ecosystem has three trophic levels, as illustrated in figure 6.4. *Hyalella montezuma* filter-feeds on attached and floating microscopic algae (epiphyton-metaphyton) in the littoral vegetation and on phytoplankton and neuston (microscopic surface film) in the open-water column. The flow of algal biomass (grams of dry weight of carbon) and energy (kilojoules) into *H. montezuma* is illustrated by the bold arrows from each algal source in figure 6.4. The annual production of these microscopic algae in the Well is phenomenal because of the near-constant warm water and high light and nutrient availability throughout the year. These microscopic plants produce about thirteen metric tons of dry organic matter per year in the 0.76-hectare area of the Well; this is among the highest rates of algal production measured in any aquatic ecosystem. The highly eutrophic, saline Humboldt Lake in Saskatchewan, Canada, is one of the few lakes with a comparable annual rate of production (613 grams of carbon per square meter; Haynes and Hammer 1978) to Montezuma Well (602 grams of carbon per square meter).

Only about 4.7 percent of the combined energy from the two algal food sources (epiphyton/metaphyton and phytoplankton/neuston) is incorporated into *H. montezuma* animal tissue. Some algal material is consumed by the tiny copepod *T. prasinus mexicanus*, but this crustacean also forages on early instars of *H. montezuma* and cannibalizes its own young. The proportions of plant and animal material consumed by this copepod are unknown. Also, the role of *T. prasinus mexicanus* in the Montezuma Well food web is unknown. Laboratory experiments have shown that the leech is unable to capture copepods because of their quick response to the suction pressure by predators, but my colleagues and I (Blinn et al. 1993) have shown that insects in the weed bed occasionally feed on copepods.

As with algae, large amounts of *H. montezuma* biomass are produced in Montezuma Well each year. However, as is true of all food webs, the total mass is considerably reduced at higher trophic levels (fig. 6.4). Estimates indicate that over 1.6 metric tons of *Hyalella* tissue (by dry weight) are produced annually. Again, these primary consumer production rates are some of the highest reported for a freshwater amphipod (Dehdashti and Blinn 1991).

All of the predators in the top trophic level in the littoral and limnetic zones feed primarily on the endemic amphipod (Blinn et al. 1987, 1993; Runck and Blinn 1992; Runck 1993; Runck and Blinn 1994, 1995). The dotted lines in figure 6.4 indicate that only limited feeding occurs between top predators, with most of their energy derived from *H. montezuma*.

The leech has the lowest ecological efficiency (i.e., predator energy/prey energy) of energy transfer from prey to predator, at 1.4 percent, while the damselfly *T. salva* has the highest, at 5.1 percent, among the top predators. The low efficiency for *M. montezuma* may result from its atypical feeding strategy from other leech species, that is, capturing most of its prey in the water column. Ecological efficiencies from primary consumers (herbivores) to secondary consumers (predators) range from 6 to 13 percent in other spring ecosystems (Odum 1957; Teal 1957; Tilly 1968). The ecological efficiency of energy transfer from one trophic level to the next is often used as an indicator of ecosystem function (Ricklefs 1997). The lower efficiencies reported in the Well may result from the intense competition between multiple invertebrate predators for one primary prey, *H. montezuma*.

As is the case for lower trophic levels, production rates for the top predators in the Well are some of the highest recorded for aquatic insects. In fact, annual rates of production for *T. salva* alone are equal to multiple species of insect predators in other aquatic ecosystems (Runck and Blinn 1993, 1994). The damselfly nymphs, which average less than five millimeters in size throughout the year, produce about sixty kilograms of dry animal tissue each year in the Well. The overall productivity and ecological efficiency in the Well is likely higher than in other aquatic ecosystems because of the near-constant environmental conditions. Because of the highly predictable conditions, organisms display finely tuned behavioral features to avoid predators and forage on prey.

Why Study Spring Ecosystems?

Because of their near-constant and sometimes extreme environmental conditions, springs provide valuable models to better understand the driving variables in aquatic ecosystems. They provide useful "outdoor chemostats" for behavioral, ecological, and physiological studies under semicontrolled conditions that are applicable to other aquatic ecosystems. The following summary provides a more focused rationale for the continued study of spring ecosystems.

- Spring ecosystems are analogous to tropical rain forests, in that potentially valuable information on biological systems may be lost forever if the organisms and ecological processes within these ecosystems are not better understood. Therefore, the continued loss of springs through anthropological events, and global climatic change, sets a race for knowledge on these valuable natural resources.
- Many spring organisms thrive under extreme conditions such as high temperature, low pH, and elevated dissolved CO_2 and mineral concentrations. Similar extreme conditions may develop under continued anthropological manipulations and global climatic change. Therefore, spring habitats provide potential gene pools for future modified environments.
- Springs serve as functional models to evaluate threshold tolerances of organisms subjected to changing climatic conditions. Relatively minor changes may yield dramatic responses in reproductive and metabolic processes in spring organisms because they have evolved under relatively narrow environmental conditions. Therefore, spring organisms may provide cues for global changes in environmental conditions.
- Many springs have fewer trophic levels in their food webs with fewer taxa at each level than most aquatic ecosystems. Therefore, ecologists can study energy flow in a more simple aquatic ecosystem in the absence of complex food webs.
- Desert springs provide isolated habitats that may be used to study processes of island biogeography and invasion/extinction.
- Desert springs provide irreplaceable natural oases for terrestrial insects, birds, and mammals to water, forage, and reproduce. These aquatic systems provide valuable habitats within which researchers can study linkages between terrestrial and aquatic systems.
- Springs provide potential units for long-term monitoring for anthropogenic effects and climatic change. Many snowpack-driven springs

respond to climatic conditions. Therefore, monitoring water depth and springs discharge may give early warning signals for climatic and anthropogenic changes.

All of these issues apply to Montezuma Well, where a nearly constant aquatic environment has allowed a rich diversity of endemic organisms to coexist and a highly structured ecosystem to develop. Continued monitoring and conservation are required to keep this spring ecosystem intact for the future.

ACKNOWLEDGMENTS

This chapter is dedicated to David B. Czarnecki. His recent untimely death left us without one of the avid admirers of the many facets of Montezuma Well. I thank the National Park Service—particularly Jack Beckman, Jim Coleman, Don Montgomery, Tom Ferrell, and Glen Henderson—for their assistance with the logistics and collection of information included in this document. I also appreciate the enthusiasm and assistance of my students who have spent many hours, during the day and night, collecting information on this intriguing system, including Paul Boucher, David Czarnecki, Behrooz Dehdashti, Fredric Govedich, David Johnson, Chris Pinney, Gaye Oberlin, Chris O'Brien, Kirsten Rowell, Clay Runck, Perry Thomas, and Vincent Wagner. Thanks are also extended to Patricia Ellsworth for assistance in the field and the laboratory and to Ronald Davies for many interesting discussions on leech biology. Various specialists—including Ian Bayly (copepods), Rick Forester (ostracods), J. Norman Grim (protozoans), Patricia Ellsworth (rotifers), and Valerie Behan-Pelletier (water mites)—verified and/or identified taxa listed within the text. I would also like to thank the National Park Service, Whitehall Foundation, and Northern Arizona University for their financial assistance over the years of study. I also thank Lawrence E. Stevens and Vicky J. Meretsky as well as Don Montgomery, Sandra Blinn, and several anonymous reviewers for their helpful comments on this manuscript. Finally, I would like to express my gratitude to the late Professor Gerald A. Cole for introducing me to this marvelous system and sharing his knowledge on the processes within Montezuma Well.

Appendix 6.1

NONINSECT INVERTEBRATE ASSEMBLAGE IN THE SPRING POOL
AND CAVE SYSTEM OF MONTEZUMA WELL

The referenced literature for taxonomic groups is indicated in brackets. A single asterisk preceding the binomial indicates that the species was rarely encountered in the Well; double asterisks indicate that the species is endemic to the Well. Insects and vertebrates are given in the text.

Protozoa

Coleps sp.
Dileptus sp.
Epistylis sp.
Frontonia sp.
Halteria sp.
Keronopsis sp.
Loxodes sp.
Pseudourostyla cristata (Jerka-Dziadosz) [Grim and Manganaro 1985]
Spirostomum sp.
Stentor sp.
Vorticella sp.

Rotifera

*Lecane sp.
*Monostyla sp.
*Testudinella sp.
*undetermined bdelloid

Bryozoa

Plumatella repens L. [Dehdashti and Blinn 1986]

Oligochaeta [Wetzel et al. 1999]

Chaetogaster diaphanus Gruithuisen
Haplotaxis cf. *gordioides* Hartmann
Limnodrilus hoffmeisteri Claparède
Limnodrilus profundicola (Verrill)
Limnodrilus udekemianus Claparède
Nais elinguis Müller
Pristina sp.
Pristinella jenkinae Stephenson

Quistadrilus multisetosus Smith
Tubifex tubifex Müller
Varichaetadrilus angustipenis (Brinkhurst and Cook)
undetermined Enchytraeidae species
undetermined Lumbricidae species
undetermined Ocnerodrilidae species

Hirudinoidea

Helobdella elongata (Castle) [Blinn and Davies 1989b]
Helobdella stagnalis (Linnaeus) [Blinn and Davies 1989b]
Helobdella triserialis (Blanchard) [Blinn and Davies 1989b]
**Motobdella montezumensis* [Govedich et al. 1998]

Cladocera

**Alona* sp.

Copepoda

**Eucyclops* sp.
Macrocyclops albidus Jurine [Ellsworth and Blinn 2003]
**Orthocyclops modestus* (Herrick)
**Paracyclops* sp.
Tropocyclops prasinus mexicanus Kiefer [Ellsworth and Blinn 2003]

Ostracoda

Candona albicans Brady
Cypridopsis vidua (Mueller)
Darwinula stevensoni (Brady and Robertson)
Herpetocypris reptans (Baird)
Physocypria sp.

Amphipoda

Hyalella azteca (Saussure) [Rowell and Blinn 2003]
**Hyalella montezuma* Cole and Watkins [1977]

Arachnida

Dolomedes triton (Walckenaer)
Hydrozetes sp.

Gastropoda

Fossaria sp.
Physa virgata Gould
**Pyrgulopsis montezumensis* Hershler [Hershler and Landye 1988; O'Brien and Blinn 1999]

Pelecypoda

Pisidium sp.

Appendix 6.2

ADDITIONAL REFERENCES

Ecosystem Dynamics

Barnett and Brazel 1975
Batchelder and Cole 1978
Brazel 1976
Damon et al. 1964
Haynes et al. 1966
Runck and Blinn 1991

Ecosystem Features

Cole 1983
Henderson 1933
Konieczki and Leake 1997

Animal Species

Boucher 1980
Dehdashti 1987
English et al. 1986
Govedich 1996
Grossnickle et al. 1985
Haury 1982
Johnson 1982
Kalarani et al. 1993
McLoughlin et al. 1999
Oberlin 1995
Runck 1989
Southwest Parks and Monuments 1985
Starkweather and Blinn 1986a, 1986b
Thomas 1995
Witt et al. 2003

Plant Species

Clark and Burgess 1966
McDougall and Haskell 1960
Spangel and Sutton 1949

Other

Anonymous 1948
Cole 1965
Schroeder 1948
Sutton 1954
Wagner and Blinn 1987b

Chapter 7

Combining Ecological Research and Conservation

A Case Study in Cuatro Ciénegas, Coahuila, Mexico

DEAN A. HENDRICKSON, JANE C. MARKS,
ANGELA B. MOLINE, ERIC DINGER, AND
ADAM E. COHEN

Desert springs, ciénegas (marshes), and rivers are high-priority sites in global and regional conservation assessments because of their ecological importance, high levels of endemism (McCoy 1984; Ponder 1986; Minckley and Deacon 1991; Shepard 1993), and vulnerability to anthropogenic disturbance (Minckley and Deacon 1968; Deacon and Minckley 1974; Williams et al. 1989; Rolston 1991; Sigler and Sigler 1992; Hubbs 1995; Grimm et al. 1996; Abell et al. 1998, 2000; Dinerstein et al. 2000). Desert springs are particularly threatened because arid zones are developing at unusually fast rates (Brune 1981; Hendrickson and Minckley 1984; Ponder 1986; Contreras-B. and Lozano-V. 1994; Crowe and Sharp 1997; Minckley and Unmack 2000). Water extraction is a primary threat to desert springs, but invasions by non-native species also pose serious threats (Courtenay and Stauffer 1984; Spencer et al. 1991; Kaufman 1992; Richter et al. 1997), particularly in desert springs, where high rates of endemism are attributed to isolation from natural invasions and long periods of evolution in simple, highly coevolved communities (Moyle 1986, 1995; Moyle et al. 1986). As a result of long periods of isolation, spring endemics often lack natural refugia, dispersal mechanisms, and behaviors necessary to avoid predation or to interact in other ways with new invasive species, which have the potential to drive local endemics to extinction (Meffe and Snelson 1989; Minckley and Deacon 1991; Minckley et al. 1991; Kaufman 1992; Seehausen et al. 1997a, 1997b). Neither reducing water extraction nor stopping introductions of non-native species can be accomplished easily. Because water issues in arid environments are politically charged, conservationists must have solid data behind their predictions of the effects of water extraction to secure water rights for natural ecosystems. Similarly, non-native species are difficult and expensive to eradicate but often might be more easily and inexpensively controlled. Managers need comprehensive data on the relative severity of problems together with measures of probable effectiveness of management options to evaluate how to best focus limited resources.

In this chapter, we describe a study conducted to identify and characterize critical threats to the endangered and highly endemic biota of Cuatro Ciénegas, Coahuila, Mexico (fig. 7.1), focusing on non-native species and water extraction. Developed in close consultation with managers of the Cuatro Ciénegas National Protected Area, this project includes surveys to document the extent of invasions of non-native species and of hydrologically altered habitats, as well as focused experiments to test hypotheses about ecological mechanisms of perceived threats. Survey results and management recommendations are based on preliminary findings, as a result of which, one non-native fish has been identified as a key threat to the reserve's biota. Our approach represents a possible model for incorporating basic ecological research into protected area management, with long-term monitoring as a critical component.

Conservation Status and Site Description

The diverse endemic aquatic biota and diverse threats to it make Cuatro Ciénegas unambiguously a high-priority conservation ecoregion (Abell et al. 1998). Virtually all of the Cuatro Ciénegas aquatic biota is listed by the federal government as endangered (Secretaría de Desarrollo Social 1994b) or is under review for such listing (CONABIO 1998). In 1994, the entire valley was declared a National Protected Area (Secretaría de Desarrollo Social 1994a). Managers now have a basic management plan (Secretaría del Medio Ambiente, Recursos Naturales y Pesca 1999) and guidelines (Secretaría del Medio Ambiente, Recursos Naturales y Pesca 2000) but need better information through which to understand and mitigate the identified threats to the aquatic biota: water extraction and non-native species.

Though less than 1,500 square kilometers in area, the basin of Cuatro Ciénegas contains well over two hundred springs, ciénegas, pozas (spring-fed pools), rivers, and other water bodies. These differ in discharge and water quality but share a common biogeographic history (Hershler 1984,

1985; Hershler and Minckley 1986; Smith and Miller 1986). The springs of this basin display the physical and chemical diversity that might be found in a sample of springs drawn from a much larger region and provide a level of control for biogeographic history that would be impossible to obtain in studies done over larger areas. Furthermore, these springs demonstrate the full range of perturbations seen in modern desert springs. In Cuatro Ciénegas, introductions of non-native species and water extraction affect some pools, but adjacent, similar pools remain pristine, thereby providing a natural laboratory for analyses.

Biota

The basin's fish fauna consists of sixteen native species, at least eight of which are endemic to the basin. Further systematic study of several additional taxa will likely determine that they too are endemic at the subspecific or higher level.

A major influence on trophic structure in most pools is an abundant trophic polymorph, Cuatro Ciénegas cichlid (*Herichthys minckleyi*), a fish species with individuals that consume detritus (detritivore morph), native snails (molluscivore morph), and even small fishes (piscivore morph) (Kornfield and Koehn 1975; Sage and Selander 1975; Kornfield et al. 1982; Smith 1982; Kornfield and Taylor 1983; Liem and Kaufman 1984). The two most abundant morphs (detritivore and molluscivore) are externally indistinguishable. The Cuatro Ciénegas cichlid is widely distributed and abundant in most of the medium to large springs in the basin. With its high abundance and unusually broad diet, this single species may constitute the majority of biomass at the second, third, and fourth trophic levels in undisturbed food webs, and it can function in the food web as up to four ecological species (three morphs and juveniles). Thus, it merits particular focus in conservation efforts because changes in its abundance, or in the relative abundance of the distinct morphs, are likely to have far-reaching, cascading effects throughout these communities. Other potentially important community members include herbivorous fishes such as *Cyprinodon atrorus*, *C. bifasciatus* (Miller 1968), and *Dionda episcopa* (Minckley 1969) and the diverse native hydrobiid snails (Hershler 1984, 1985; Hershler and Minckley 1986; Hershler and Hayek 1988), which feed on epiphytic and epilithic algae and algal mats,

c

d

Figure 7.1. Poza Chiqueros (Laguna Garabatal [Winsborough 1990]), in the upper Río Garabatal system. Though this poza dried seasonally in the 1970s and perhaps into the 1980s, it has been continually dry since at least the mid 1990s. Distinctive stromatolites at an adjacent poza in 1983 (*c*) and 1999 (*d*). (*a* and *c*: 1983 photos by Barbara Winsborough; *b* and *d*: 1999 photos by DAH)

some of which produce a diversity of stromatolites unrivaled anywhere else in the world (Winsborough 1990; Winsborough et al. 1994).

Aquatic insects are present in pools with lower salinity, consuming algae, detritus, and small insect larvae. Largemouth bass (*Micropterus salmoides*) and the piscivorous morph of *H. minckleyi*, as well as flathead catfish (*Pylodictus olivaris*—relatively rare) consume smaller fishes but only in the larger sites that provide sufficient habitat for them. Aquatic birds are relatively rare but may play a role, at least seasonally, in some areas. Aquatic snakes of the genera *Nerodia* and *Thamnophis* are abundant in many areas and may be the primary fish predators. Four species of aquatic turtles, including the endemics *Terrapene coahuila* and *Trionyx ater*, also are important species in many areas.

The spatial heterogeneity of fish and other taxa is striking, with many small pools being fishless and others having only one or two species, while other habitats have as many as fourteen fish species. Waters in the basin often flow underground for substantial distances, but subterranean systems are relatively inaccessible and virtually unstudied. At least three troglobitic (cave-adapted) isopods are found in spring outlets (Cole 1984), and many fishes (especially the native catfish, *Ictalurus lupus*, and other species) are sometimes abundant in subterranean streams.

Water Resources

Prehistorically, no surface waters flowed out of the valley. If water left the valley it was only via subterranean outlets, and thus surface aquatic habitats were isolated from surrounding river basins. In the 1890s, however, a canal was constructed to carry Cuatro Ciénegas surface water into the adjacent headwaters of the Río Salado de los Nadadores for delivery to water users up to one hundred kilometers downstream. Additional canal construction quickly proliferated in the basin, connecting many other springs to the new outlet and diverting others to agriculture within the Cuatro Ciénegas basin, mostly near town. These canals decreased local water tables and substantially altered spring discharge and water level and also facilitated movements of species among habitats via new aquatic connections. Water levels of many pools in the basin have decreased as water extraction increased (Minckley 1969; Cole 1984; Contreras-B. 1984; Cohen et al. 2001).

There were 1,690 hectares under irrigated agriculture in the municipality of Cuatro Ciénegas in 1930, and 9,321 hectares in 1994 (Secretaría del Medio Ambiente, Recursos Naturales y Pesca 1999). Within the basin (municipality), extraction from groundwater is 49 million cubic meters per year, of which 48 million cubic meters per year is used to irrigate animal forage products (alfalfa, sorghum, and oats), although some is used to irrigate human food products, primarily corn and wheat (Secretaría del Medio Ambiente Recursos Naturales y Pesca 1999). If preliminary estimates of groundwater recharge at a rate of 25 million cubic meters per year (HRS Water Consultants 1998) are accurate, the water table is declining. Estimates of total water flow through the six main canals range from 1,730 liters per second to 2,620 liters per second, with approximately 50 percent of that leaving the basin through two canals (Fondo Mundial para la Naturaleza and Protección de la Fauna Mexicana 1999). Canal flow rates do not include evaporation or leakage, estimated to range from 10 percent to 80 percent of the water drained from the pozas. Loss rates are high because most canals are neither covered nor lined (Fondo Mundial para la Naturaleza and Protección de la Fauna Mexicana 1999). Details of downstream water rights and socioeconomic issues surrounding them have only recently come under study (Cotera 2001; Medellín and Cotera 2001; Medellín et al. 2001) to help managers devise sustainable plans.

Changes in water levels and flow rates have serious implications. Clearly, water extraction will diminish the areal extent of aquatic habitats, particularly the shallow marginal habitats. Fish community diversity and structure in the basin has been shown to be positively correlated with habitat size (Minckley 1984), so reduced water levels in some of the larger shallow pools may threaten the top predators there (e.g., *H. minckleyi* piscivore morph and largemouth bass, *Micropterus salmoides*). Discharge changes attributable to water extraction may also severely impact these ecosystems by, for example, changing water turnover rate in pools, and thus water chemistry and temperature. Continued water extraction could alter native food webs by shrinking habitats and altering abiotic conditions beyond species' natural tolerance ranges. Beyond water extraction, resident humans directly impact aquatic ecosystems in many other ways. The city's sewage water is impounded in evaporation ponds that are susceptible to overflow during flood events. Additionally, many houses, including all of those in the surrounding *ejidos*, and all recreational sites are not connected to city water and either use septic tanks or latrines or lack sanitation facilities (Secretaría del Medio Ambiente Recursos Naturales y Pesca 1999). Outfalls from these are obvious in some places, presenting threats to both human and ecosystem health. Although chemicals are used in local agriculture, whether there is significant runoff into the pools is unclear.

Cumulative effects of human impacts on aquatic habitats of the basin are not well documented. Historic spring discharges and water quality data are limited. In an attempt to more rigorously examine long-term habitat changes, we have been collecting historic photos from earlier researchers, local residents, and various archives. Most historic photos are of often-visited, popular areas, and comparisons between these and more recent photos help document changes. These paired photos, many interviews with long-time residents, and the sparse documentation that predates them clearly indicate that the basin contains far less surface water today than it used to and that changes are occurring at a relatively rapid rate. Documentation of more recent changes, however, has not improved markedly. Though more individuals visit more sites now on a more regular basis, quantitative, consistent data on water levels and discharges are still not available, and the source of the groundwater that feeds the surface aquatic ecosystems remains unknown. To produce such large volumes, recharge areas very likely include areas outside the surface drainage to the valley. Potential recharge areas are experiencing large-scale water extraction for agriculture, and additional large agricultural projects that would increase groundwater extraction are proposed. In the valley immediately to the north, Valle de las Calaveras, large-scale alfalfa production based on pumped groundwater has expanded rapidly in the past decade, dramatically dropping the water table in that valley. Similar development is proposed to the south in the El Hundido valley. Hydrologic studies to determine the origin of groundwater feeding the pools, rivers, and marshes of Cuatro Ciénegas are urgently needed.

Water Survey Results

We sampled over thirty sites for water quality (table 7.1). Water quality varies among sites but has not changed dramatically since the 1980s (table 7.1), indicating no serious eutrophication or other major recent shifts in quality. As mentioned above, however, comparisons of current and historic photos provide striking evidence of significant decreases in surface-water quantity (levels), especially at higher elevations of the valley floor. In headwaters of the Río Garabatal, for example, we duplicated photos from the 1980s that show that several connected pozas dried in the late 1980s or early 1990s (fig. 7.1) and have been continually dry since, whereas they dried only seasonally in the 1980s (Winsborough 1990). Because there is no direct water extraction from this site, the cause of dropping water levels here is unclear, but it may be related to diversions from large spring pools nearby, such as Poza de la Becerra. We also retook photographs at twelve other sites originally photographed between 1920 and 1980 (Cohen et al. 2001). Changes in many cases are dramatic, again providing unambiguous graphic evidence of lowered water tables, especially in habitats on the perimeter of the valley. We found small springs and seeps near the tip of the Sierra San Marcos, where Hershler (Hershler 1984, 1985; Hershler and Hayek 1988) collected snails in 1984 and 1985, dry and lacking snails in August 2001. Continued monitoring will determine whether these habitats are now permanently dry.

Historical Changes

In the company of William S. Brown, we returned in the summer of 2002 to his *Terrapene coahuila* study sites of the late 1960s and early 1970s (Brown 1971, 1974) and found that all of his localities for this endemic, aquatic box turtle surrounding the tip of Sierra San Marcos were now dry or very nearly so, and lacking turtles.

Regional newspapers have also frequently carried stories regarding decreasing water levels in pozas in Cuatro Ciénegas (e.g., Anonymous 1999; Recio 1999). Older written accounts often contain general references indicating the former presence of far more water, such as George F. Weeks's (1918) mention that the operator of the train (who had long traveled the length of the valley) often stopped in the basin in the early 1900s to allow passengers to shoot ducks. Ducks are no longer common, and there is no place along the tracks today where significant wetlands are anywhere near shotgun range. Walter Scott Adkins (1920) described decreases in water levels attributable to water diversions from pozas when he visited in 1920, and he pointed out the nonsustainability of continuing to extract water in this way. None-

TABLE 7.1. Water quality comparisons between 1983 and 1999–2001

Site and year	Temperature (°C)	pH	Conductivity (μS/cm)	Dissolved oxygen (mg/L)	Alkalinity (mg/L)
Anteojo					
1983	29.7	6.40	—	—	137
1999	30.5	7.00	1762	3.9	166
Churince					
1983	30.4	7.05	—	—	163
2001	29.7	7.12	2538	5.1	156
Escobedo					
1983	35.0	6.99	2700	—	167
2001	34.7	7.08	2651	3.3	184
Juan Santos					
1983	26.9	7.21	—	—	142
2000	29.5	7.52	2913	5.6	176
Laguna Grande					
1983	25.7	8.35	—	—	116
1999	31.0	8.44	5373	7.4	107
Mojarral Oeste					
1983	31.9	7.20	2800	5.3	199
2001	33.2	7.04	2659	3.3	184
Poza Azul					
1983	31.0	7.50	2900	4.8	170
2000	28.9	7.64	2789	4.88	181
Río Mesquites					
1983	30.4	6.70	—	—	150
1999	28.8	7.77	2935	5.1	178
Santa Tecla					
1983	30.3	7.07	—	—	226
1999	30.2	7.34	1394	4.1	186
Tío Cándido					
1983	29.7	6.75	—	—	142
2001	31.8	7.19	2425	5.4	154

Sources: For 1983 data, see Winsborough 1990; data for 1999–2000 are from the current study.
Note: $N = 2$ for 1999 and 2000; $N = 3$ for 2001. When we had multiple years for the same habitat, we used the most current data available. Sample size and standard error for 1983 data are unavailable. Alkalinity is expressed as CO_3. ND = not detected. A dash indicates that data were not available. Numbers in parentheses denote standard error.

Nutrients and ions (mg/L)						
NO_3^-	PO_4^{-3}	SO_4^{-2}	Ca^{+2}	Mg^{+2}	Na^+	Cl^-
2.8	0.004	—	280	58	37	25
—	—	—	292 (2.5)	68 (0.5)	23 (0.0)	—
6.4	—	—	350	102	145	110
1.3 (0.01)	0.04 (0.0)	1169 (4.2)	287 (2.6)	95 (0.5)	140 (0.9)	105 (0.7)
—	—	—	340	110	151	124
1.4 (0.0)	0.03 (0.0)	1224 (20.5)	298 (9.8)	97 (2.1)	139 (3.5)	99 (0.8)
4.3	0.009	—	400	114	150	127
1.26 (0.06)	0 (0.01)	1132 (210.1)	314 (1.5)	100 (0.5)	149 (0.5)	121 (32.3)
0.1	0.02	—	780	580	900	666
0.08 (0.075)	ND	3697 (461.0)	545 (60.0)	340 (10.0)	472 (32.0)	375 (50.0)
6.8	0.003	—	370	110	140	103
1.5 (0.0)	0.04 (0.0)	1245 (2.0)	304 (1.0)	100 (0.8)	142 (1.5)	101 (1.5)
—	—	1374	360	120	170	120
1.14 (0.1)	ND	1282 (292.7)	305 (1.5)	102 (1.5)	154 (0.5)	121 (24.2)
5.6	0.006	1480	380	120	170	140
1.6 (0.03)	ND	1470 (15.5)	379 (9.0)	130 (2.5)	163 (1.0)	107 (1.0)
8.61	0.015	—	110	35	36	32
1.61 (0.03)	ND	461 (15.0)	160 (1.5)	53 (0.5)	46 (0.0)	36 (1.5)
5.6	0.009	—	340	102	140	110
1.26 (0.0)	0.04 (0.0)	1116 (6.2)	279 (2.7)	93 (0)	131 (0.3)	93 (0.3)

theless, canal construction and modifications, such as the deepening and lining of existing canals, continued well into the early 1970s (Minckley 1969; Contreras-B. 1984; Minckley 1992; Contreras-B. 2000) and continues to be proposed and, to some extent, carried out (Cristino Villareal and Arturo Contreras-A., reserve staff, pers. comm., 2000–2001). Compared to at least three sites just outside the valley, however, many sites in Cuatro Ciénegas still remain relatively pristine (Cohen et al. 2001).

Biological Surveys of Native Species

The unique biota of Cuatro Ciénegas was extensively described in the 1960s and early 1970s (see bibliography by Minckley [1994]), providing a useful baseline, but few field studies were conducted between 1975 and 1999. We initiated new surveys in 1999 to determine whether distributions of native and endemic species had changed, any extinctions had occurred, or invasions of non-native species had expanded since the early surveys. From summer 1999 to spring 2002, 16,013 fish specimens of twenty-one species (table 7.2) were collected from ninety-one sites in the basin of Cuatro Ciénegas and downstream in Río Salado de los Nadadores and its tributaries. Selected voucher specimens from these collections were preserved and are temporarily archived at the University of Texas at Austin, Texas Natural History Collection, for analysis and subsequent distribution to Instituto de Biología, Universidad Autónoma de México, and Universidad Autónoma de Nuevo León, per permit stipulations.

All native fish species previously recorded were still present in the basin. At most sites, species collected in our surveys were previously recorded or were a subset of those previously reported. We found some earlier fish collection sites now dry, but most of these had been previously reported as dry by Minckley (Minckley 1969, 1992). At sites where we found fewer species than had been previously reported, we consider that our efforts may have been inadequate to detect rare species, and we will continue to resample before declaring local extirpations. Many important former fish habitats had been lost or substantially altered by the time of the first fish collections, and most of the historic fish collections were taken from the larger and more accessible habitats fed by springs from deep aquifers. These habitats, for the most part (but with important exceptions), have suffered less alteration in

TABLE 7.2. Fish species collected between the summer of 1999 and spring of 2002

Family/scientific name	Status	Location	Abundance
Cyprinidae			
Cyprinella rutilus	N/E	RSN	Present
Cyprinella xanthicara	N/E	CC	Common basinwide, especially in larger systems with strong flow
Cyprinus carpio	NN	CC/RSN	Extirpated? (see text)
Dionda episcopa	N	CC/RSN	Common basinwide
Ictaluridae			
Ictalarus lupus	N	CC/RSN	Common basinwide; attempted *I. punctatus* introduction was stopped in the mid-1990s
Pylodictus olivaris	N	CC/RSN	Rare basinwide in larger habitats
Poeciliidae			
Gambusia longispinis	N/E	CC	Common basinwide in shallow marshes
Gambusia marshi	N	CC/RSN	Common basinwide in spring pools and rivers
Poecilia reticulata	NN	RSN	Collected only from springs at San Blas
Xiphophorus gordoni	N/E	CC	Common in SE area
Cyprinodontidae			
Cyprinodon atrorus	N/E	CC	Abundant in shallow, highly dynamic marshes
Cyprinodon bifasciatus	N/E	CC	Abundant in spring-fed pozas and stable rivers
Lucania interioris	N/E	CC	Common in densely vegetated marshes
Centrarchidae			
Lepomis megalotis	N	CC/RSN	Common basinwide in larger habitats
Micropterus salmoides	N	CC/RSN	Common basinwide in larger habitats (see text regarding *M.s. floridianus*)
Cichlidae			
Hemichromis guttatus	NN	CC	Common at sites indicated in fig. 7.3
Herichthys cyanoguttatus	N?	CC/RSN	Common in eastern parts of CC basin, decreasing to rare in western portions
Herichthys minckleyi	N/E	CC	Common basinwide to rare in far SE portions of valley
Oreochromis aureus	NN	CC	Common in Rio Mesquites; abundant in Las Teclitas in far SE
Percidae			
Etheostoma lugoi	N/E	CC	Locally common in strong flows; occasional in quiet pozas
Etheostoma segrex	N/E	RSN	Rare and apparently restricted to short reaches of appropriate habitat in Celemania area

Note: N = native; N/E = native/endemic; NN = non-native; CC = Cuatro Ciénegas valley; RSN = Río Salado de los Nadadores, downstream from Cuatro Ciénegas valley.

the three to four decades since biologists first collected in them than have the shallow marshlands, but they are far from pristine. A prime example is Mojarral Oeste, where early biologists sampled extensively. It retains its warm, constant-temperature inflow, and water levels do not vary more than about eight to ten centimeters, but the cool waterfall inflow (Arnold 1971) apparently dried in the 1970s. This tributary was fed by springs and seeps farther south, near Sierra San Marcos, that were mentioned in numerous publications (e.g., Taylor 1966; Taylor and Minckley 1966; Minckley 1969; Brown 1974; Cole 1981, 1984; Hershler 1984) and which, as reported above, are now mostly dry. Poza Escobedo during our survey (until September 2001, when it was dammed as part of a habitat restoration project) was also very much as biologists had found it in the late 1960s and 1970s, with its water level perhaps two meters or more lower than it had been near the end of the nineteenth century, when it was lowered by canal construction.

Probably the largest single loss of habitat that occurred just prior to most fish collections was the result of diversion of the outflow of Poza de la Becerra. Until late December 1972 (W. L. Minckley and W. S. Brown, pers. comm. 2001–2002), this outflow fed extensive marshes and a large river system that inundated much of the now mostly dry area west of Sierra San Marcos (fig. 7.2). The original primary canal caused the water level of the poza to drop an estimated two meters, drying the river and extensive marshlands that extended at least west, south, and northwest of the poza (Benigno Vázquez, Beto Lugo, and W. L. Minckley, pers. comm. 1999–2002; DAH, pers. obs. 2002). The formerly higher water level of the poza probably maintained a very large, near-surface aquifer and associated wetland that fed not only the river that flowed southwest from the poza, but also the headwaters of the Río Garabatal system to the northwest, including the pozas featured in figure 7.1. The extent and rapidity of decline in the local water table maintained by the La Becerra system were obviously increased by the many tributary drainage ditches that are still obvious (albeit now mostly dry) today west and northwest of the poza.

As mentioned, extirpations of the fauna at habitats that have not dried completely have not yet been rigorously documented, but we fear that such documentation may be imminent. The endemic darters (Norris and Minckley 1997), for example, should be carefully watched. Although still present at historic localities, some populations, such as that of *Etheostoma*

a

b

Figure 7.2. *a*, The outlet of Poza de la Becerra as photographed in December 1972 by W. S. Brown, reportedly two days after water began flowing down the newly opened canal visible in the lower right. *b*, The same view in the summer of 2002. The canal visible in the photos no longer carries water and has since been replaced by a buried pipeline that drains water from the poza, carrying it to the open canal visible along the road above and to the right of center in *b*. The house visible slightly right of lower center in *a* still stands but is partially hidden by trees. All the large trees near the poza in *b* are *Tamarix aphylla*, and the dense band of vegetation extending from lower left toward the poza, as well as the less-dense vegetation along the dirt road, is predominantly mesquite (*Prosopis* sp.). Note considerable tourism infrastructure at this now heavily utilized public bathing and picnic area.

lugoi in the Río Garabatal (where discharge has decreased tremendously from conditions seen in the 1960s), are restricted compared to what we suspect their distribution and abundance must have been in that system three to four decades ago, and they are found in habitats atypical for the species. Similarly, we observed one specimen of *Pylodictus olivaris* in the upper Río Garabatal, in habitat that is exceptionally small for this large predator. These findings lead us to conclude that the fish fauna of the Río Garabatal today still reflects the river's formerly much greater discharge — a situation that we expect will not last much longer unless water management in the area is altered to increase the discharge. Downstream, in the Río Salado de los Nadadores, where discharge is but a small remnant of historically far greater flows (Contreras-B. 2002), *E. segrex* remains at and near its type locality but is not common. It is apparently highly restricted to short reaches in what is now a very different habitat from that described by R. R. Miller in the early 1960s (Norris and Minckley 1997).

Native Invertebrates

Prior to 1999, there had been no broad-scale surveys of nongastropod invertebrates in the basin. Samples of invertebrates collected from twenty habitats in 1999 and 2000 revealed a diverse assemblage of aquatic invertebrates that varies among habitat types (pozas, lagunas, rivers, canals, and marshes), indicating that preservation of all of these habitat types will be necessary if the entire fauna of the basin is to be conserved. Our faunal surveys include at least 136 taxa, of which 118 are insects (Dinger 2001).

Taxonomic distribution of aquatic insects in Cuatro Ciénegas is typical of other desert systems. The most diverse and widespread predators are Odonata (14 genera) and Hemiptera (15 genera). Coleoptera (29 genera) and Ephemoptera (11 genera) are found in cooler, flowing habitats. Different groups of Trichoptera (13 genera) are found throughout the basin. Among the Diptera (14+ genera), Chironomidae are ubiquitous, but ceratopogonids, tabanids, and others are also present (Dinger 2001).

Other nongastropod invertebrates include the amphipod *Hyalella* spp., which occur in all habitat types and are often the most abundant, nongastropod invertebrate. Cole (1984) suggested on the basis of coloration that there might be as many as seven distinct species of *Hyalella*. The glass shrimp, *Palaemonetes suttkusi*, is also abundant throughout most of the basin (Rodriguez-A. et al. 1997; Dinger 2001) and is widespread throughout the state of Coahuila (Smalley 1964). Isopods (four species) have been well described in other publications (e.g., Cole 1984). Flatworms, leeches, ostracods, and oligochaetes are generally rare and found only in isolated habitats (Dinger 2001).

Although aquatic invertebrate diversity is typical of other desert systems, densities are typically far lower than elsewhere. High fish abundance and diversity may limit insect abundance. Invertebrate densities and species abundances are highest in flowing water, littoral vegetation, and on stromatolites, and lowest in stagnant waters and sediments. However, stagnant waters and waters with physicochemical extremes offer habitat for certain specialists (e.g., brine fly, Hydrometridae; *Artemia* sp.). Life-history characteristics, such as time of emergence and number of generations per year, are almost completely unknown, but seasonal collections do not indicate any large changes in species composition at any sites surveyed.

Non-native Species

We have surveyed throughout the basin for fishes and aquatic invertebrates, as well as plants (although our botanical surveys have not been systematic or as comprehensive as have those for aquatic biota). We have documented the presence of eleven non-native aquatic/semiaquatic species in or near the Cuatro Ciénegas basin (figs. 7.3 and 7.4).

NON-NATIVE FISH. A single specimen of the spotted jewelfish, or *mojarra joya* (*Hemichromis guttatus*), an African cichlid (Loiselle 1979) popular in the aquarium industry, was first observed (by DAH and Matt Stephens) in Poza Churince in 1995, but it could not be captured. The species was abundant there by 1998 and is now found in at least five sites: Poza Churince, Poza Juan Santos, Río Mesquites, Mojarral Este, and a small pool and canal at Rancho San Juan de Anteojo (fig. 7.3).

At Rancho San Juan de Anteojo, *H. guttatus* occurs with only one other fish, the native *Astyanax mexicanus*. The ranch owners indicate that the native *mojarra* (*Herichthys mincklyi*) had formerly occurred there but had disappeared some years after the introduction of *H. guttatus*, which they

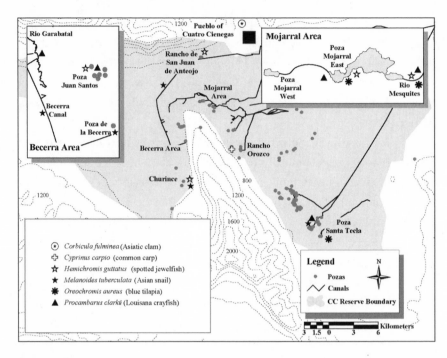

Figure 7.3. The distribution of non-native fishes and invertebrates in the Cuatro Ciénegas valley.

indicate may have been as early as 1990. This isolated site thus may have been the first habitat in the basin to receive *Hemichromis* introductions. Extirpation of *H. minckleyi* following the introduction of *Hemichromis* suggests competitive exclusion; however, physical habitat quality for *H. minckleyi* was subjectively considered marginal at best at the time the *H. guttatus* population was discovered. Extensive modification of this small habitat for irrigation may have caused extirpation of *H. minckleyi*, with or without the non-native *Hemichromis*.

Growth of the *H. guttatus* population in Poza Churince appears to have been paralleled by a reduction in the detritivore morph of *H. minckleyi*, from about 75 percent of the individuals before the introduction of *H. guttatus* to about 42 percent in 1998, by which time *H. guttatus* had come to constitute a large proportion of the total fish biomass (DAH, unpubl. data). This suggests that the diet of *H. guttatus*, which has pharyngeal teeth similar to those of the detritivore morph of *H. minckleyi*, likely overlaps with the

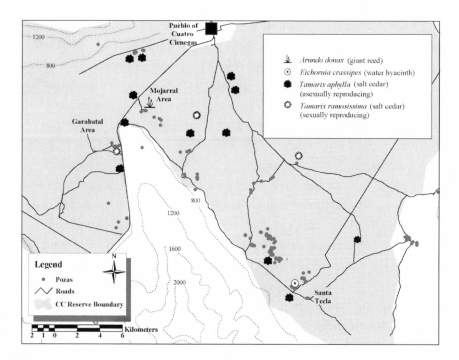

Figure 7.4. The distribution of selected non-native plants in the Cuatro Ciénegas valley. The *Tamarix ramosissima* indicated to the north of Santa Tecla is at Los Gatos.

detritivore morph of *H. minckleyi* and that the *H. guttatus* invasion may have caused the depletion of a key member of the Laguna Churince food web. Our ongoing research is combining stable isotopes and manipulative experiments to help predict the outcome of interactions between *H. guttatus* and native fishes. Preliminary results indicate that *H. guttatus* competes more strongly with juvenile than with adult *H. minckleyi*, and with *Cyprinodon bifasciatus*. Managers are implementing a *Hemichromis* eradication program. Although the species is highly susceptible to the trapping methods being used, our anecdotal observations indicate that populations rebound quickly. We are confident that intensive trapping could reduce populations of *Hemichromis*, but trapping is very unlikely to ever extirpate them in any habitat in the basin. Trapping of *Hemichromis* may be necessary to lessen their impacts in selected habitats, but for the moment, especially in light of the broad distribution of this exotic and its easy access to most of the basin, efforts to control *Hemichromis* populations should be limited to habitats that are important for juvenile *H. minckleyi*, those where it has been newly introduced, and valuable human resources used elsewhere. Control of other exotics, such as *Tamarix ramosissima* (below), with potentially very severe impacts but for which proven control technologies are available, should be implemented.

Tilapia (*Oreochromis aureus*) were introduced to the Río Mesquites at least ten years ago, and they persist upstream to Mojarral Este, although they are restricted in distribution and generally rare. This population remains small, and our anecdotal observations indicate that it has not increased substantially in the Río Mesquites in the past six to eight years. However, from this site, the species can easily access many other connected habitats. Additional populations were established about the same time in multiple ponds in the Santa Tecla area in the southeastern part of the valley, where the species is well established and abundant. Overall, however, Cuatro Ciénegas is unusual when compared to many other drainages in North American deserts (especially in Mexico; DAH, pers. obs.) where this species has rapidly expanded and established dense populations (Minckley 1973; Moyle 1976). What might limit expansion and population growth in the Río Mesquites is unclear, though we suspect that it may have something to do with the still relatively intact and diverse native fish community, including the native polymorphic cichlid.

The common carp (*Cyprinus carpio*) occurred in 1993 in a small, artificial pond at Rancho Orozco (specimens at IB-UNAM-P). Our recent surveys indicated that it is no longer present, and more recently (summer 2002), the artificial pond in which it had been found was purposefully dried by the landowner. The species (or possibly another carp species) was observed in 1998 at Las Playitas by reserve staff (SEMARNAT employees, pers. comm.). We have been unable to capture specimens there despite repeated efforts, but we have found large scales typical of this species along the shorelines of this habitat.

Historic collections do not allow us to determine with certainty whether the Texas cichlid (*Herichthys cyanoguttatus*) is native to Cuatro Ciénegas or whether, prior to the first collections, it may have invaded from the Río Salado de los Nadadores (where it is native; Guerra 1952) via the earliest canals. Its natural occurrence upstream in Ocampo would, however, seem to support the hypothesis that it was native to the Cuatro Ciénegas valley. In Cuatro Ciénegas, it is found most commonly and is most abundant in the eastern half of the basin, near canal outlets to the Río Salado de los Nadadores, but we found it rarely elsewhere, including locations in the far western parts of the basin, such as Río Garabatal. Our westernmost records, however, are from disturbed habitats, where its presence might reflect recent invasion. It was not recorded in the western parts of the basin in early surveys. Many *Herichthys* specimens from eastern, and some central, parts of the basin appear to display hybridlike characters, suggesting that these two similar and probably closely related species are hybridizing. In some areas, such as Canal de Escobedo, *H. cyanoguttatus* was found in marginal and fluctuating habitats immediately adjacent to the relatively stable habitats occupied by *H. minckleyi*, but it has not been taken in syntopy with *H. minckleyi* in this system. As habitats continue to change and become more variable, the relative abundance of *H. cyanoguttatus* may increase.

Florida largemouth bass (*Micropterus salmoides floridianus*) has been widely introduced for sport fisheries throughout much of the United States and Mexico, but we were surprised to detect evidence of its introduction in Cuatro Ciénegas. In collaboration with Francisco García de León of the Instituto Tecnológico de Ciudad Victoria and his student Rocio Inéz Rodríguez Martínez, we detected allozymic markers of this subspecies indicating its past introduction in the basin. It has hybridized with the native subspecies, which has been considered to represent an endemic form (Minckley 1969, 1974; Contreras-B. 1977). Genetic markers for Florida bass were found at some but not all sites sampled. Rodríguez is continuing her studies and will report results independently.

NON-NATIVE INVERTEBRATES. *Melanoides tuberculata*, a non-native snail introduced around the globe through the aquarium trade and for biological control of schistosomes, has had devastating impacts on snail communities in other aquatic ecosystems (Abbott 1973; Madsen and Frandsen 1989; Pointier et al. 1993, 1994). The species was first found in Cuatro Ciénegas in 1994 (Contreras-Arquieta 1998) and is now abundant in the Becerra canal and maintains high densities in the littoral vegetation of Poza Churince, as well as low densities in Pozas de la Becerra and Santa Tecla. It is also abundant at sites outside the basin (e.g., Río Salado de los Nadadores and La Mota at Ocampo, north of Cuatro Ciénegas). We hypothesize that it may represent a threat to the diverse assemblage of native hydrobiid snails and coevolved native fishes that feed on them. The turban snail (*Thiara granifera*) also was found in the Río Salado de Nadadores but has not been observed in the Cuatro Ciénegas basin (Contreras-A. 1998).

Louisiana crayfish (*Procambarus clarkii*), long known from the basin but previously rare, is now abundant in the now-fluctuating habitats of the partly desiccated Río Garabatal and nearby Laguna Juan Santos. It is also found in the Río Mesquites up to at least Mojarral Este. Because the Río Mesquites traverses much of the basin, crayfish will probably colonize more habitats. This species' ability to withstand periods of dryness may give it a competitive advantage over other aquatic biota during and following dry spells. As more habitats are altered by water extraction and declining water tables, crayfish may become a more significant threat. This crayfish may have invaded the basin by ascending canals from the Río Salado de los Nadadores, to which it may be native, but early surveys in Cuatro Ciénegas were sufficiently thorough that we feel confident that it was not part of the basin's fauna in the 1960s. In other regional aquatic systems, crayfish have rapidly invaded and become extremely abundant (Mather and Stein 1993; Rodriguez-A. and Campos 1994; Hill and Lodge 1995; Gamradt and Kats 1996).

Asiatic clam (*Corbicula fulminea*) is now widely distributed throughout North America and the region of interest here (Hillis and Mayden 1985) but had not previously been reported from Cuatro Ciénegas. It was discovered recently living immediately upstream from town in the canal from Río El Cañon. Shells, but no live specimens, of this species have also been found at Poza de la Becerra and Mojarral West. If restricted to El Cañon, the municipal water supply, this non-native species may not spread quickly, but it could have impacts on water quality in town and the distribution system by clogging plumbing systems. If, as reported anecdotally, the sewage treatment plant sometimes spills into Canal de la Becerra, this non-native clam could spread throughout the basin. Furthermore, extreme flood events carry water down the normally dry Río El Cañon through town and into the adjacent valley, where individual clams may reach permanent surface aquatic habitats. There are no native filter feeders in Cuatro Ciénegas, and thus establishment of *Corbicula* could have far-reaching impacts on natural food webs. The species is also found in the Río Salado de Nadadores (unpubl. data), into which the waters of Cuatro Ciénegas now flow through canals, and it could move upstream into the basin from there.

NON-NATIVE AQUATIC MACROPHYTES AND RIPARIAN PLANTS. Water hyacinth (*Eichornia crassipes*) has had devastating effects on other aquatic ecosystems (Muli 1996) and is locally abundant in the Santa Tecla and El Venado area (fig. 7.4). A program to control this pest has been at least partially successful (L. Lozano-V., UANL, pers. comm.).

Salt-cedar (*Tamarix ramosissima*), a native of Eurasia, is reported here for the first time in the published literature as established in Cuatro Ciénegas, where we feel it poses a potentially severe new threat to aquatic ecosystems. This invasive, now widely distributed throughout the western United States and northern Mexico, has prompted considerable research (Sala et al. 1996; DeLoach et al. 2000). It is not yet abundant in the Cuatro Ciénegas basin except in a few well-established stands in the upper Río Garabatal, near Las Salinas, and at Los Gatos (fig. 7.4). We expect that it will soon be found elsewhere, because its seeds are aerially dispersed and we have seen flowering and seed-dispersal occurring at all sites, as well as in dense stands along the Río Salado de Los Nadadores not far from Cuatro Ciénegas. In the Río Garabatal area, rings of young plants around shorelines of now-dry pools indicate that seeds had germinated on recently exposed pond-bottom sediments. The same phenomenon is obvious at Los Gatos where water levels vary annually; at Las Salinas, older trees clearly established as water levels were manipulated by humans. Several generations of plants at Las Salinas extend generally westward from the largest specimens found near a dam. This species can affect channel morphology, decrease discharge through high evapotranspiration, and decrease overall riparian diversity. Local managers have already made good progress with physical control of this pest at Río Garabatal, and we removed all that we could find in summer 2002 at Los Gatos, but we recommend immediate implementation of a more comprehensive and intensive basinwide detection and control project. Much research has gone into the development of effective, albeit sometimes expensive, control methods that could be applied (e.g., Larmer 1998; Colorado State University Cooperative Extension Tri River Area 2002; Nature Conservancy Wildland Invasive Species Team 2002).

Giant reed (*Arundo donax*), another highly invasive species for which very expensive control projects have been implemented elsewhere (Team Arundo del Norte 2002), is well established below the valley along the Río Salado de los Nadadores and also at Las Tortugas (the pool adjacent to the visitor center in the middle of the valley), along the Canal de Tío Julio about five to six kilometers east of Las Tortugas, and on the banks of the canal from El Cañon above the town of Cuatro Ciénegas (fig. 7.4). On the Río Salado de los Nadadores, *Arundo* has nearly completely replaced *Phragmites* since 1968, when an apparently monospecific stand of *Phragmites* was photographed there by Donald Pinkava (1984). Compared to the probably native *Phragmites* sp., which is widely established throughout the basin, the much more dense, faster growing, and taller stands of *Arundo* are capable of greater evapotranspiration. The stand at Las Tortugas is rapidly expanding, and control projects should be considered. Luckily, this plant propagates only asexually, and so, especially in the floor of this basin where scouring flooding is virtually nonexistent, it can be expected to spread relatively slowly compared to its rather rapid spread on major rivers elsewhere on the continent.

THE THREAT OF NON-NATIVE SPECIES. In summary, non-native species pose potentially severe threats to native and endemic species of Cuatro Ciénegas. Despite some indications that extreme abiotic conditions in Cuatro Ciénegas springs and the unique and highly coevolved biota have allowed the basin to resist invasions, we now see evidence that non-natives are invading disturbed habitats and that disturbances are becoming more widespread. The modes of disturbance that we see, such as dewatering and invasions by non-natives, are not easy to control. Whether the ranges of these non-native species will be affected by either abiotic or biotic conditions or whether they just have not had sufficient time to spread is unclear. However, as abiotic conditions in the basin continue to change because of water extraction and development, the environment may become more conducive to the spread of some of these species. Further research on each of these species will help managers predict which species are most likely to spread and/or have detrimental effects on native taxa.

Many other less obvious and less studied non-native organisms may have adverse impacts on natural ecosystems in the valley. The common morning glory (*Convolvulus* sp.), for example, now occurs in moist areas throughout the basin. If it is impacting the native vegetation, either it is doing so in subtle ways or we are naïve in this perception. Athel, another salt-cedar (*Tamarix aphylla*) is common but localized at human settlements, where it is planted for shade and as a windbreak. Because it reproduces asexually, it has apparently not expanded rapidly, but an occasional individual is found far from human settlements, sometimes appearing as though it got there by humans using branches as fence posts. The potential of this species for spreading or localized impacts should not be overlooked. Far larger than *T. ramosissima*, it must transpire large quantities of water. A great diversity of other cultivars can be found at most ranchos and ejidos, and the possibility of introduction and establishment of new pests via horticultural trade should not be overlooked. Domestic animals also pose a threat. Once abundant, goats surely had impacts (Minckley 1969, 1992), but today they are rare. Cattle do not fare well in the saline conditions prevalent on the valley floor, but horses do, and many hundreds, if not thousands, now graze there and up into the foothills. Their impacts on riparian areas are ubiquitous in the form of trampling and bank erosion, as well as obvious nutrient enrichment in the form of manure. They play a role in the dispersal of mesquite seed, and we suspect that the apparent increases in mesquite coverage of the valley floor that are evident in many matched historic and recent photos of habitats throughout the basin may be attributable to grazing.

Finally, humans are having direct impacts in many areas not only through water extraction but also by tourism. Cuatro Ciénegas has become a nationally and internationally famous tourist destination. Year-round tourism has increased in recent years, especially following publication in 1995 of a beautifully illustrated article in *National Geographic* magazine (Grall 1995). During Easter week, thousands of individuals crowd public bathing areas such as Poza de la Becerra, Río Mesquites, and more recently the much more remote Las Playitas. Their impacts include physical trampling of bottom substrates (including easily broken stromatolites), nutrient enrichment (especially obvious at Río Mesquites, where outfalls from public latrines filter quickly into the shallow aquifer and adjacent river), trash deposition, and the risk of introduction of new non-native species or dispersal of those already found in the basin. *Melanoides* is especially resistant to drying and can easily be transported in shoes or other clothing from one habitat to another. *Corbicula* shells are attractive to bathers, and bathing suit pockets could be containers unwittingly used by humans to disperse this potential pest over great distances. The spotted jewelfish (*Hemichromis guttatus*) is attractive by anyone's standards; although this species is somewhat more difficult to move from place to place than are snails or clams, the lack of natural connections between the major water bodies that it now occupies in the basin is evidence of human introduction from one habitat to another.

Effects of *Hemichromis guttatus* on Natural Food Webs

Our surveys found that populations of *H. guttatus* became established rapidly in five distinct habitats over a six-year period. In one habitat (Rancho San Juan de Anteojo), introduction of this species was possibly associated with local extirpation of the endemic *H. minckleyi* (ranch owners, pers. comm.); in another (Poza Churince), the frequencies of morphotypes of *H. minckleyi* apparently shifted, perhaps in response to *H. guttatus* invasion. To better determine the severity of the threat posed by *H. guttatus*, we carried out focused observations and experiments.

Stable Isotope Studies of Trophic Position

Stable isotope analysis can describe energy flow through different food webs (Peterson et al. 1986; Peterson and Fry 1987; Fry 1988, 1991) and help identify important trophic interactions not obvious through direct observation. In part because adequate control sites are generally not available, stable isotopes have only infrequently been used to compare historical changes in energy flow as a result of anthropogenic disturbance, and to compare trophic structure between pristine and disturbed habitats (Vander Zanden et al. 1999). Cuatro Ciénegas offers an unusual opportunity for such research because of extensive and well-documented museum collections of native fishes and snails, as well as the presence of replicate control and treatment springs, with and without non-native species.

We analyzed isotopes in fin clip samples from fishes from two habitats, Pozas Churince and Juan Santos, both invaded by the African cichlid *H. guttatus*. The invasive African species overlaps in isotope values (and therefore diet) with juvenile native *H. minckleyi* and with the native pupfish *Cyprinodon bifasciatus*. Stable isotope values of *H. guttatus* differ significantly from those of adult *H. minckleyi* (in both papilliform and molariform morphs), suggesting that dietary interactions between *H. guttatus* and juvenile *H. minckleyi* are more intense than are interactions between *H. guttatus* and adult *H. minckleyi* (Marks et al. unpubl. data).

The Experimental Verification of Stable Isotope Patterns

We tested for competitive effects between *H. minckleyi* and *H. guttatus* using two large field-competition experiments. The first experiment tested for effects of *H. guttatus* on growth rates of juvenile *H. minckleyi* using three experimental treatments: *H. minckleyi* at low densities (four per enclosure), *H. minckleyi* at high densities (eight per enclosure), and a combination of *H. minckleyi* and *H. guttatus* (four of each species per enclosure), thus allowing quantification and comparison of intraspecific and interspecific effects. The second experiment used an identical design to test for effects of *H. guttatus* on adult *H. minckleyi*. Weighed and measured fish were placed in enclosures in Mojarral Este (a protected habitat that had recently been invaded by *H. guttatus*) in late June 2000 and removed and weighed and measured again three months later. Results demonstrated an interspecific effect of *H. guttatus* on the native cichlid that was greater for juveniles than for adults, and interspecific effects that were greater than intraspecific effects for juveniles than for adults.

These results are consistent with the stable isotope data, which also indicated that *C. bifasciatus* could be sustaining competitive pressure from *H. guttatus*; future experiments will examine this possibility. Our results indicate that *H. guttatus* may detrimentally affect both endemics, *H. minckleyi* and *C. bifasciatus*. Though elimination of *H. guttatus* is unlikely, efforts to control it are under way. In light of our studies, we recommend that control measures focus on habitats where populations of either *H. minckleyi* or *C. bifasciatus* are low or otherwise sensitive and on habitats determined to be important sources of juvenile *H. minckleyi*.

A Geographic Information System for Cuatro Ciénegas

Long-term monitoring is essential for evaluating the status of native species populations and for documenting the spread of non-native species. Our survey efforts highlight the importance of developing an accessible database to facilitate future monitoring by our team or other researchers. One challenge that we confronted in our survey work was identifying and locating historic sample sites, many of which have more than one name, while locations of others were ambiguously described, in part because there are few definitive landmarks on the valley floor. We often had to rely on anecdotal information passed among researchers and local naturalists. Though much of this older information is accurate, conservation efforts would benefit from a more systematic assessment of species distributions and habitat quality.

We are developing a geographic information system (GIS) for the reserve that builds on earlier work by Profauna (Fondo Mundial para la Naturaleza and Protección de la Fauna Mexicana 1999) and Instituto Tecnológico de Estudios Superiores de Monterrey (Vela-C. and Lozano Garcia 2000). There are currently no high-resolution maps available for the site. The Instituto Nacional de Estadística Geográfica e Informática (National Institute of Statistics, Geography, and Informatics) maps for the basin are 1:50,000 scale (INEGI 1974a, 1974b), which is not fine enough for identifying most aquatic habitats.

Our goal is to incorporate field data into a high resolution GIS database that will include habitat locations (UTM coordinates), standardized place-names, water-quality data, and historic (museum) and recent species distribution and abundance data, as well as habitat photos. This format will enable future researchers to retrieve historic data to facilitate evaluations of habitat and distribution changes over time. Many pozas, roads, canals, and rivers have been mapped and incorporated into an ArcView database, together with all public-access remote-sensing images and GIS coverages, such as watersheds, vegetation maps, and climatic data. Aerial photos obtained in 1968 by W. L. Minckley, and provided by him to us, were georeferenced to the GIS database and now provide an important snapshot of past land use and vegetative cover.

A detailed map of Poza Escobedo was created that includes fish and snail densities in the poza and at various sites along the canal draining it. Less than two months after the mapping effort was completed, the canal was dammed to restore upstream aquatic habitat, resulting in a significant alteration of system hydrology. The GIS database provides a concrete description of fish and snail distributions and abundance before the alteration, in addition to a comprehensive prerestoration snapshot of the system for future evaluation of the success of the restoration project. Many small springs and seeps in the basin are nearly impossible to locate when hiking owing to a lack of definitive landmarks. An accurate map with coordinates will ensure that future workers can return to specific sites for monitoring or follow-up research. Additionally, landmarks such as fences, caves, and roads plotted on the habitat maps help researchers orient and reduce reliance on the global positioning system (GPS), which can sometimes be unreliable because of signal blockage by surrounding mountains and poor satellite availability or geometry.

Field-collected GPS data and satellite images are being used to map aquatic habitats and surface flows among them. Many small habitats are not found on existing maps (e.g., INEGI 1974a, 1974b) but can be identified in high-resolution satellite images or aerial photos or can be physically mapped by being walked with a GPS receiver. Visual displays of the basin output from the GIS will help educate visitors and local residents regarding the global importance of the area and help them appreciate changes and threats to biodiversity. Providing maps for tourists that identify visitor areas and prohibited areas will help manage tourism-related impacts.

Defining Needs, Conserving, and Monitoring

As has long been pointed out, Cuatro Ciénegas is truly an exceptional field laboratory with a long history of evolution of persistent and apparently stable aquatic islands isolated in a sea of desert. Its highly diverse and endemic biota has evolved fascinating specializations that have long intrigued evolutionary biologists; these still stand to provide valuable insights into evolutionary and ecological processes.

This remarkable fauna has not yet suffered extinctions of macro-invertebrates or vertebrates, but distributions of some species have already been decreased and fragmented by lowering water levels. The fauna is gravely threatened by continuing human development in the region and continued expansion of non-natives and is desperately in need of conservation action. Non-native control efforts are urgently needed, and hydrologic studies will be necessary to develop a regional water budget and determine recharge areas for the aquifer so that sustainable regional water-development plans can be developed. Studies are also needed to place current meteorologic conditions in a long-term perspective, so that long-term climate history and future can be considered in hydrologic modeling and planning. The current regional drought, generally perceived as especially long and severe, may be relatively mild when compared to earlier, prehistoric events (e.g., Acuna-Soto et al. 2002), and managers should be aware of such extreme events when developing long-term plans.

Finally, consistent long-term monitoring of fauna and flora is critically needed. Perhaps one of the greatest contributions of our own studies, which came after more than twenty years of relative inactivity in the area by researchers, has been the additional time spent in the field by knowledgeable biologists who perceived and documented habitat and faunal changes both within and beyond the scope of our specific focal projects. Future monitoring, however, will have to be carefully planned, quantitative, standardized, continual, and long-term to ensure that it will support the rigorous demands that will undoubtedly be placed on monitoring and all other studies in the future. With such actions, we hope that the remarkable biota of Cuatro Ciénegas can be preserved in harmony with sustainable regional human development for the benefit of future human generations.

ACKNOWLEDGMENTS

We acknowledge the assistance of countless individuals who have helped us in the field and in many other ways. Space limitations allow us to mention only a very small proportion of them here. The late W. L. Minckley provided many hours of discussion about the habitats, fauna, and flora of Cuatro Ciénegas. We would have been lost without the many hours DAH spent with him being introduced to the diversity of the basin's habitats. The staff of the Cuatro Ciénegas Protected Area, especially Director Susana Moncada, Pepe Dávila, Arturo Contreras, Cristino Villareal, and Luis de la Garza, among others, have gone far out of their way to help us with all aspects of the study, including help with everything from general logistics for our sometimes large groups to entertainment. Reserve volunteers Sigilrey and Leo contributed many hours of help in the field. Francisco García de León has collaborated with us in many ways and has devoted considerable effort to helping us obtain the permits required to carry out our research (Comisión Nacional de Acuacultura y Pesca, permiso de fomento 180202-

613-03-050 and amendments). His student, Aldo Guevara, provided valuable field assistance, as did our students (Jennifer Howeth, Heidi Kloeppel, Jean Krejca, Jaina Moan, Matt Stephens, Brook Swanson, and Chris Williamson). Countless residents of Cuatro Ciénegas, especially Roberto Arredondo, Pepe Lugo, Beto Lugo, Prof. Héctor ("Tito") Mendez, and Benigno Vázquez, provided valuable information and general assistance. Funding for our efforts was provided by the Nature Conservancy's Ecosystem Research Program. Additional funding came from Earthwatch Institute's Student Challenge Award Program (to DAH) and a National Science Foundation POWRE award (to JCM).

Chapter 8

The Desert Fan Palm Oasis

JAMES W. CORNETT

The Colorado Desert of Alta and Baja California is second only to Death Valley as the hottest and driest place in North America. The mean daily maximum temperature in July reaches 43°C, with an average annual precipitation of just eight centimeters (Zabriskie 1979). With concomitant high evaporation rates and little rainfall, vegetation is predictably scant; plants are generally small, and perennials often lifeless in appearance. Given these conditions, the Colorado Desert might seem the last place to find dense groves of large, water-loving palm trees.

Yet palms were what General Stephen Kearny and his exhausted soldiers encountered after weeks of trekking across the bleak landscapes of western Arizona and southeastern California. In the heart of the Colorado Desert, they were surprised and relieved when, on 29 November 1846, they spotted green foliage at the base of a distant hillside. Lieutenant William Emory, a member of Kearny's contingent, wrote of the encounter, "A few miles from a spring called Oro Grande . . . several scattered objects were seen projecting against cliffs, hailed by the Florida campaigners . . . as old friends. They were cabbage trees, and marked the locale of a spring and small patch of grass" (quoted in Henderson 1961:148). The so-called cabbage trees were not the *Sabal* palms with which the Florida campaigners were familiar. Instead, they had encountered a genus and species new to science. Thirty-three years later, the German botanist Herman A. Wendland (1879) would give the palm the scientific name of *Washingtonia filifera* in honor of America's first president.

In comparison to the surrounding desert, an oasis dominated by *W. filifera* appears as a verdant paradise beckoning weary travelers. The dense palm crowns give shade and create a cooler environment. Plant life is both abundant and frequently diverse and, in earlier times, provided moist green forage for livestock and a variety of food resources for desert Indians

(Cornett 1987a). Most important is that drinking water was at the surface or could be easily reached by digging.

Approximately 160 desert oases support one or more mature individuals of *W. filifera*, popularly known as the desert fan palm. Although dozens of perennial species are associated with desert springs, when the visually dominant *W. filifera* is present the site is usually referred to as a palm oasis. This designation reveals the importance humans place on plant height regardless of the ecological significance of a species.

Distribution of Desert Fan Palm Oases

Although *W. filifera* is one of the most widely planted ornamental trees in the world, wild populations are primarily restricted to the Colorado Desert subdivision of the Sonoran Desert (fig. 8.1). In this region, the combination of relatively mild winters and hot summers provides a temperature regime conducive for this species of obvious tropical affinities. More important, however, is the existence of two geological features in the Colorado Desert that provide an abundance of seeps, springs, and streams in an otherwise extremely arid environment. These two features are the Peninsular Ranges and the San Andreas Fault. When considered together, these provide for the water requirements of over 90 percent of wild desert fan palms. The remaining 10 percent of wild desert fan palms are scattered about isolated springs in California, Nevada, and Arizona. The only nondesert locality occurs in the chaparral community of Valle de los Palmas, Baja California (Moran 1977).

Damp soil at the surface is an absolute requirement for the germination of desert fan palm seeds. Permanent water within approximately four meters of the surface is required for the survival of mature trees. These requirements are only met where (1) the runoff from high mountains is continuous; or (2) where earth faults allow groundwater to rise to the surface, forming permanent springs. The Peninsular Ranges of Alta and Baja California, specifically the northern one-third of these ranges, provide the lofty peaks and extensive watersheds necessary for continuous runoff. The highest peaks reach more than three thousand meters in elevation and are the first obstacle to winter storms traveling east off the Pacific Ocean. Snow typi-

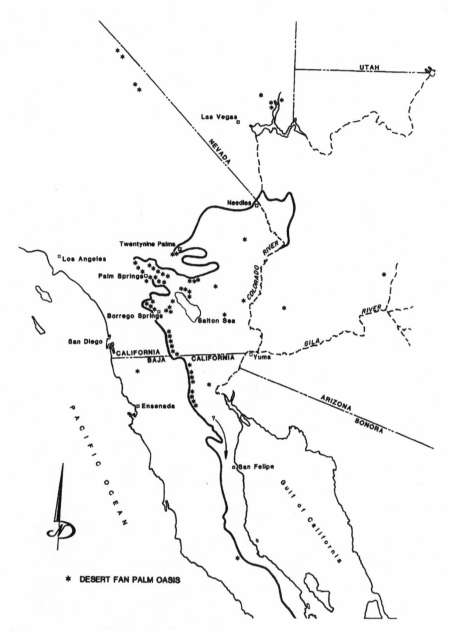

Figure 8.1. The distribution of desert fan palm oases.

cally blankets the summits from December through March. The Peninsular Ranges are also the first obstacle to occasional summer storms traveling north off the Gulf of California. The peaks receive occasional showers during this season as well. The bimodal precipitation regime ensures surface or at least subsurface runoff down the major canyons throughout the year. Thus, it is within the canyons along the eastern desert slopes where the majority of wild desert fan palms are found. Indeed, one of the largest undisturbed palm oases in the world, with 2,511 mature trees, is located in Palm Canyon, at the intersection of the San Jacinto and Santa Rosa mountains of the Peninsular Ranges (Cornett 1986a).

The second area of desert fan palm concentration lies along the San Andreas Fault, the longest and most extensive active fault system in the Southwest. At the San Andreas, groundwater flowing from high to low topography and perpendicular to the fault is prevented from advancing by an impermeable layer of fine soil referred to as fault gouge. At more than two dozen sites in the Colorado Desert, there is sufficient flow for the impounded water to reach the surface. This has allowed the establishment of numerous oases with some seeps and springs supporting several hundred desert fan palms (Vogl and McHargue 1966; Cornett 1993).

In addition to being restricted in its distribution by the availability of water, *W. filifera* is also limited by the severity of winter cold. The desert fan palm has never been found growing wild above 1,250 meters or, until very recently, at latitudes greater than 37° N. Presumably, the degree and duration of minimum temperatures on winter nights are the factors that contribute to the desert fan palm's altitudinal and latitudinal limits. That being said, the desert fan palm is one of the most cold-tolerant palms in the world. Seedling palms can survive temperatures of −21°C for one hour, and mature palms are known to withstand temperatures down to −11°C (Cornett 1987b).

There do not appear to be any significant soil limitations on the distribution of *W. filifera*. The species thrives in the coarse volcanic soils of the Kofa Mountains in western Arizona, the granitic soils at the base of the Peninsular Ranges, and the fine alkaline silts along the San Andreas Fault. Near Dos Palmas Spring, east of the Salton Sea, palms are found growing out of soil surfaces that are white with alkaline deposits left from upwelling groundwater.

The range of the desert fan palm is currently expanding, with new records from Death Valley National Park (Cornett 1988a), southern Nevada (Cornett 1988b), and several localities in the Colorado Desert (Cornett 1985a, 1989a, 1991). Contrary to the conclusion reached by Brown and colleagues in 1976, the palm oases at Castle Creek, Arizona, probably represent recent establishments of *W. filifera*. The absence of old palms, downed trunks, and exit holes of the giant boring beetle (*Dinapate wrightii*) at Castle Creek are characteristic of recent establishments by this species. Furthermore, desert fan palms are present as ornamentals at a nearby ranch and hotel and could have rapidly established at natural springs in the area after their introduction. Interestingly, most of the new palm populations have been found north of the species' known range and in environments that experience winter temperatures colder than those found in the Colorado Desert. Examples include Grapevine and Travertine springs in Death Valley National Park (Cornett 1988a) and Warm and Juanita springs in southern Nevada (Cornett 1988b).

The Origin of the Desert Fan Palm

The desert fan palm has been referred to as a relict species by Axelrod (1950). The tropical affinities of its family (Arecaceae), relative scarcity, and disjunct distribution were cited as evidence of its relict status. In addition, Axelrod stated that *Washingtonia* fossils of Miocene and Pliocene age had been found in the present-day Mojave Desert, suggesting a much broader distribution than at present. Numerous authors continued Axelrod's thesis, stating that the fan palm is "a relict species dating back to the Miocene and Pliocene" (Vogl and McHargue 1966) and describing it as "a relict species now limited to sites with a permanent water supply" (Burk 1977), "a semi-tropical plant that was once more widespread than at present" (Olin 1977), and "a holdover from Miocene and Pliocene times" (Schwenkmeyer 1986).

Relict species are those that "in the past were widely distributed, were affected by climatic changes and survive now only in a few islands of favorable climate" (Cox et al. 1976). All available evidence, however, indicates that *W. filifera* has never been more abundant or widely distributed than it is today. Adult desert fan palms now number nearly 24,000, as compared with 17,600 just fifty years ago (Cornett 1989b). Additionally, Axelrod misidentified his fossil palms; therefore, no known fossils exist. Furthermore, genetic heterogeneity between scattered oases is nonexistent (McClenaghan and Beauchamp 1986), and as described earlier, the range of the desert fan palm has expanded with colonizations at many new localities. This body of evidence suggests that, as a species, the desert fan palm is more like a recently evolved, invasive taxon than a relict.

The probable geographic origin of *W. filifera* is the Baja Peninsula. The genus is represented by just two species, *W. filifera* and *W. robusta* (Mexican fan palm). Both are native to the Baja Peninsula and do not occur naturally in adjacent Sonora, Mexico (Shreve and Wiggins 1964). Only the more cold-tolerant *W. filifera* occurs north and east of Baja California. Barring evidence to the contrary, biogeographers assume that the origin of a taxon is most likely the region harboring the greatest number of surviving representatives (Cox et al. 1976). Today, the only genus of palm that occurs in both mainland Sonora and the Baja Peninsula is *Brahea*, the blue fan palm. *Brahea* is thought to be closely related to the genus *Washingtonia* (Uhl and Dransfield 1987). *Washingtonia* may thus be speculated to have evolved from a *Brahea* population isolated on the Baja Peninsula at the time of the peninsula's creation some 12–14 million years ago (Gastil et al. 1983). Both genera are generally restricted to isolated springs and seeps on the peninsula; this factor combined with the region's steep topography could facilitate the isolation of populations and, ultimately, speciation.

The same evidence mentioned previously—the lack of fossils, increase in numbers, range expansion, and close genetic affinities between desert fan palm populations—suggests a recent arrival from Baja California into what is today the United States. Perhaps *W. filifera* spread northward into the United States sometime after the warming of the Southwest at the close of the Wisconsin ice age.

Characteristics of the Desert Fan Palm

The desert fan palm is one of the tallest native plants of the Sonoran Desert and the largest palm in North America. Mature individuals are known to reach twenty-five meters in height and attain a trunk diameter of up to one meter. Its most distinctive characteristic is the skirt of dead leaves, or "petticoat," that can completely obscure the trunk. Unlike most other palm species, the leaves of the desert fan palm adhere to the trunk throughout its life. Such a "virgin" palm appears even larger because the dead hanging leaves add another two meters to the tree's diameter. The woody trunk is revealed only when the petticoat has been burned away or, in rare instances, if unusually high winds or floods have torn it off (fig. 8.2).

The desert fan palm is considered a fast-growing species, adding up to thirty centimeters of trunk height annually for the first twenty years (Cornett unpubl. data). After that, the growth rate steadily decreases, with trees over fifteen meters in height not adding more than a few centimeters per year. At sites where water supplies are limited (such as at Pygmy Grove in Anza-Borrego Desert State Park) or where climate conditions are relatively cold (as at Grapevine Springs in Death Valley National Park), even young palms may add just five or six centimeters per year.

The desert fan palm increases the height of its trunk by producing new leaves at its growing tip, or apical meristem. New leaves emerge vertically from the meristem and are then pushed aside as more leaves emerge. The trunk is made taller by this process. The apical meristem may be considered the most important part of the desert fan palm. The trunk, leaves, and roots can sustain appreciable amounts of damage, but if the growing tip is injured, the palm dies. Although date palms (*Phoenix* spp.) can grow sprouts from their root crowns, the desert fan palm has no such capacity. Like most monocots, palms have fibrous root systems composed of thousands of pencil-width rootlets that are rarely more than five meters in length. (The longest desert fan palm root measured was ten meters.) The root mass is so dense that the roots of other competing plant species seldom, if ever, penetrate the soil space occupied by the palm.

The world's approximately 2,800 palm species belong in the plant family known as Arecaceae. The desert fan palm is also a member of a subgroup termed the coryphoid palms, a division that includes the genera *Brahea*, *Livistonia*, *Pritchardia*, and *Sabal*. A key feature of the subgroup is the fan-shaped or palmate leaf, which is best displayed in *Washingtonia*. Desert fan palm leaves, or "blades," can be nearly two meters long, with a comparable width, and resemble a huge fan with forty to sixty accordion-like folds. Typically, each leaf remains functional for about one year, turns brown, and then becomes part of the petticoat. The flattened petiole, or leaf stalk,

Figure 8.2. Desert fan palms (*Washingtonia filifera*) in Magnesia Springs Canyon, Santa Rosa–San Jacinto Mountain National Monument, Riverside County, California. Unburned palms show full skirts, a unique feature of this palm species.

can be up to two meters long and, in palms under ten meters in height, is armed with rows of dangerously sharp spines on the two edges.

Flower stalks, called spadices, emerge from a palm crown in late spring. Counts of flower stalks on several hundred palms (and counts of flowers on a few large spadices) indicate that healthy palms produce up to fourteen spadices, each with approximately 400,000 flowers (Cornett 1989c). Thus, an extremely robust palm could produce approximately 6 million flowers. Fertilized ovaries develop into fruits through the summer and ripen into purplish black, pea-sized dates in fall. Fruit removal from several palms indicates a maximum production of approximately 350,000 mature seeds.

Palms and Fire

Once in a great while, perhaps every few decades, lightning can be expected to strike a tall oasis palm. The fronds are highly flammable. Chris Moser (pers. comm.) observed a lightning bolt strike a *Washingtonia* palm. He related how the top of the palm literally exploded when struck, sending flaming leaves in all directions. Such an event would likely set an entire oasis on fire. Most oasis fires today are started by humans—either accidentally or as acts of vandalism. There are many accounts of vandalism-started fires. The Dry Falls fire of 1980, for example, was started by two young boys playing with matches near Palm Springs, California. The palm oasis at Corn Springs in Riverside County's Chuckwalla Mountains burned in 1979 when campers built a campfire under the dry skirt of one of the trees. The palms at Hidden Spring in the Mecca Hills were torched in 1977 by several young men—just for fun. Of the approximately 160 desert fan palm oases, all but 14 show individuals with charred trunks indicating that a fire has occurred.

Although many palm oases are located far out on the desert and are surrounded by sparsely vegetated terrain, most large palm oases and most wild desert fan palms grow in canyons at the eastern base of the Peninsular Ranges. The palms and other oasis-associated plant species—such as cottonwoods (*Populus* spp.), willows (*Salix* spp.), and California sycamore (*Platanus racemosa*)—are abundant at the lower reaches of the canyons. White alder (*Alnus rhombifolia*), pines (*Pinus* spp.) and species of willows are often abundant at the upper elevations, forming a continuous fuel corridor up and down the canyon. A fire that starts anywhere in the high country, or many miles away on the western slopes, may eventually reach the lower canyon bottoms along the eastern sides of the mountains. Insofar as the oases are concerned, these may be referred to as contact fires. The fire does not originate in the palm oasis but rather spreads there as a result of the oasis coming in contact with other plant communities subject to burning.

Regardless of the origin of a fire, the desert fan palm shows a remarkable ability to survive a conflagration. When a palm oasis burns, adult palms typically lose their leaf skirts, trunks are charred, and green crowns are destroyed. Within months, however, the trees reveal that they are alive when they produce an entire new rosette of leaves. The ability of the desert fan palm to survive fire is related to its taxonomy. Many, if not most, members of the palm family are adapted to a fire regime. Because palms are monocots, the sensitive vascular tissue is arrayed throughout the moist trunk, not confined to a narrow cambium layer beneath the bark as is the case with conifers and dicots. Many fires are sufficiently hot to destroy the vascular tissue of a conifer or dicot, but rarely is a fire hot enough to kill the vascular tissue in the center of a palm.

Not only does *W. filifera* survive fire, but it is also the only desert community that thrives under a fire regime. Hidden Palms, near the floor of the Coachella Valley in Riverside County, California, burned three times between 1939 and 1979: the greatest fire frequency of any palm oasis (Cornett 1989c). Yet Hidden Palms had the highest ratio of young to mature palms of all the oases examined. By 1983 there were 264 young palms and 191 adult palms, a ratio of seven to five. Fire promotes the establishment and growth of young palms by removing competing species.

Fire affects the plant species composition of a palm oasis. It favors palms over cottonwoods, willows, sycamores, and most other perennials because the latter (or at least their aboveground portions) are destroyed by fire. An absence of fire allows other plant species to gain a foothold and compete for dominance with the palms. Thus, the Fremont cottonwood (*Populus fremontii*) and species of willow are large and relatively abundant in oases that have not burned for several decades and are scarce or even absent in oases that burn more frequently (Cornett and Zabriskie 1981).

Fire also increases fruit production. In one study (Cornett 1986b), mature palms that had been burned recently were likely to grow eleven fruit stalks (spadices) compared with seven fruit stalks for unburned palms. (The removal of competing vegetation and resultant increase in soil moisture for the surviving palms was the likely explanation for the observed increase.) Thus, fire not only promotes the growth of young palms but also increases seed production.

Synecology

The Vegetation Composition of Palm Oases

Although the desert fan palm may be the most conspicuous plant in the oasis environment, it is usually not the dominant species. Ground cover is the single criteria most often used by ecologists to assess plant species dominance. Using this criterion, *W. filifera* is not the dominant species in even half of all oasis environments; stated another way, palm oases are most often dominated by a plant species other than the desert fan palm.

Vegetation analyses were conducted on 122 of the approximately 160 desert springs where *W. filifera* was known to occur. For the purposes of the vegetation analysis, a palm oasis was defined as everything enclosed within an imaginary line drawn ten meters around an individual palm or the outermost individuals of a discrete cluster of adult palms. (Common and scientific nomenclature for plant species in the subsequent discussion follows Hickman 1993.)

In total, 86 perennial plant species were associated with desert fan palm oases (appendix 8.1). Honey mesquite (*Prosopis glandulosa*) occurred in 74 of 122 palm oases and was therefore the most frequent associate of *W. filifera* (table 8.1). Tamarisk (*Tamarix ramosissima* × *chinensis*) was second with 47, and squaw waterweed (*Baccharis sergiloides*) and salt grass (*Distichlis spicata*) were third with 38 occurrences each.

At 55 of the locations (45 percent), *W. filifera* was estimated to be the dominant plant species using ground cover criteria (table 8.2). Other plants that dominated the vegetation at palm oases included *P. glandulosa* (8 percent of those oases evaluated), species of *Baccharis* (7 percent), and *Tamarix ramosissima* × *chinensis* (5 percent).

Nine exotic perennials have succeeded in establishing in desert fan palm oases (see appendix 8.1). Three are the dominant species in the palm oases where they occur. Only *Tamarix*, however, is the dominant exotic plant

TABLE 8.1. Ten most frequent perennial plant species found in palm oases

Plant species	Oasis occurrences (N)	(%)
Washingtonia filifera	122	100
Prosopis glandulosa	74	61
Tamarix ramosissima × chinensis	47	39
Baccharis sergiloides	38	31
Distichlis spicata	38	31
Pluchea sericea	37	30
Salix spp.	32	26
Acacia greggii	31	25
Typha domingensis	30	25
Sporobolus airoides	26	21

Note: The total number of oases is 122.

in more than one oasis. Despite the arrival of *Tamarix* and several other exotic plant species, the numbers of desert fan palms have increased dramatically within the past fifty years.

Other than the desert fan palm, there are no perennial species that are restricted in their distribution to desert fan palm oases. All of the species listed in tables 8.1 and 8.2 and appendix 8.1 are also found in environments other than a desert fan palm oasis.

Faunal Associates of the Palm Oasis

Just as the palm oasis has served as a verdant paradise beckoning weary human travelers, so too has it been an important respite and sanctuary for many animals. Some species may only pause at an oasis for shade or food during a migratory journey. Others may take up seasonal residency. Still others depend on the palm oasis for survival.

ARTHROPOD POLLINATORS. Although some species of palm are wind pollinated, others are known to rely on insects for pollination (Blombery and Rodd 1982). Researchers have not determined which of these two systems is operative for *W. filifera*. Lepesme (1947) listed twenty insect

TABLE 8.2. Plant species dominating vegetation at two or more desert springs supporting *Washingtonia filifera*

Plant species	Springs N	%
Washingtonia filifera	55	45
Prosopis glandulosa	10	8
Baccharis spp.	9	7
Tamarix ramosissima × *chinensis*	6	5
Atriplex spp.	5	5
Salix spp.	5	5
Vitis girdiana	4	3
Platanus racemosa	3	2
Distichlis spicata	2	2
Juncus acutus	2	2
Pennisetum setaceum	2	2
Pluchea sericea	2	2
Quercus turbinella	2	2
Sporobolus airoides	2	2

Note: Vegetation evaluated in 122 palm oases.

species associated with *W. filifera* but, with the exception of the giant palm boring beetle (*Dinapate wrightii*), did not describe the relationship between the insects and the palms. A first step in ascertaining the mechanism of pollination is to determine which, if any, insects visit *W. filifera* flowers.

Insects visiting palm inflorescences have been observed and collected on three occasions (Cornett 1986c). Over seventy-five insects per hour may visit inflorescences on calm days. The most frequent visitor is the introduced honeybee (*Apis mellifera*), followed by paper wasps (*Polistes* spp.) and the California carpenter bee (*Xylocopa californica*). The latter insect often destroys the flower as it feeds. Most insects visit five or more flowers on a single inflorescence and then fly from the palm. Whether they immediately visit a second palm has not been determined. Many of these species may be pollinators of *W. filifera*, and the diversity and abundance of insects in general suggest that the flowers of this palm species are insect pollinated. Insects and other arthropods found on desert fan palm inflorescences are listed in appendix 8.2 in decreasing frequency of occurrence. None of the species collected appeared in Lepesme's (1947) list.

GIANT PALM BORING BEETLE. The animal most dependent on *W. filifera* is the giant palm boring beetle. With a huge head bearing power-ful jaws and a length of up to five centimeters, it is an intimidating insect nearly twice as large as any of the other wood-boring bostrichid beetles. The species is associated exclusively with the desert fan palm, whether in natural or ornamental situations.

The giant palm boring beetle was unknown to science until 1886, when the beetle and larva were described by G. H. Horn. However, not until 1899 (Hubbard) was a paper published describing the habits and life history of the beetle. Henry Hubbard of Los Angeles had found a specimen of *D. wrightii* beneath a desert fan palm near Palm Springs, California. Upon further investigation, he discovered that the large, pale yellow larvae lived only in this species of palm and survived by eating extensive tunnels through the trunks. Just how many years *D. wrightii* larvae spend in their palm nests is unclear. Baker (1971) suggested that the time period from egg to adult beetle lasted from three to five years. However, a carved human figure made from a dead *W. filifera* log was seven years old when an adult beetle emerged from its dime-sized exit hole.

When the larva is ready to pupate, in April or May, it constructs a larval chamber about one inch from the exterior surface of the trunk. This chamber is a wider-than-normal tunnel that provides sufficient space for the beetle to turn around after it chews an exit hole. About seven weeks elapse between pupation and emergence. The adult chews its way out of the trunk. Baker (1971) found that males and females show no interest in each other upon emergence. Not until a few days later, when the female begins to tunnel into the crown of a palm, do the sexes become receptive. The female enters the crown by chewing into a leaf base and constructs her nest chamber at the end of a tunnel a few centimeters in length. Several males may pursue one female into the nest chamber, and mating appears to take place there. A female lays from 400 to 500 white eggs, each measuring about two milli-meters long by one millimeter wide.

Of special interest is the impact of *D. wrightii* on the health and lon-gevity of desert fan palms. There are four obvious causes of palm mortality: extended drought, flood, fire, and beetle damage. Drying-out of the soil is probably the most significant cause of death for seedlings, and numerous mature palms in Wentworth Canyon near Palm Springs died as a result of a thirty-year drought that ended in 1976. A single flash flood has been known to uproot hundreds of palms in such places as Cloudburst Canyon, which drains the Sierra Juarez of Baja California. Each palm oasis experiences fire, but the mortality of adult palms is low: less than 1 percent were lost in the fire that occurred in the canyons surrounding Palm Springs, California, in 1980. About 45 percent of palms over thirty-five feet in height (and less than 1 percent of fruit-bearing trees under twenty feet) harbor *D. wrightii* holes. It is obvious that the beetles damage and often kill the palms by chewing through the vascular tissue and interrupting the flow of water and nutrients through the trunks. In taller palms, there may be several hundred exit holes, and the trunks become nothing more than columns of sawdust.

The exact percentage of palms killed by *D. wrightii* would be difficult to determine since a fallen palm log with exit holes does not necessarily indicate beetle-caused death. However, after examining hundreds of dead standing trees, I have calculated that in oases not subject to severe floods, 70 percent of the adult palms are killed outright by beetles, and many remaining trees are killed by fire made more harmful because of beetle exit holes. The percentage can vary tremendously depending on the location of the palm oasis and the frequency of fire. Although many trees have exit holes, a few seem to have unusually large concentrations, giving the impression that some trees are more attractive to egg-laying females than are others. Such trees are usually the tallest and presumably the oldest palms in the groves. Because of their age, the older trees have had the greatest opportunity of being the recipients of ovipositing females. Whereas most living palms do not have numerous beetle holes, most tall dead palms do, reinforcing the idea that very old trees are more likely to be attacked than young ones. These latter observations suggest that the palms have some kind of defense against the beetles but that this defense deteriorates with age. Because palms are monocots, they do not have annual growth rings to help determine age, as do conifers and dicots. Thus, the age of an individual palm can be surmised only with knowledge of growth rate, condition of the tree, and in a small number of cases, a review of historical photographs.

One of the interesting aspects of the beetle's life history is its apparent ability to disperse from one isolated desert fan palm oasis to another, sometimes flying distances of several miles. Giant palm boring beetle exit holes have been found in the vast majority of palm oases in southeastern

California and northeastern Baja California, and some are more than twenty kilometers apart. The two likely explanations for this distribution are that the beetles (1) have been in the vicinity since the groves first came into existence or (2) arrived later after the palms became established. If the beetles arrived after the establishment of a grove, then random chance or scent attraction are the likely explanations for their arrival. Sight must be excluded since the beetles are practically blind (Baker 1971).

What are the chances that both a male and a female beetle would accidentally fly into an unoccupied grove within a few days of each other and initiate a new colony of palm boring beetles? Undoubtedly, this could have happened once or twice, but this scenario is not likely to explain the beetle's presence in every palm grove in which it occurs. Scent attraction is a tempting explanation. The ability of salmon to "smell" their way home through thousands of miles of ocean waters to the stream where they were born is well documented (Orr 1971). Many moths and other insects are also known to produce pheromones that attract mates from miles away (Alcock 1975). Perhaps desert fan palms have an odor that can be detected by the beetles.

Recently the presence or absence of *D. wrightii* has assisted in determining the relative age of certain desert fan palm oases. Of particular interest in this regard was the age of groves far removed from the center of the palm's distribution—along the San Andreas Fault and in the canyons draining the desert slopes of the Peninsular Ranges. As has been mentioned previously, some naturalists believe that palm oases are relict communities that have been present since the warmer, moister Pliocene epoch, when palms were more widespread than they are today. If this were so, there should have been ample opportunity for the beetles to fly the presumed shorter distances necessary to reach neighboring groves. If, however, the groves are not relicts but are of recent origin, then they might be unoccupied because they became established after desert climatic conditions appeared some 11,000 years ago; these recent palm communities would have always been too isolated to be reached by *D. wrightii*. In fact, nearly all of the isolated palm oases lack *D. wrightii* populations, suggesting that they are of recent rather than ancient origin.

The only known predators of *D. wrightii* are birds, specifically the northern flicker (*Colaptes auratus*) and ladder-backed (*Picoides scalaris*) and Gila (*Melanerpes uropygialis*) woodpeckers. All three locate larvae tunneling close to the exterior, apparently by listening to them chew. Evidence of a capture site often shows up as a doughnutlike pattern with a circle of small holes and a large hole in the center where the grub was located and withdrawn. The pecking of such small test holes may cause a larva to reveal its precise location when it attempts to escape by chewing deeper into the interior.

Not only does *D. wrightii* provide food for woodpeckers, but it also facilitates the construction of their nests. Living desert fan palm trunks are too resistant for woodpeckers to peck out cavities. Hence, their nests are almost always restricted to dead standing palms—and only those that show numerous *D. wrightii* exit holes. These trees are soft enough to attract breeding woodpeckers intent on constructing a new nest cavity. Of course, in later years, these cavities may also be used by other avian species such as the house finch (*Carpodacus mexicanus*), American kestrel (*Falco sparverius*), western screech-owl (*Otus kennicottii*), and introduced European starling (*Sturnus vulgaris*). Another animal that benefits from *D. wrightii* exit holes is the southern California carpenter bee (Cornett 1985b). The females of this insect often lay their eggs inside the empty exit tunnels and may be seen almost every spring day entering the beetle holes to provide food for their developing young. In some areas, these may be the only suitable nesting sites for the bees, as they cannot bore their own holes in living palm trunks.

The giant palm boring beetle is an excellent example of the often complex and subtle relationships between an organism and its environment. As just described, the presence of many other animals in the palm oasis is facilitated by, and perhaps dependent on, *D. wrightii*. The vitality of the groves is based on a continuous replacement of older palms by younger palms—a cycle that is enhanced if not perpetuated by the activities of this insect.

AMPHIBIANS. A single amphibian species is restricted to a palm oasis: the desert slender salamander (*Batrachoseps aridus*; Brame 1970). This species is found only at Mayhew Palms in the Santa Rosa Mountains of the Peninsular Ranges province. The habits of this salamander do not suggest that it is dependent on the desert fan palm. Rather, both species rely on the moisture provided by the spring that occurs at the bottom of the ravine in which they live. The desert slender salamander has been listed as an endangered species by both the state and the federal government. The only other salamander known to occur in a palm oasis is the garden slender salamander (*Batrachoseps pacificus major*; Cornett 1981). This species is found in the Andreas Canyon palm oasis along the eastern base of the San Jacinto Mountains.

The most widespread amphibian found in palm oases is the red-spotted toad (*Bufo punctatus*). This species is associated with almost every *W. filifera* grove that lies at the eastern base of the Peninsular Ranges. The only other toad that has been reported from a palm oasis is the western toad (*Bufo boreas*).

In those palm oases located at the eastern base of the Peninsular Ranges, where permanent or near-permanent surface water is present, the California treefrog (*Pseudacris cadaverina*) is the most encountered amphibian. The Pacific treefrog (*Hyla regilla*) has also been recorded at Push-walla Palms, Riverside County, California (Stebbins 1985). The only true frog known to be present in a palm oasis is the bullfrog (*Rana catesbeiana*), a species introduced at ponds located in Thousand Palms Oasis, Riverside County, California.

REPTILES. The most noticeable reptiles in the oasis are members of the genus *Sceloporus*, commonly referred to as spiny lizards. These lizards are often seen basking or climbing on palms whose leaf skirts have been burned off by a wildfire, have been worn off by floodwaters, or have fallen away from prone or standing dead palms. In palm oases along the eastern base of the Peninsular Ranges the granite spiny lizard (*S. orcutti*) is the species usually observed. Away from the Peninsular Ranges, along the San Andreas Fault and at isolated desert oases, the desert spiny lizard (*S. magister*) is usually encountered.

The most frequently observed serpent in groves of *W. filifera* is the common kingsnake (*Lampropeltis getula*). This species seems attracted to moist habitats in desert environments, a factor that may explain the frequency with which it is seen in palm oases. The two-striped garter snake (*Thamnophis hammondii*) is a surprising but relatively common inhabitant in palm oases at the eastern base of the Peninsular Ranges where permanent surface water is present. It enters the water freely to escape predators and to feed on tadpoles and adult California treefrogs and red-spotted toads. The small and secretive ringneck snake (*Diadophis punctatus*) is also an inhabitant of this type of palm oasis environment.

BIRDS. Intensive yearlong avian surveys were conducted in two large palm oases located in the Colorado Desert subdivision of the Sonoran Desert of Riverside County, California (Cornett unpubl. data). The vegetation of both oases was dominated by *W. filifera* and contained hundreds of adult trees. Thousand Palms oasis is situated on the San Andreas Fault and in 1989 consisted of 517 adult trees. Palm Canyon is located at the eastern base of the San Jacinto Mountains, the northernmost extension of the Peninsular Ranges geological province. As mentioned previously, Palm Canyon contains 2,511 adult trees, making it the largest desert fan palm oasis in existence. Both oases have year-round standing water.

I observed that birds used four different approaches in utilizing the palm oasis environment (for species lists in each category, see appendix 8.3). The first category is migratory birds that appeared in the spring and fall. They were often observed to drink at waterholes or feed before moving on. Several species were known to have spent the night in the crowns of tightly packed palms before continuing on their migratory journey. A second group of migratory species spent all or a portion of the winter occupying the palm oasis environment but left in spring to breed elsewhere. The third category is those few species that arrived in spring to breed and then departed in summer or early fall. Fourth, many birds were considered permanent residents and were known to breed in the oasis.

Although all bird species listed in the appendix have been observed in one of the two palm oases, their categorization is not as discrete as the list might suggest. The American robin (*Turdus migratorius*), for example, is essentially a migrant (category 1) but also occasionally spends the winter (category 2) and might even breed (category 3) at northern or high-altitude palm oases.

The hooded oriole (*Icterus cucullatus*) is the only avian species that may be said to depend on the palm oasis, or at least palm trees, for breeding within the Colorado Desert. These beautiful birds utilize the fibers from the margins of the palm trees to weave their basketlike nests. They usually hang the nests from the underside of a living frond. By late spring, any palm oasis with ten or more adult palms can be expected to support at least one pair of breeding hooded orioles. The males arrive first, typically by April, to establish a breeding territory. Their melodious song attracts the much drabber-colored female. By June the pair has constructed the nest, the female has laid her eggs, and the young have hatched and are being fed. Both males and females use the palms not only for nesting but also as display perches and as feeding sites. Hooded orioles frequently visit the palm flowers to feed on the insects that are attracted to the nectar. By late August, many hooded orioles have left the palm oases and presumably have headed south into Mexico. With the widespread ornamental planting of desert and Mexican fan palms, as well as many other palm species, in the past century, hooded orioles have been able to expand their numbers and range dramatically. They are now common breeders in coastal southern California, the central valley of northern California, southern Nevada, and southern Arizona. In most (though not all) instances, their presence is associated with palms in either natural or ornamental situations.

MAMMALS. The only mammalian species that seems dependent on the desert fan palm oasis is the southern yellow bat (*Lasiurus ega*). By North American standards it is considered medium-sized, weighing 12–19 grams with a total length of 109–124 millimeters (Jameson and Peeters 1988). Its common name reflects the yellowish brown fur that covers all but the wings, ears, and feet. In palm oases, *L. ega* is often the most abundant bat and can be easily captured in mist nets as it comes to water holes. In the author's experience, the western yellow bat is typically the only species captured in large, dense palm groves, and no individuals of this species have ever been taken outside of palm oases. Although it has been observed to drink by touching the water while in flight, it also is very likely attracted to water holes by the abundant flying insects that provide its food. During the day, these bats prefer roosting in the dead leaf skirts that characterize the desert fan palm. They have been found hiding among the dead fronds of *W. filifera* in both natural and ornamental environments. Dead or injured individuals have been discovered in sections of palm skirts that had been downed by high winds or pruning. Oasis fires, by removing the leaf skirts, temporarily eliminate an important habitat for these bats.

The most abundant terrestrial mammals in palm oases are rodents. In a survey of sixteen palm oases in California and Baja California, eight rodent species were recorded, though only two species were considered common: the cactus mouse (*Peromyscus eremicus*) and the spiny pocket mouse (*Chaetodipus spinatus*; Cornett 1989d). Both species also occurred, albeit in lesser numbers, in nearby scrub vegetation of mesquite and creosote bush.

Many large desert mammals have been observed in or near palm oases. Most notable is the coyote (*Canis latrans*), whose presence is detected most often by its scats. In the fall, coyotes consume large numbers of palm fruits that have fallen to the ground. Though these canids easily digest the sweet skin and flesh of the fruit, they cannot digest the extremely hard seeds. In the fall, some scats are composed almost entirely of fan palm seeds.

Once consumed and passed by coyotes, palm seeds germinate readily if deposited on moist soil (Cornett 1985c). This fact, coupled with the coyote's known ability to travel up to fifty-eight kilometers in a single twenty-four-hour period, indicates that coyotes are excellent candidates for short- and medium-range dispersal of *W. filifera* seeds (Cornett 1984).

Conclusion

The desert fan palm is an important component in the ecology of desert oases. Along with other perennial species it forms an environment that is decidedly unique when compared to the surrounding desert scrub vegetation.

Washingtonia filifera provides critical necessities to several animal species and important nesting sites for numerous birds. Because of its ability to survive fire, the desert fan palm is a postfire stabilizing component in the oasis environment through the maintenance of shaded microhabitats and the continued provision of food and shelter for many desert animals.

A wide variety of evidence suggests that the desert fan palm is not a relict species but a recently evolved invasive species that has been forming new colonies around desert springs throughout the past century. These colonies often dominate oasis vegetation.

ACKNOWLEDGMENTS

I wish to thank the Richard King Mellon Foundation of Pittsburgh, Pennsylvania, the Palm Springs Desert Museum, and the Joshua Tree National Park Association for providing the financial support necessary for ongoing studies on the ecology of desert fan palm oases. For fifteen years, the Agua Caliente Band of Cahuilla Indians provided unrestricted access to several palm oases within the Indian Canyons Tribal Park. My appreciation is extended to all members of the band as well. Last, I wish to extend my gratitude to Thomas R. Van Devender of the Arizona-Sonora Desert Museum, Vicky J. Meretsky of Indiana University, and Larry Stevens for their thoughtful review of the manuscript and their many helpful suggestions.

Appendix 8.1

PERENNIAL PLANT SPECIES OCCURRING
IN DESERT FAN PALM OASES

An asterisk indicates an introduced species. Scientific names are given first, followed by common names.

Acacia greggii, catclaw
Adiantum capillus-veneris, southern maiden-hair fern
Allenrolfea occidentalis, iodine bush
Alnus rhombifolia, white alder
Ambrosia dumosa, burro-weed
Amorpha fruticosa, false indigo
Artemisia dracunculus, tarragon
*Arundo donax, giant reed
Atriplex canescens, fourwing saltbush
Atriplex hymenelytra, desert holly
Atriplex lentiformis, quail brush
Atriplex polycarpa, cattle spinach
Baccharis emoryi, Emory baccharis
Baccharis salicifolia, mule fat
Baccharis sergiloides, squaw waterweed
Bebbia juncea, sweetbush
Brahea armata, blue fan palm
Brandegea bigelovii, brandegea
Capparis isomeris, bladder-pod bush
Chilopsis linearis, desert willow
*Cynodon dactylon, Bermuda grass
*Cyperus involucratus, umbrella plant
Datura discolor, jimsonweed
Datura wrightii, jimsonweed
Distichlis spicata, salt grass
Encelia farinosa, brittlebush

Epilobium canum, California fuchsia
Epipactis gigantea, stream orchid
Eriodictyon trichocalyx, yerba santa
Eriogonum inflatum, desert trumpet
Fraxinus velutina, velvet ash
Hibiscus denudatus, desert hibiscus
Hyptis albida, desert lavender
Isocoma acradenia, alkali goldenbush
Juncus acutus, rush
Juncus mexicanus, Mexican rush
Juncus xiphoides, rush
Juniperus californica, California juniper
Justicia californica, chuparosa
Larrea divaricata, creosote bush
Lotus rigidus, deerweed
Lycium fremontii, Fremont's thornbush
Mimulus bigelovii, wishbone bush
Muhlenbergia rigens, deer grass
*Nerium oleander, oleander
*Nicotiana glauca, tree tobacco
Nolina parryi, Parry's nolina
Nolina wolfii, wolf nolina
Parkinsonia florida, blue palo verde
*Pennisetum setaceum, fountain grass
Penstemon clevelandii, beardtongue
*Phoenix dactylifera, date palm
Phoradendron californicum, desert mistletoe
Phragmites australis, common reed
Pinus monophylla, single-leaf pinyon
Platanus racemosa, California sycamore
Pluchea sericea, arrowweed
Populus fremontii, Fremont's cottonwood
Prosopis glandulosa, honey mesquite
Prunus fasciculata, desert almond
Prunus fremontii, desert apricot
Psorothamnus emoryi, Emory's dalea
Psorothamnus schottii, indigo bush
Psorothamnus spinosus, smoke tree
Pteridium aquilinum, bracken fern
Quercus turbinella, turbinella oak
Rhus ovata, sugar bush
Ribes tortuosum, gooseberry
Salix exigua, narrow-leaved willow
Salix laevigata, red willow

Salvia apiana, white sage
Scirpus americanus, bulrush
Sesuvium verrucosum, western sea-purslane
Simmondsia chinensis, jojoba
Sphaeralcea ambigua, apricot mallow
Sporobolus airoides, alkali sacaton
Suaeda moquinii, bush seepweed
* *Tamarix aphylla*, athel
* *Tamarix ramosissima* × *chinensis*, tamarisk
Tidestromia oblongifolia, tidestromia
Trixis californica, trixis
Typha domingensis, southern cattail
Typha latifolia, broad-leaved cattail
Viguiera parishii, goldeneye
Vitis girdiana, desert wild grape
Washingtonia filifera, desert fan palm

Appendix 8.2

ARTHROPODS OBSERVED OR COLLECTED ON INFLORESCENCES
OF *WASHINGTONIA FILIFERA*

Listed in decreasing frequency of occurrence, with the name of the order in parentheses after the scientific and common names. Several arthropods could be identified only to the level of the family.

Apis mellifera, honeybee (Hymenoptera)
Polistes spp., including *P. major*, *P. apachus*, and *P. dorsalis*, paper wasps (Hymenoptera)
Xylocopa californica, California carpenter bee (Hymenoptera)
Dermestidae, hide beetles (Coleoptera)
Forelius foetidus, ant — no common name (Hymenoptera)
Stratiomyidae, soldier flies (Diptera)
Pepsis sp., tarantula hawks (Hymenoptera)
Tripoxylon xantianum, mud dauber wasp (Hymenoptera)
Alleculidae, comb-clawed beetles (Coleoptera)
Litoprosopus coachella, palm moth (Lepidoptera)
Tachytes sp., sand wasps (Hymenoptera)
Prionyx parkeri, sand wasp (Hymenoptera)
Megachile sp., leafcutter bees (Hymenoptera)
Salticidae, jumping spiders (Araneae)

Appendix 8.3

BIRD SPECIES OBSERVED AT DESERT FAN PALM OASES

Species are listed in alphabetical order by common name.

CATEGORY 1: Migratory birds seeking temporary refuge

Turdus migratorius, American robin
Hirundo rustica, barn swallow
Ceryle alcyon, belted kingfisher
Pheucticus melanocephalus, black-headed grosbeak
Dendroica nigrescens, black-throated gray warbler
Vireo solitarius, blue-headed vireo
Passerella iliaca, fox sparrow
Empidonax hammondii, Hammond's flycatcher
Passerina amoena, lazuli bunting
Oporornis tolmiei, MacGillivray's warbler
Vermivora ruficapilla, Nashville warbler
Contopus cooperi, olive-sided flycatcher
Vermivora celata, orange-crowned warbler
Empidonax difficilis, Pacific-slope flycatcher
Pheucticus ludovicianus, rose-breasted grosbeak
Selasphorus rufus, rufous hummingbird
Oreoscoptes montanus, sage thrasher
Catharus ustulatus, Swainson's thrush
Dendroica townsendi, Townsend's warbler
Cathartes aura, turkey vulture
Tachycineta thalassina, violet-green swallow
Vireo gilvus, warbling vireo
Piranga ludoviciana, western tanager
Empidonax traillii, willow flycatcher
Wilsonia pusilla, Wilson's warbler
Dendroica petechia, yellow warbler

CATEGORY 2: Winter-visiting birds

Carduelis tristis, American goldfinch
Bombycilla cedrorum, cedar waxwing
Spizella passerina, chipping sparrow
Junco hyemalis, dark-eyed junco
Catharus guttatus, hermit thrush
Melospiza lincolnii, Lincoln's sparrow
Sitta canadensis, red-breasted nuthatch
Regulus calendula, ruby-crowned kinglet

Amphispiza belli, sage sparrow
Icterus parisorum, Scott's oriole
Sialia mexicana, western bluebird
Zonotrichia leucophrys, white-crowned sparrow

CATEGORY 3: Spring- and summer-breeding birds

Icterus galbula, Baltimore oriole
Vireo bellii, Bell's vireo
Archilochus alexandri, black-chinned hummingbird
Icterus cucullatus, hooded oriole
Carduelis lawrencei, Lawrence's goldfinch
Piranga rubra, summer tanager
Tyrannus verticalis, western kingbird
Zenaida asiatica, white-winged dove

CATEGORY 4: Year-round residents

Pipilo aberti, Abert's towhee
Falco sparverius, American kestrel
Calypte anna, Anna's hummingbird
Tyto alba, barn owl
Thryomanes bewickii, Bewick's wren
Sayornis nigricans, black phoebe
Polioptila melanura, black-tailed gnatcatcher
Amphispiza bilineata, black-throated sparrow
Molothrus ater, brown-headed cowbird
Campylorhynchus brunneicapillus, cactus wren
Catherpes mexicanus, canyon wren
Corvus corax, common raven
Accipiter cooperii, Cooper's hawk
Calypte costae, Costa's hummingbird
Sturnus vulgaris, European starling
Callipepla gambelii, Gambel's quail
Geococcyx californianus, greater roadrunner
Bubo virginianus, great horned owl
Carpodacus mexicanus, house finch
Passer domesticus, house sparrow
Troglodytes aedon, house wren
Charadrius vociferus, killdeer
Picoides scalaris, ladder-backed woodpecker
Carduelis psaltria, lesser goldfinch
Lanius ludovicianus, loggerhead shrike
Zenaida macroura, mourning dove

Colaptes auratus, northern flicker
Mimus polyglottos, northern mockingbird
Stelgidopteryx serripennis, northern rough-winged swallow
Phainopepla nitens, phainopepla (some individuals leave during summer)
Buteo jamaicensis, red-tailed hawk
Salpinctes obsoletus, rock wren
Sayornis saya, Say's phoebe
Auriparus flaviceps, verdin
Otus kennicottii, western screech-owl

Chapter 9

Spring-Supported Vegetation along the Colorado River on the Colorado Plateau

Floristics, Vegetation Structure, and Environment

JOHN R. SPENCE

Around on the rocks in the cave-like chamber are set beautiful ferns, with delicate fronds and enameled stocks. . . . This delicate foliage covers the rocks all about the fountain, and give[s] the chamber great beauty.
—J. W. Powell, 1895

Despite extensive research on the distribution, ecology, and management of riparian ecosystems in the arid American Southwest, spring-supported ecosystems remain poorly studied. One region of the Southwest, the Colorado Plateau (fig. 9.1), is characterized by numerous springs that support a rich diversity of wetland flora and vegetation in an otherwise arid to semiarid climate. Many of these sites are isolated by large expanses of desert vegetation and support unusual and distinctive plant assemblages, such as hanging gardens (Welsh 1989a, 1989b). The Colorado Plateau is a large region that until recently has seen relatively little biological research (West 1988; Harper et al. 1994); only a handful of published studies exist on the flora and vegetation of its springs (e.g., Malanson 1980; Malanson and Kay 1980; Welsh and Toft 1981; Malanson 1982; Welsh 1989a; Romme et al. 1993; Spence and Henderson 1993; Fowler et al. 1995; Graham 1997). These ecosystems represent ideal study sites in which to examine evolution in isolation, the varying roles of dispersal, extinction and disturbance in the structuring of community patterns, community assembly rules, and the effects of future global warming and related changes in regional weather patterns.

The results of a two-year project designed to inventory the biotic and physical components of selected springs along the Colorado River permits a more precise exploration of general relationships between flora, vege-

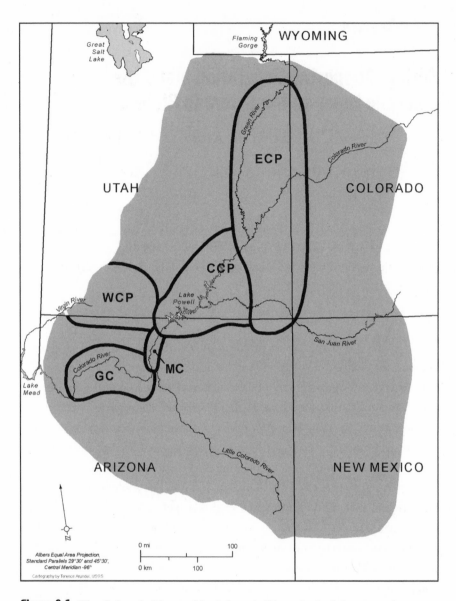

Figure 9.1. The Colorado Plateau (shaded area). The principal rivers are shown, and the five regions discussed in the text are outlined: eastern Colorado Plateau (ECP), central Colorado Plateau (CCP), western Colorado Plateau (WCP), Marble Canyon (MC), and lower Grand Canyon (GC).

tation, and physical factors than has previously been available for this area. Principal physical and geographic factors can thereby be correlated with floristic diversity, species distribution, and vegetation structure. Yet this is only a preliminary effort; the results of this analysis, as well as a review of previous work, suggest potential future research questions on spring ecosystems on the plateau.

Background and Review

The area under consideration consists of the central and northern third of the Colorado Plateau, dominated by the Colorado River and numerous associated tributaries such as the Green, San Juan, Escalante, Paria, and Little Colorado rivers, as well as hundreds of shorter side canyons (see fig. 9.1). This portion of the Colorado Plateau is relatively low in elevation, with the Colorado River ranging from 1,215 meters at Moab, Utah, to 370 meters where the river reaches Lake Mead. Most of the surrounding landscape consists of mesas and cliffs of either sandstone, limestone, or crystalline bedrock. A distinct break in geology occurs at Lees Ferry. Below this break, older Paleozoic limestones are interbedded with shales along the Colorado River; above this break, younger Mesozoic sandstones and shales are found. The entire study area lies below 2,000 meters in elevation in an arid climate, with hot summers, relatively mild winters, and less than twenty-five centimeters of precipitation per year (Spence 2001). That part of the study area in Marble Canyon below RK65 (measured in kilometers downstream from Lees Ferry), at and below approximately 890 meters, is Sonoran Desert in climate, with Colorado River riparian vegetation dominated by mesquite (*Prosopis glandulosa*) and catclaw (*Acacia greggii*). Above that point, Colorado River riparian communities are (or were, through Glen Canyon) characterized by a mix of Apache plume (*Fallugia paradoxa*), live oak (*Quercus turbinella*), New Mexico olive (*Forestiera pubescens*), and poison ivy (*Toxicodendron rydbergii*). Tamarisk (*Tamarix ramosissima*), arrowweed (*Tessaria sericea*), seepwillow (*Baccharis emoryi*), and coyote willow (*Salix exigua*) are abundant throughout the study area. Netleaf hackberry (*Celtis reticulata*), Goodding willow (*Salix gooddingii*), and western redbud (*Cercis occidentalis*) are less common but still relatively widespread.

The earliest descriptions of springs in the region come from Powell in his descents of the Colorado River (Powell 1895). Eastwood (1896) collected plants in the area around Bluff, Utah, and commented on the unusual character of the hanging gardens in the vicinity. Woodbury (1933) briefly described primary succession at seeps in Zion National Park. Clover and Jotter (1944) described the principal plant species at springs along the Colorado and Green rivers, from Green River in Utah to Separation Canyon in the lower Grand Canyon.

More recent vegetation studies in the study area have concentrated on hanging gardens (Welsh 1989a). These are unique plant communities that develop under certain geologic and climatic features in arid to semi-arid climates. They are fed by groundwater aquifers either in fine-grained sandstones or in limestones. They exist on cliff faces or in undercut alcoves formed by spalling and groundwater sapping at aquifer/aquiclude contacts that create one or more springs or horizontal seep lines. At least two distinct habitats occur at most hanging gardens: the seeping, more or less vertical, backwall; and a colluvial-detritus slope at the base. Often a plunge-pool basin develops at the foot of the colluvial slope where water erodes a basin as it falls during flood events from the cliffs above. These pools often retain water for extended periods of time and can support wetland communities. Principal regional aquifers include the Cedar Mesa, Entrada, and Navajo Sandstones at higher elevations and the Coconino, Redwall and Muav Limestones in Marble and Grand Canyons. The development and physical features of the alcoves supporting hanging gardens have been discussed in detail elsewhere (Welsh and Toft 1981; Laity and Malin 1985; May et al. 1995).

Some descriptive work has been done on the floristics and geographic distribution of hanging gardens (e.g., Loope 1977; Welsh and Toft 1981; Rushforth and Merkley 1988 [algae]; Tuhy and MacMahon 1988; Welsh 1989a, 1989b; Van Pelt et al. 1991; Romme et al. 1993; Fowler 1995; Keate 1996; Wong 1999). The most characteristic species of the gardens is the widespread tropical-subtropical maidenhair fern *Adiantum capillus-veneris*, found in most gardens on the plateau. Hanging gardens support mixtures of species of different phytogeographic elements, including species representative of tallgrass prairies, boreal-montane climates, and the Colorado Plateau (Welsh 1989b). Malanson (1980, 1982; Malanson and Kay 1980) examined a series of gardens in the Virgin River Narrows of Zion National Park. Those studies indicated that flooding and soil development strongly controlled the floristics and dispersal spectra of the species occurring in gardens. Flood-prone gardens tended to have fewer species and more species with adaptations to long-distance dispersal. He termed this the fugitive guild. In less flood-prone gardens, there were more species and a larger array of dispersal types represented. Floristic differences between gardens appeared to be more related to chance events, such as flooding, rather than to specific characteristics of the habitat. Spence and Henderson (1993) examined the floristics and dispersal ecology of *tinajas* (water-filled rock tanks) and hanging gardens in the Waterpocket Fold of Capitol Reef National Park. They noted that the garden floristics were much less similar to each other than were tinajas, suggesting either greater site-to-site habitat differences or a larger random component related to flooding or chance dispersal. The species in these gardens included more species adapted for wind dispersal compared with tinaja floras.

Fowler (1995) examined vegetation and invertebrates in a variety of hanging gardens across the northern and central Colorado Plateau, testing his data against species-area relationships and the core-satellite hypothesis of Hanski (1982). He documented differences in vegetation and floristics across the region and noted that species that were locally abundant tended also to be geographically widespread. Fowler found a weak positive relationship between species richness and garden area but failed to differentiate widespread riparian species that occur in adjacent riparian zones from species restricted to gardens (Fowler et al. 1995). He did not find any support for the core-satellite hypothesis.

Keate (1996) analyzed the distribution of hanging garden species around Moab, Utah. She realized that for some plant species, gardens were not islands, so she restricted her analyses to narrow garden endemics that did not occur in adjacent riparian zones. Keate detected patterns in the distribution of endemics at several scales, including regional and local between-garden and within-garden scales. Important factors related to the presence of endemics included the extent of garden seep-line development and distance above the adjacent riparian channel, a proxy variable for flooding. Gardens with more endemic species were higher above adjacent flood zones, had complex microhabitats, and were exposed to low levels of solar radiation. Species richness of endemics was only weakly correlated with garden area.

Finally, Wong (1999) examined hanging garden vegetation in Zion National Park. She found that floristic and vegetation differences between gardens could not be attributed to measured site differences. Wong also analyzed floristic relationships in gardens from across the Colorado Plateau and found that significant differences in floristic composition occurred from region to region. Three floristic regions were noted: the eastern Colorado Plateau, western Colorado Plateau, and Grand Canyon regions. Distinct assemblages of endemic and wetland-riparian species occurred in each region.

Little is known about other kinds of spring-supported vegetation in the region. Phillips and coworkers (1980, 1987) and Warren and coworkers (1982) described spring-supported vegetation in the Grand Canyon region. Most springs they studied supported stands of riparian-like woodlands, dominated by widespread riparian species such as *Fraxinus pennsylvanica*, *Populus fremontii*, and *Salix gooddingii*. Elsewhere I have described and classified the vegetation associated with springs in twenty-five side canyons around Lake Powell and in the Escalante River basin (Spence 1996). Several unusual types occurred, including mixed deciduous woodlands dominated by birchleaf buckthorn (*Rhamnus betulifolia*) and thicket creeper (*Parthenocissus vitacea*), as well as types dominated by poison ivy (*Toxicodendron rydbergii*), common reedgrass (*Phragmites australis*), and spikerush (*Eleocharis palustris*).

About twenty endemic plant species and varieties occur at springs below 2,000 meters in elevation on the north and central two-thirds of the Colorado Plateau; these species represent approximately 15 percent of the endemic flora of this part of the plateau. Table 9.1 summarizes the current habitats, distributions, and closest congeners of these species. Most are strongly associated with hanging gardens rather than riparian and other wetland communities. All but one (*Phacelia indecora*) are perennials, and most are herbaceous rather than woody. Some species are essentially restricted to the gardens and often dominate the vegetation (e.g., *Flaveria mcdougalii*), while other species are more common in the drier margins around the gardens. Eight species are also typically found in riparian zones as well. Most of these endemic species appear to be derived from widespread boreal-temperate elements, with relatively fewer species of Madrean (southwestern United States and Mexico) or autochthonous origins (western North America). None of the endemics is widespread on the Colorado Plateau. Most species occur in

TABLE 9.1. Endemic plant species associated with springs on the northern and central Colorado Plateau

Species	Ecology	Distribution	Congeners	Probable origin
Aquilegia micrantha	hgw	cCP, eCP	*A. caerulea?*	BT
Aster welshii	hgw, rip	wCP	*A. hesperius*	BT
Carex curatorum	hgw, rip	all except wCP	*C. scirpoidea*	BT
C. haysii	hgw	wCP	*C. scirpoidea*	BT
C. specuicola	hgw	cCP, eCP, MC	*C. aurea*	BT
Cirsium rydbergii	hgw	cCP, eCP	*C. calcareum?*	BT
C. phillipsi	hgw	MC, GC	*C. virginense?*	BT
Dodecatheon pulchellum var. *zionense*	hgw	wCP	*D. pulchellum*	BT
Erigeron kachinensis	hgw	eCP	*E. pumilus*	BT
E. sionis	hgw, rip	wCP	*E. flagellaris*	BT
Flaveria mcdougallii	hgw	GC	*F. sonorensis?*	Madrean
Jamesia americana var. *zionis*	hgw, rip	wCP	*J. americana*	BT
Mimulus eastwoodiae	hgw	eCP	*M. guttatus?*	BT
Perityle specuicola	hgd, rip	eCP	*P. tenella*	Madrean
Phacelia indecora	hgd	eCP	*P. incana*	wNA
Platanthera zothecina	hgw, rip	cCP, eCP	*P. sparsiflora*	BT
Primula specuicola	hgw	all except wCP	*P. mistassinica*	BT
Sphaeromeria ruthiae	hgd, rip	wCP	*S. diversifolia*	BT
Viola clauseniana	hgw	wCP	*V. nephrophylla*	BT
Yucca toftiae	hgd, rip	cCP	*Y. angustissima*	Madrean
Zigadenus vaginatus	hgw	cCP, eCP	*Z. volcanicus*	Madrean

Sources: Holmgren et al. 1976; McArthur et al. 1988; Kelso 1991; Welsh et al. 1993; Stone 1998.
Note: Abbreviations for ecology, distribution, and probable origin columns are as follows: hgw = hanging garden wet, hgd = hanging garden dry margins, rip = riparian; c = central, e = eastern, w = western; CP = Colorado Plateau, MC = Marble Canyon, GC = lower Grand Canyon; BT = boreal-temperate, wNA=western North American, Madrean = southwestern United States and Mexico. A question mark beside the name of the congener (presumed closest related species based on available evidence) indicates that more work is needed.

more than one of the five regions in Arizona and Utah: western Colorado Plateau (Zion Canyon–Virgin River area), central Colorado Plateau (Glen Canyon and Escalante River), eastern Colorado Plateau (Canyonlands–San Juan River–Green River area), Marble Canyon, or lower Grand Canyon.

The ecology, reproductive biology, and phylogenetic relationships and origins of most of these endemic plant species remain poorly studied. Allphin and Harper (1994) described the habitat requirements of *Erigeron kachinensis*, a species essentially restricted to hanging garden environments of the eastern plateau. Kelso (1991) discussed the probable origin of the endemic *Primula speculicola* from a widespread boreal-temperate *P. mistassinica*-like species during and following Pleistocene glaciation. Holmgren and coworkers (1976) indicated that *Sphaeromeria* is a western U.S. genus closely related to *Tanacetum*, which is boreal-temperate in origin. This relationship has been recently confirmed by McArthur and coworkers (1998). Hudson and coworkers (2000) provided preliminary morphological and reproductive data on a population of *Platanthera zothecina* from Betatakin Canyon, Navajo National Monument. This species is distributed on the eastern side of the Colorado Plateau, with a disjunct population known from Oak Creek Canyon on the Mogollon Rim south of Flagstaff (Ron Coleman, pers. comm. 2000). Flanagan and coworkers (1997) examined carbon isotope ratios from several garden species, including *Aquilegia micrantha*, *Cirsium rydbergii*, *Mimulus eastwoodiae*, and *Primula speculicola*. No detailed ecological and phylogenetic work has been published on the remaining endemics.

Methods

Data were collected in 1997–98 on a National Park Service project to sample and characterize the flora, vegetation structure, water quality and chemistry, and aquatic invertebrates of springs along the Colorado River from Arches National Park through Grand Canyon National Park. Specific details on field and statistical methods are described in the final report (Spence 2002). The flora of each spring was sampled for approximately two hours, and canopy cover was estimated visually for each species using a modified Daubenmire scale. Each species was classified by current distribution, presumed origin (based on methods of Raven and Axelrod 1978; Stebbins 1982; Axelrod and Raven 1985), and dispersal ecology (Spence and Henderson 1993). Association tests were used to examine the relationship between distribution, origin, and dispersal ecology and selected groupings of the flora.

Each sampled site was classified into one or more of four principal geomorphic types (henceforth termed habitats): backwall, colluvial slope, wetland (often associated with a plunge pool), or simple spring. Each habitat was sampled in its entirety for flora and species canopy cover unless it was larger than 0.1 hectare, in which case it was sampled by a circular 0.1-hectare plot centered on the type. Floristic lists were relatively complete for each site and habitat, but vegetation structure was not completely described for habitats larger than 0.1 hectare. Analyses of species accumulation rates suggest that a single site visit for a two-hour duration would find approximately 90 percent of the species at the site (Spence, unpubl. data).

The relationship between geography and other physical factors and vegetation and floristic structure was explored through the use of ordination, cluster analysis, and TWINSPAN. Spring vegetation was grouped by cluster analysis and TWINSPAN. Ward's flexible sorting method was used with Euclidean distance on a matrix of springs by cover estimates of species. For the TWINSPAN, pseudospecies were defined by the levels of the Daubenmire scale. For both classification techniques, PC-ORD was used (McCune and Mefford 1995), with default methods chosen unless otherwise specified. Relationships between vegetation structure as revealed by classification techniques were then compared with specific physical variables using nonmetric multidimensional scaling ordinations (NMS). The number of dimensions chosen was based on preliminary solutions where reductions in final stress were small. Pearson-Kendall correlations were then computed between site scores and site physical variables. Separate ordinations were computed for the groups defined by classification. Default methods were used for the NMS ordinations as specified in McCune and Mefford 1995.

Species richness estimates were computed for the vegetation types derived by classification analysis using first-order jackknife estimation, based on the species-area algorithm in PC-ORD. Although there are concerns about the assumptions of the basic jackknife techniques (e.g., Boulinier et al. 1998), they do provide a rough baseline of estimated species richness against which to measure field observations. On the basis of data from Spence 2002, seventy-six species that were mostly or entirely restricted to hanging garden habitats were grouped into two categories: those with long-distance dispersal mechanisms (megawind, fleshy, sticktight, floater) and those lacking them (miniwind, smooth). Simple linear regression and rank sum tests using the Mann-Whitney statistic, and association tests, were used to examine the relationship between abundance and distribution.

To permit examination of the pattern of geographic distribution of endemic plant species, a parsimony analysis of endemicity (PAE) was done following the method as first described in Myers and Giller 1988. The PAE used the current distributions of the twenty endemic species in five regions on the Colorado Plateau. The five regions were defined primarily by a combination of geology and elevation. For each species, presence in each of the five regions was determined based on herbarium specimens and the literature.

Flora and Vegetation Patterns along the Colorado River

Sixty springs were sampled along the Colorado River during the project. In all, 125 species of native obligate phreatophytes (see Campbell and Green 1968) were documented, representing about 50 percent of the total known phreatophyte flora in the study area. In addition, 18 species of exotic phreatophytes were also found. The current distributions, presumed origins, and dispersal spectra for the native flora are listed in table 9.2, divided among three regions: Canyonlands (eastern Colorado Plateau, centered on Arches and Canyonlands national parks), Glen Canyon (central Colorado Plateau, Glen Canyon National Recreation Area), and Grand Canyon (both Grand Canyon and Marble Canyon in Grand Canyon National Park). The majority of species were either widespread temperate North American or southwestern North American in distribution, with origins primarily with temperate, Madrean, or boreal taxa. Overall, the majority of the species lack specialized adaptations for long-distance dispersal. Tests of association reveal that for dispersal type ($\chi^2 = 3.5$, $p = 0.967$) and presumed origin ($\chi^2 = 5.33$, $p = 0.868$), the frequency distributions among the three regions are homogeneous. However, current distributions show a significant departure from expectation ($\chi^2 = 19.0$, $p = 0.040$), with the Grand Canyon region having significantly more southwestern species and significantly fewer boreal-temperate species than the other two regions. In addition, the Glen Canyon region supports significantly more boreal-temperate species than could be expected by chance.

Ordination and classification techniques using sites and species cover estimates revealed four main groupings of vegetation: hanging garden backwalls, colluvial slopes, wetland sites dominated by species such as *Phragmites australis* and *Typha domingensis*, and riparian-like woodlands. Considerable overlap in floristics and vegetation structure occurred between the wetlands and woodlands. Different geomorphic types within a site grouped with the

TABLE 9.2. Current distribution, presumed origins, and dispersal spectra of phreatophytes found in sixty springs along the Colorado River on the Colorado Plateau

	Region			
Flora category	Glen Canyon (cCP)	Canyonlands (eCP)	Grand Canyon (MC/GC)	Entire flora
Current Distribution				
Boreal–temperate	14	8	6	14
Widespread temperate	27	14	21	30
Widespread in North America	17	11	10	20
Colorado Plateau endemic	9	5	4	12
Interior–Rocky Mountain	4	3	4	5
Southwestern U.S.–Mexico	18	9	37	40
Total	89	50	82	121[a]
Presumed Origin[b]				
Boreal	21	11	10	23
Cosmopolitan	7	2	6	8
Madrean	11	5	20	25
Temperate	37	22	32	49
Tropical	9	6	11	12
Western North America	5	4	6	8
Total	90	50	85	125
Dispersal Type[c]				
Fleshy fruit	12	7	8	16
Floater	4	1	4	5
Megawind	11	9	9	12
Miniwind	16	9	18	25
Smooth	37	18	35	55
Sticktight	10	6	11	12
Total	90	50	85	125

Note: The flora is grouped into three regions: Glen Canyon (central Colorado Plateau), Canyonlands (eastern Colorado Plateau), and Grand Canyon (including Marble Canyon).
[a] The distribution of four taxa could not be determined.
[b] Methods follow those of Raven and Axelrod (1978) and Stebbins (1982).
[c] Based on the classification of Spence and Henderson (1993).

same geomorphic types at other sites rather than grouping with the different types within a site. There were particularly strong vegetative distinctions within sites between backwalls and other geomorphic types.

Figure 9.2 summarizes the principal results for vegetation structure using TWINSPAN and floristic similarities (fig. 9.2A) versus cluster analysis

A

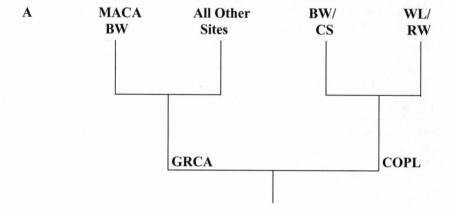

B

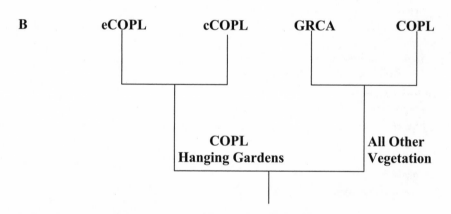

Figure 9.2. Dendrograms of the first two principal branches in vegetation structure and floristics. *A*, Vegetation structure based on cover estimates, using TWINSPAN. *B*, Floristic structure based on species presence/absence, using cluster analysis. Abbreviations: COPL = Colorado Plateau, GRCA = Grand Canyon, MACA = Marble Canyon, c = central, e = eastern, BW = backwalls, CS = colluvial slopes, WL = wetlands, RW = riparian-like spring woodlands.

and Sorenson's coefficient of similarity (fig. 9.2B). The vegetation and geographic correlates with structure are identified through the first two levels of dendrogram branching based on TWINSPAN. For vegetation structure (fig. 9.2A), the first division separates sites by elevation and region, with the lower-elevation Grand Canyon–Marble Canyon sites distinct from the higher-elevation Colorado Plateau sites. Another factor correlated with this division is geology, with the majority of Grand Canyon sites on limestone and the majority of Colorado Plateau sites on sandstone. Within the Grand Canyon region, the second primary division separates Marble Canyon backwalls from all other Grand Canyon and Marble Canyon sites. Within the Colorado Plateau group, the second primary division separates hanging garden backwall and colluvial slope sites from wetland, plunge-pool, and spring woodland sites.

However, when floristic structure is based on simple presence and absence of species, the results differ significantly from vegetation structure (fig. 9.2B). The first division separates Colorado Plateau hanging garden backwall and colluvial slopes from all other vegetation. For Colorado Plateau gardens, the second division separates eastern (Canyonlands) sites from central (Glen Canyon) sites. For other sites, the second division separates Grand Canyon sites from the remaining non–hanging garden plateau sites.

Table 9.3 summarizes the results of NMS ordinations for the following habitats: backwalls, colluvial slopes, and combined plunge pool/ woodlands. The three variables most strongly correlated with the first two ordination axes are listed for each vegetation group. For all groups, elevation is the most strongly correlated factor on the first axis. For backwalls and colluvial slopes, geology and electrical conductivity (in μS) are additional strongly correlated variables. For the woodlands and wetlands, summer solar radiation is also strongly correlated with vegetation structure. For the second major axis, aspect and solar radiation are generally the factors most strongly correlated with structure. Water temperature is most strongly correlated with colluvial slopes.

Table 9.4 summarizes observed species richness at sites, estimated species richness, and the ratio of the two. Species richness is greatest on colluvial slopes and least in woodlands. Overall, the first-order jackknife results suggest that the observed number of species is 73–80 percent of the estimated number of species likely to occur in each habitat. Estimated number

TABLE 9.3. Summary results of Kendall-Pearson correlations between vegetation types and physical site characteristics

Habitat	NMS Axis I	NMS Axis II
Backwalls	Elevation Geology Conductivity	Summer solar radiation Aspect
Colluvial slopes	Elevation Geology Conductivity	Winter solar radiation Water temperature
Woodlands and wetlands	Elevation Summer solar radiation	Winter solar radiation Aspect

Note: Kendall-Pearson correlations are used following a nonmetric multidimensional scaling (NMS) ordination. Only the three variables showing the strongest correlation with the first two ordination axes are listed.

of species in the study area for each habitat ranges from 64 species in woodlands to 121 species on colluvial slopes.

The relationship between species richness and physical site factors was explored using regression analysis for the four habitat groups (backwalls, colluvial slopes, wetlands, and woodlands). Richness is not strongly correlated with elevation or geology for any of the habitats. For backwalls, only summer solar radiation is strongly correlated with richness: richness increases with increasing summer solar radiation. However, for endemic species richness of backwalls, the relationship with summer radiation is negative, with more species in shaded summer sites (see Flanagan et al. 1997). For the other three habitats, richness is most strongly correlated with discharge rates and solar radiation. For colluvial slopes, richness increases in sites with larger discharge rates. For woodlands, richness increases with higher discharge rates but decreases with higher June solar radiation inputs. For wetlands, richness increases with both higher discharge rates and higher autumn (September–November) solar radiation inputs.

Among the 60 springs sampled in the study area, 18 species of exotic phreatophytes were found, representing 13 percent of the total flora. Among the most widespread of these exotics were *Agrostis semiverticillata* (10 sites), *A. stolonifera* (10 sites), *Tamarix ramosissima* (22 sites), and *Sonchus asper* (11 sites). None of the exotics dominated any site, and most were relatively rare and local. Among regions, Glen Canyon had the most exotics (14 species),

TABLE 9.4. Species richness estimates for vegetation types found at sixty springs along the Colorado River

Vegetation type	Observed species (N)	Estimated species (N)	Observed/estimated (%)
Backwalls	58	79	73
Colluvial slopes	97	121	80
Wetlands	59	81	73
Woodlands	49	64	77

Note: Species richness is based on the first-order jackknife estimate. For each vegetation type, the observed number of species, first-order jackknife estimate, and their difference are listed.

compared with Canyonlands (11 species) and Grand Canyon (7 species). Exotics were fewer in number and abundance in the relatively inaccessible Grand Canyon sites compared with the relatively more accessible Canyonlands and Glen Canyon sites.

Discussion

General Patterns

Several general points emerge based on the analyses presented above. The principal points are as follows:

1. Little is known about most spring-supported vegetation in the study area other than hanging gardens.
2. Hanging garden vegetation is structurally and floristically distinct from other spring-supported vegetation in the U.S. Southwest.
3. Previous research suggests that flooding disturbance and amount of solar radiation are important factors in controlling species assemblages and the distribution of plateau endemics in hanging gardens.
4. Elevation, geology, and solar radiation are important factors in structuring spring-supported floras and vegetation at the regional (Colorado Plateau) level.
5. Species richness is relatively high in spring sites and is most strongly correlated with amount of water and solar radiation.

6. There is no evidence for the core-satellite hypothesis but some evidence for the hypothesis that widespread species are locally more abundant than less widespread species.

7. A small but distinctive endemic wetland flora exists on the Colorado Plateau, concentrated in hanging gardens and with primarily boreal-temperate affinities.

8. The endemic flora appears to show distinct nonrandom patterns of distribution on the Colorado Plateau related to major river systems.

9. At the regional level, similar geomorphic habitats in different sites tend to be floristically and structurally more similar to each other than they are with different geomorphic habitats in the same site.

General Hypotheses and Preliminary Tests

Of the many hypotheses emerging from the analyses above, four have been selected and are presented below and tested with available data sets. These, however, represent only a subset of the possible hypotheses on the origin, structure, and dynamics of spring-supported communities on the Colorado Plateau.

DISPERSAL: Are widespread species restricted to hanging garden sites more likely to have adaptations for long-distance dispersal?

Both Malanson and Kay (1980) and Spence and Henderson (1993) provide evidence that hanging gardens support nonrandom concentrations of species with long-distance dispersal mechanisms. These include very small seeds or spores, fruits, or seeds with plumes or structures allowing effective wind dispersal; fleshy fruits attractive to birds; and fruits with sticky or barbed structures that adhere to animals. If dispersal is restricted between isolated garden sites (nonriparian species), one would expect nonrandom patterns of dispersal types in species assemblages among these sites. One preliminary test is to determine whether species with long-distance dispersal mechanisms are more widespread (occur at more sites) than those species lacking such mechanisms. A rank sum test using the Mann-Whitney statistical test indicates that there is no significant difference in number of sites for species in the two categories ($p = 0.322$ for a two-tailed test, Mann-

Whitney statistic). There is a weak but still nonsignificant difference using a one-tailed test ($p = 0.161$), which is likely the result of the presence of maidenhair fern; this is the most widely distributed garden species on the Colorado Plateau, and it disperses via spores. Comparing the site-restricted species with the widespread riparian and wetland species for dispersal types, there is no difference between the two groups ($\chi^2 = 0.27$, $p = 0.61$). Isolated and site-restricted hanging garden species and riparian and wetland species, most of which occur commonly throughout the region, do not differ in dispersal spectra. Hence, regionally isolated garden floras are not enriched in long-distance dispersed species.

This preliminary test suggests that dispersal may not be an important factor in the composition of site floras in the study area; however, more work is clearly needed. Comparing dispersal spectra using local floras with source floras through Monte Carlo simulations would provide a more robust test of the role of dispersal. I have done this type of simulation with restricted species found in isolated low elevation stands of Douglas-fir (*Pseudotsuga menziesii*) associated with springs on the Colorado Plateau (Spence 1995). The results suggest that the presence of boreal-montane disjunct and endemic species found in these sites was more likely to be a result of persistence *in situ* rather than the result of independent long-distance dispersal events. This hypothesis is examined in more detail later in this chapter.

ABUNDANCE AND DISTRIBUTION: Do locally abundant species also have wider geographic distributions?

To test this hypothesis, regressions between site abundance (canopy cover estimates) and the number of sites at which each species is found were computed for all species and for endemic species. For both groups, there was no relationship between the two variables (for all species, $r^2 = 0.01$, $F = 0.89$, $p = 0.34$; for endemics, $r^2 = 0.07$, $F = 1.57$, $p = 0.20$). Brown (1984, 1995) points out, however, that the abundance-distribution relationship is most likely to occur between congeners or species with similar habitat requirements. Because of the specialized nature of hanging garden backwall habitats, those colonists are much more likely to be ecologically similar than are co-occurring species in other habitats. With canopy cover estimates serving as the dependent variable, the relationship between backwall species abundance and distribution is shown in figure 9.3. The regression is highly sig-

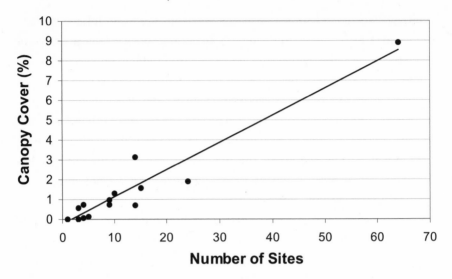

Figure 9.3. The relationship between local abundance (canopy cover) and regional distribution (number of sites) in the study area for species found on hanging garden backwall environments.

nificant ($r^2 = 0.93$, df = 17, $F = 198.67$, $p < .00001$). Because of the strength of the relationship, it is possible to predict the geographic distribution of a backwall species based on local site abundance. Species that tend to dominate backwalls in terms of canopy cover, such as maidenhair fern or columbine (*Aquilegia micrantha*), also tend to occur at more sites along the Colorado River compared with locally rarer species.

The summary of the probable or known origins of the endemic wetland species on the Colorado Plateau indicates that the majority (75 percent) are derived from more northern boreal-temperate taxa (table 9.1). Currently, most non-endemic species with these affinities tend to occur at an elevation above about 2,500 meters, in the upper montane, subalpine, and alpine zones. Endemics with boreal-temperate affinities currently occur at elevations between 1,000 and 2,000 meters in arid and semiarid habitats. During the late Wisconsin glaciation, many species of boreal-temperate affinities were widespread at lower elevations along riparian zones on the Colorado Plateau. Among those that are reported to have occurred in the study area (Betancourt 1984; Betancourt et al. 1990; Cole 1990; Cole and Murray 1997; Dryer 1994; Sharpe 1991, 1993; Withers and Mead 1993) are subalpine and white firs (*Abies bifolia*, *A. concolor*), bigtooth and Rocky Mountain maples (*Acer grandidentatum*, *A. glabrum*), Engelmann and blue spruce (*Picea engelmannii*, *P. pungens*), limber pine (*Pinus flexilis*), and Douglas-fir (*Pseudotsuga menziesii*). A variety of herbaceous taxa also are likely to have occurred in these riparian communities. Most of these boreal-temperate species disappeared at lower elevations circa 11,000–10,000 BP. Remarkably, several stands of bigtooth maple still exist at elevations of 1,150–1,200 meters in Cow and Fence canyons on the Escalante Arm of Lake Powell (Spence 1996). Other boreal-temperate species with scattered populations in the region include *Amalanchier alnifolia*, *Cystopteris utahensis*, *Rosa woodsii*, and *Smilacina stellata*. Other possible survivors of the late Wisconsin include the low-elevation Douglas-fir stands described in Spence 1995.

One plausible scenario for the origin of the endemic species is that they evolved during earlier glacial periods on the Colorado Plateau, when conditions were cooler and probably somewhat wetter than in the subsequent Holocene. Hanging garden and spring environments were probably more widespread during glacial periods. With the onset of the warmer and drier climates of the Holocene, boreal-temperate species dispersed to higher elevations or became locally extinct. Only populations of endemics and boreal-temperate species that occurred in the cool, wet, and shaded microclimates of the hanging gardens have survived to the present. Many typical high-elevation boreal-temperate species also occur in the gardens. For example, the subalpine *Calamagrostis scopulorum* is a common and widespread species on north-facing or shaded desert garden backwalls and colluvial slopes. In all, approximately forty high-elevation species are disjunct in low-elevation hanging gardens. Areas where concentrations of these high-elevation species occur include the lower parts of the Escalante River and Zion Canyon, which Betancourt and coworkers (1990) suggest may have functioned as "megarefugia" during the Holocene, with its cool shaded and well-watered habitats. Canyons off the Escalante Arm of Lake Powell, in particular Cow and Fence canyons, also may have functioned as refugia (Spence 1996).

Overall, the available evidence suggests that these endemic and boreal-temperate populations represent a vicariant pattern (*sensu* Nekola 1999) rather than one of recent dispersal. Hence, the current scattered distribution of endemic species with poor dispersal abilities is best explained by local extinction at the end of the Wisconsin and during the early and middle

Holocene. Most of the endemic species do not appear to have good dispersal abilities. Only *Platanthera zothecina*, which occurs across a wide elevation range on the Colorado Plateau (1,200–2,700 meters), has good long-distance dispersal abilities through its tiny seeds. A second garden endemic, *Erigeron kachinensis*, is known from high-elevation sandstone sites in southeastern Utah, but these populations are genetically distinct from those in typical garden environments (Allphin et al. 1996). If extinction since the late Wisconsin were occurring, one would expect that extinction rates in boreal-temperate (mostly montane-disjunct) species would be highest at lower and warmer elevations and lowest at higher and cooler elevations. Species richness of boreal-temperate relicts is positively correlated with elevation ($r^2 = 0.26$, $F = 19.5$, $p = 0.0001$), although there is a lot of scatter in the data. High-elevation springs tend to have more boreal-temperate species compared with low-elevation springs in the study area. Future tests of these hypotheses should include genetic work on isolated populations of garden-restricted species, searches for packrat midden remains from the late Wisconsin and Holocene, and more rigorous statistical tests of dispersal ecology.

ENDEMIC DISTRIBUTION PATTERNS: Are the endemic species in the study area distributed in a geographically random pattern?

The preliminary answer to this is no, as there appears to be a pattern in the distribution of the species (fig. 9.4). There is no overlap in distributions between Zion Canyon and the Colorado River corridor in Arizona and Utah. Seven species and subspecies are endemic to upper Zion Canyon near the headwaters of the Virgin River, while the other thirteen species are distributed along the Colorado River. Two species occurring widely along the Colorado River are *Carex curatorum* and *Primula specuicola*. One species, *Flaveria mcdougalii*, is endemic to the lower Grand Canyon, while a second species, *Cirsium rydbergii*, is found only at springs in Paria Canyon and Marble Canyon. Four species are widespread in the center and eastern portions of the plateau in Utah: these are *Aquilegia micrantha*, *Cirsium rydbergii*, *Platanthera zothecina*, and *Zigadenus vaginatus*. One species, *Yucca toftiae*, is endemic to Glen Canyon. Another three species — *Erigeron kachinensis*, *Perityle specuicola*, and *Phacelia indecora* — are endemic to the eastern portions of the plateau. The Navajo sedge, *Carex specuicola*, is found both on the eastern plateau and in Marble Canyon, suggesting either a dispersal event or per-

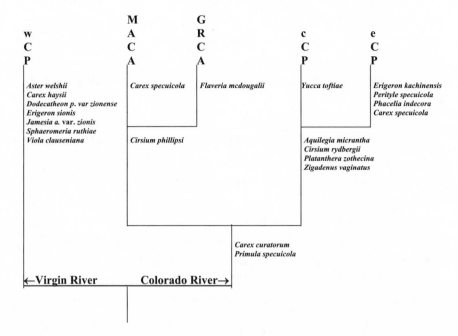

W
C
P

M
A
C
A

G
R
C
A

c
C
P

e
C
P

| Aster welshii Carex haysii Dodecatheon p. var zionense Erigeron sionis Jamesia a. var. zionis Sphaeromeria ruthiae Viola clauseniana | Carex specuicola | Flaveria mcdougalii | Yucca toftiae | Erigeron kachinensis Perityle specuicola Phacelia indecora Carex specuicola |

Cirsium phillipsi

Aquilegia micrantha
Cirsium rydbergii
Platanthera zothecina
Zigadenus vaginatus

Carex curatorum
Primula specuicola

←Virgin River   Colorado River→

Figure 9.4. Parsimony analysis of endemicity (PAE) of five principal regions along two rivers of the Colorado Plateau: the Colorado River and Virgin River. The distributions of the twenty endemic plant species are mapped onto the dendrogram. The five principal regions are eastern Colorado Plateau (eCP), central Colorado Plateau (cCP), western Colorado Plateau (wCP), Marble Canyon (MACA), and Grand Canyon (GRCA).

haps extinction in intervening areas. However, the taxonomy of this species and its allies in the region, *Carex aurea* and *C. hassei*, is confused, and the plants from Marble Canyon may be reassigned to a different taxon following revision.

The distribution of the endemic species is clearly not random. The primary factor determining their current distributions appears to be the presence of the Virgin and Colorado rivers. Currently, the confluence of these rivers is under the waters of Lake Mead. Prior to the construction of Hoover Dam, the confluence was in an extremely low-elevation arid portion of the Mojave Desert, which may have acted as a barrier to dispersal of species between the two river corridors. Interestingly, the Zion Canyon endemics are not found downstream along the Virgin River. All are restricted to cool shaded gardens and adjacent sandstone walls in Zion Canyon and its principal tributary canyons. Downstream in the Mojave Desert regions of the Virgin River, a single endemic spring-associated species, *Cirsium virginense*, is known. Within the Colorado River corridor, the number of endemic species increases from Grand Canyon upstream to the eastern Colorado Plateau. In the area from Arches National Park south through eastern Utah to the San Juan River, twelve endemic species occur, compared with seven in Glen Canyon, four in Marble Canyon, and four in the lower Grand Canyon.

If the scattered and isolated distributions of these species are the result of extinction, then additional populations may be found in areas not yet explored. One major area where fieldwork is needed includes those spring systems between the Colorado River corridor and the canyon rims through Marble and Grand canyons, which spans an elevation range of 2,000 meters. Examination of the ecology and distributions of boreal-temperate species in low-elevation sites also may provide additional information on the relative contributions of extinction and dispersal processes to community assemblages.

COMMUNITY ASSEMBLY: Dispersal or vicariance?

Earlier I suggested that the current patterns of distribution of endemic and boreal-temperate species in the study area might be the result of extinction rather than dispersal. In a broader sense, both vicariance and dispersal are important factors in structuring communities (Brown 1995; Rosenzweig 1995; Nekola 1999). However, as pointed out by Nekola (1999), the varying contributions of these processes are often difficult to distinguish. He described two categories of refugia: neorefugia and paleorefugia. Neorefugia consist of habitats that formed more recently than the surrounding matrix of vegetation, with dispersal processes being relatively more important than extinction. Paleorefugia are older than the surrounding matrix, with extinction processes being more important than dispersal. Some of Nekola's (1999) eight predictions on assemblage structure and history are more testable than others, and Nekola tested four of these predictions with landsnail data from Iowa fens and algific talus slopes.

Nekola (1999) predicted that neorefugia should be enriched in species with long-distance dispersal mechanisms. For the site-restricted species (including endemics) of isolated hanging garden environments, there was no support for this hypothesis at the regional level. No significant differences were found in dispersal types between site-restricted species compared with other groups of riparian and wetland species. Overall, a majority of restricted species lacked any mechanism for long-distance dispersal. A second prediction is that the slope of the species-area curve should be greater for paleorefugia compared with neorefugia. However, earlier work, as well as my own work, indicates that the species-area relationship is weak in these isolated gardens. Rather, as pointed out by Keate (1996) as well as in this chapter, species richness seems to be most strongly correlated with springs discharge. Springs with larger discharges are more likely to provide long-term stability to wetland communities, compared with springs with smaller discharges. During periods of extensive drought, as well as during the mid-Holocene thermal maximum, only the larger springs may have existed to support wetland species. Hence, the test of the species-area hypothesis may not be valid for these communities.

Dispersal is likely to contribute to the species assemblages of spring-supported communities along the Colorado River. The Colorado River has provided a continuous corridor for the dispersal of many plant species from the lower Grand Canyon regions as well as from higher-elevation regions in eastern Utah and western Colorado. Many of the typical wetland and riparian species in genera such as *Baccharis, Carex, Eleocharis, Equisetum, Juncus, Scirpus*, and *Salix* are widely distributed in the interconnected riparian zones throughout the region. Other primarily southwestern species clearly have spread up the river corridor from the lower Grand Canyon; these species include *Chloracantha spinosa, Cladium californicum, Imperata brevifolia, Parthenocissus vitacea, Rhamnus betulifolia*, and *Tessaria sericea*. In recent times, tamarisk (*Tamarix ramosissima*) also spread rapidly up the Colorado River into Utah and western Colorado from the lower basin between 1890 and 1930 (Graf 1978). Even for hanging gardens, local dispersal appears to play a role in recolonization following disturbance (Malanson and Kay 1980; Malanson 1982). A small "fugitive" guild of species, including the fern *Adiantum capillus-veneris*, orchids such as *Epipactis gigantea*, and bryophytes are readily dispersed via spores and dust seeds. Thus, both dispersal and persistence since the late Wisconsin are likely to have contributed to patterns of species richness and community structure in spring-supported vegetation in the study area.

Conclusions and Promising Areas of Future Research

The springs along the Colorado River on the Colorado Plateau support a rich diversity of plant species and vegetation communities. However, as yet, relatively little research has been conducted on their structure, dynamics, and origins. A variety of interesting research programs could be established on the isolated spring biotas of the region; some topics of focus for future research are suggested below.

Although climate appears to be the principal controlling factor at the regional level, it is not strongly correlated with species richness patterns. Regressions indicate that richness is most strongly correlated primarily with water discharge rates and secondarily with solar radiation. Species-rich sites have higher discharge rates than do species-poor sites. Solar radiation shows a different kind of pattern depending on the habitat type. Radiation is not strongly related to backwall and colluvial slope richness. Woodland sites support more species where summer radiation is low, while wetland and plunge-pool vegetation richness is higher in sites with higher levels of autumn radiation. Clearly, these relationships are preliminary, and additional work is required to determine the role of solar radiation in controlling vegetation structure and species richness patterns.

Significant evidence suggests that vicariance best explains the current isolated distribution patterns of two groups of species associated with springs along the Colorado River: Colorado Plateau endemics and boreal-temperates. If this is correct, then several predictions can be made. Nekola (1999) provided eight predictions for isolated communities based on the contrasting effects of vicariance and dispersal (one of which is tested in the earlier discussion). Another approach is to study genetic change in isolated populations. If isolated populations have persisted since the Pleistocene, then a variety of genetic changes could occur. Populations that have persisted for a long period of time and have descended from more than one original genotype are likely to exhibit greater genetic variability and have more unique alleles than does a population that is the result of a long-distance dispersal event. The only studies to date that have looked at genetic changes include Allphin and coworkers (1996) on *Erigeron kachinensis* and Kimberling and coworkers (1996) and Miller and coworkers (1999) on northern leopard frogs (*Rana pipiens*). Allphin and coworkers noted that a high–elevation population of the endemic *Erigeron kachinensis* was genetically and morphologically distinct from hanging garden populations. Kimberling and coworkers showed that populations of leopard frogs in northern Arizona and in isolated canyons off of Lake Powell had genetically diverged from one another to varying degrees. The species was originally widespread along the Colorado River through Glen Canyon, but current remnant populations in side canyons are genetically different from one another. These changes may have occurred in the past thirty years, since the isolation of side-canyon populations resulting from the filling of Lake Powell.

Another promising area of research on the origins of these communities and the roles of extinction and dispersal is the examination of associated phytophagous invertebrate assemblages on endemic and boreal-temperate plant species. The presence of groups of closely associated boreal-temperate or montane invertebrates, such as some groups of beetles or true bugs with poor dispersal abilities, would suggest a vicariant pattern. A plant species that dispersed recently into a spring is likely to have a more depauperate invertebrate fauna associated with it compared with one that has persisted in situ since the end of the Pleistocene. I am not aware of any such studies that have been conducted in the study area on this hypothesis.

Springs in the region are fed primarily by aquifers that are recharged by local precipitation events. Currently, about 50–70 percent of regional precipitation occurs during the cooler winter and early spring months, with the rest from either the summer monsoonal rains or from autumn storms from the Pacific (Petersen 1994; Spence 2001). Global warming predictions indicate that substantial changes in precipitation patterns could occur. The principal models currently predict weaker summer monsoon and increased winter precipitation. The relevance of these changes in precipitation patterns to springs discharge rates is unclear for several reasons. First, the water recharging the springs varies considerably in estimated age. Kimball and Christensen (1996) point out that the water discharging from springs in Zion Canyon varies in age from essentially modern to four thousand years old. Second, we know little about recharge and depletion rates for most aquifers in the region. Third, the relative contributions of winter and summer precipitation to regional aquifers have yet to be determined with certainty. The vicariance hypothesis for the presence of boreal-temperate and endemic species in many springs suggests that these springs have been flowing for most of the Holocene, including the thermal maximum of the mid-Holocene. Despite these uncertainties, springs on the Colorado Plateau and their associated aquatic and wetland biotas could be at risk in the future. Further research on the origin, age, and depletion rates of regional aquifers is urgently needed.

Although this chapter focuses on the Colorado River and some of its major tributaries, other rivers draining off the Colorado Plateau also support spring-related vegetation. These include the upper Green River and its tributaries in and around Dinosaur National Monument, the upper Rio Grande, the upper and middle reaches of the Little Colorado River, and streams draining off the Mogollon Rim. The riparian corridors associated with these rivers connect Colorado Plateau regions with adjacent floristic provinces, especially the southern Rocky Mountain, the Apachean, and the Chihuahuan regions (see McLaughlin 1989). Floristic and vegetation surveys in these areas would provide valuable additional information on the roles of climate, dispersal and extinction in structuring spring-related vegetation in the region.

ACKNOWLEDGMENTS

Portions of this research were supported by a grant from the Water Resources Division, National Park Service. Logistical support was supplied by Grand Canyon Monitoring and Research Center, Bureau of Reclamation, and Glen Canyon National Recreation Area. I am grateful to many people who helped with the fieldwork, including Ken Baker, Dave Baker, Kevin Berghoff, Bill Liebfried, Jerry Monks, Suzanne Rhodes, Charlie Schelz, Howard Taylor, and Eric Wilson. Steve Chubbeck and his helicopter safely got me in and out of many interesting remote springs on the Colorado Plateau. I am particularly grateful to Larry Stevens for his interest in and support of the project and to John Ritenour, chief of resource management at Glen Canyon National Recreation Area, for supporting my interest in springs research. I am grateful to Terry Arundel of the U.S. Geological Survey, who kindly supplied the study area map. Finally, I am indebted to Peter Rowlands for many stimulating discussions on the climate of the Southwest.

Chapter 10

Mechanisms of Change in Seep/Spring Plant Communities on the Southern Colorado Plateau

VICKY J. MERETSKY

Springs exist at the intersection of geological, hydrological, and biological processes: they are breeding grounds for change. Most agents of change in riparian areas may be present, along with most agents of change in uplands. To date, discussions of southwestern desert spring plant communities have tended to focus on rock-based settings such as hanging gardens (vegetation associated with water seeping or flowing from vertical or near-vertical rock; e.g., Malanson 1980; Welsh and Toft 1981; Malanson and Kay 1980; Malanson 1982; Welsh 1989a; Spence and Henderson 1993; Fowler et al. 1995; Wong 1999; see Spence, chapter 9 of this volume) and *tinajas* (eroded pockets in bedrock that collect water; Spence and Henderson 1993). Several researchers have investigated factors that affect plant community composition in hanging gardens, but these studies have dealt with composition as a static characteristic (e.g., Malanson 1980; Malanson and Kay 1980; Spence and Henderson 1993; Fowler et al. 1995). Changes in spring plant communities have been addressed for one particularly damaging agent of change: human alterations of hydrology, often channelizing or pumping. The affected plant communities in these cases are typically marsh communities that have been damaged or destroyed (e.g., Hendrickson and Minckley 1985; Perkins et al. 1984; Williams 1984; Williams et al. 1985).

The distinction between the well-studied springs (hanging gardens) and the heavily impacted springs (low-gradient, often marshlike) points to a gap in our information about spring ecosystems: the ecosystems that are most likely to be damaged are not the ones we understand best. Restoration efforts at such springs will rely primarily on standard wetland restoration information that may not reflect characteristics linked specifically to spring-fed systems. Furthermore, the range of variation that may occur in these systems is not well known, reducing our ability to differentiate variability resulting from ecosystem changes attributable to restoration activities from naturally occurring variability.

In this chapter, I describe plant community change at non–hanging garden springs in Zion and Grand Canyon national parks. Zion's hanging gardens are well studied, and the Zion flora is generally well described, but no low-gradient springs have ever been specifically studied. At Grand Canyon, spring hydrology has been of concern because of water withdrawal, and floristics are well described. Springs in national parks are rarely completely untouched by human activity, but some have had a number of years to recover from interventions and so have begun to reflect natural levels of variability in desert spring plant communities. My observations provide the beginnings of an understanding of the factors that affect these communities and the extents of their impacts.

Methods

The Study Area

Grand Canyon National Park (Grand Canyon) and Zion National Park (Zion) both encompass deep canyons on the Colorado Plateau and contain many springs. Both support desert vegetation in their lower elevations (Mojave and Great Basin in Grand Canyon, and Great Basin, alone, in Zion) and semidesert, Great Basin pinyon-juniper woodlands, and conifer forests above the canyon walls.

Grand Canyon, in northern Arizona, was carved by the Colorado River; of the two, it has a greater vocabulary of geologic strata, by virtue of being both deeper and much larger than Zion Canyon. In particular, it has several limestone and sandstone strata that serve as aquifers. Springs in limestone strata generally emerge from faults and fractures within the strata. In contrast, sandstone generally produces springs at the contact zones where an overlying, permeable sandstone stratum meets an underlying less permeable (aquitard) or impermeable (aquiclude) stratum. Many of the springs in Grand Canyon lie in high-gradient settings and support hanging gardens or plant communities on bedrock or talus slopes. However, some are in lower-gradient settings: for example, in tributary valleys or on the Tonto Plateau.

I selected two study sites in Grand Canyon National Park, both on the South Rim. Santa Maria Spring emerges at the contact of the Supai Formation and the Hermit Shale, above the Hermit Trail. The Garden Creek spring emerges in the Bright Angel Shale one drainage east of Garden Creek and the Bright Angel Trail. Both sites produce perennially flowing water: at the time of the study, Santa Maria Spring had flow only in the lower portion; the spring at the Garden Creek site had a very small stream running through it. Surface flow in both cases was less than one liter per minute. Additional site information is in table 10.1.

Zion National Park is located in southwestern Utah, immediately northeast of the confluence of the eastern and northern forks of the Virgin River, both of which flow through the park. Navajo sandstone is the primary aquifer in the park, and springs form at its base and to a lesser extent at the base of the sandstone Springdale Member of the Moenave Formation. Most of the sandstone-sourced springs are in high-gradient settings, often creating hanging gardens. Zion National Park also contains upwelling springs in very low gradient settings, but these are much rarer than the outwelling, steep-gradient springs, although the former support considerably larger wetland communities than the latter.

I selected four study sites in Zion National Park. The Temple site lies in the area called the Temple of Sinuwava, within Zion Canyon, in a flat, sand-covered area at the base of a sandstone cliff with a productive contact zone. The Parunuweap site is partway up a side canyon in Parunuweap Canyon in a flat area in the outflow of a hanging garden; a small stream with a two-meter-diameter pool borders that study area. The Potato Hollow site is in the valley called Potato Hollow on Horse Pasture Plateau, and the wetland arises from a line of seep that waters a broad swath in the lower end of the valley. The Pocket Mesa site is immediately below Pocket Mesa, near the Wildcat Trail; it comprises an area of saturated soil. Additional site details are given in table 10.1.

Field Data Collection

Long-term study areas were selected to meet two main requirements: (1) presence of non–hanging garden wetland vegetation on sand/soil substrates associated with a spring, and (2) topographic and soil conditions

TABLE 10.1. Descriptive information for study wetlands at seep/springs in Zion National Park (ZNP) and Grand Canyon National Park (GCNP) studied during 1997–2000

Location	Park	Elevation (m)	Plot size (m²)	Slope	Substrate	Context and total size	Disturbance history
Pocket Mesa	ZNP	2,134	121	<10%	Sand lens	Perched water table supports 400 × 6 m wetland bordering stream segment that is normally saturated, briefly flowing after rain and snowmelt	2 small ponds dug in basin away from main drainage before 1950; grazed until 1979
Potato Hollow	ZNP	2,103	18	<10%	Alluvial soil	Aquifer/aquitard contact supports broad front of unevenly saturated soil in lower valley, ~200 × 30 m	Grazed until late 1950s; severe natural fire in 1996
Santa Maria	GCNP	1,615	24	<40%	Soil over bedrock	Aquifer/aquitard contact supports ~16 × 4 m saturated soil interspersed with rock; very sparse flow from study portion (uppermost portion); pipe driven into rocks ~10 m below study portion	Possible signs of long-gone pipe near top; protected since 1919
Temple Upper	ZNP	1,372	20	<10%	Sand	Hanging garden outflow from aquifer/aquitard contact supports saturated sand platform ~30 × 40 m	A portion of the wall is piped for water, but not the portion watering the study areas
Temple Lower	ZNP	1,372	9	<10%	Sand		
Parunuweap	ZNP	1,341	36	<10%	Soil	Hanging garden outflow from aquifer/aquitard contact supports saturated soil along slow-flowing stream segment; 9 × 1.5 m	Apparently untouched; unreachable by stock; too close to Virgin River to have been tapped for water; no sign of any occupation or use
Garden Creek	GCNP	1,158	21	<30%	Soil and rock	Aquifer/aquitard contact supports 7 × 1 m wetland along slight but steady flow	Not recently disturbed; very small and within 200 m of a much larger water source with much more vegetation

Note: Study wetlands here are listed in descending order of elevation.

that permitted close study of vegetation without major impact to the site. Small monocultural patches of common reed (*Phragmites australis*) and other less mesophytic seep species (e.g., *Aster glaucodes*) that tended to form along barely productive seep contact zones were avoided.

One-meter-square grids were established within the wetland vegetation. At small springs (Parunuweap, Garden Creek), grids encompassed the entire wetland. Where size precluded timely work on the entire wetland, one or two grids were established within the wetland area (Santa Maria, 1; Pocket Mesa, 1; Potato Hollow, 1; Temple, 2). Within each one-square-meter cell, I visually estimated percent cover and average height of each vascular plant species as well as total ground cover of vegetation, soil, rock, and litter. Plant species were identified as closely as possible (following Welsh et al. 1993 and Hurd et al. 1998; scientific names follow Welsh et al. 1993). Identification and determination of plant species was provided by the staff at the S. L. Welsh Herbarium at Brigham Young University (notably, Stan Welsh) and the Deaver Herbarium at Northern Arizona University (notably, David Hammond and Tina Ayers). Some plants, particularly Compositae and *Epilobium*, could not be identified to species level, owing either to a lack of opportunity to collect at the appropriate time of year or to a lack of necessary parts for identification on specimens because of herbivory or other damage. Wetland affiliation of all species was determined using the PLANTS database (U.S. Department of Agriculture NRCS 2001); species with wetland status of "facultative" or higher were considered wetland species. For plants that could not be identified to species and for which some species within the genus had wetland affinities, the taxon was listed as potentially having wetland affiliation and was included in the "maximum wetland species" count for the site. Wetland species were considered common if maximum height in the grid was at least ten centimeters and if they accounted for at least 5 percent cover at a site or if they occurred in at least 5 percent of grid squares at a site. The height restriction was used because the two Temple study grids (both usually having areas of bare sand) sustained substantial germination but very low survival of sprouted plants. All wetland plant species encountered during the study readily attained heights of at least ten centimeters and typically bloomed at heights above ten centimeters. The height restriction did not eliminate any plants that accounted for at least 5 percent cover within any study grid but did eliminate some species that germinated in 5 percent or more of grid squares and never survived to grow above ten centimeters.

Because of the imbricate nature of the vegetation, plant species covers often summed to over 100 percent. Total ground cover measurements summed to 100 percent. Soil moisture, litter depth, and soil depth were measured at the corners of the one-meter squares. Long-term study areas were visited each summer, beginning in 1997. Data up to and including the 2000 study season are included here.

Visual estimates of percent cover lack the precision of most measurements. Using only a single observer should improve consistency, but I have no measurements against which to confirm estimates. However, in the course of fieldwork, miscommunication with the assistant recording data provided regular opportunities to determine repeatability. Variation in repeatability of visual estimates of cover was less than 10 percent of the estimated percent cover for the large majority of estimates above about 30 percent (e.g., a repeat estimate ranged at most from 54 to 66 for an original estimate of 60 percent). Differences as large as 100 percent occurred in small estimates (e.g., a repeat estimate of 2 percent when the original estimate was 1 percent). All springs observed during this study showed changes in total percent cover of wetland species (range over the study period) exceeding 10 percent.

In analyses of total wetland plant cover, adjustments were made to account for rock cover. To limit these analyses to wetland grid squares, I eliminated grid squares that did not, at least once in 1997–2000, have at least 25 percent cover (adjusted for rock) of wetland plants.

I used Jaccard's coefficient to assess between-site similarity. Jaccard's coefficient is calculated as follows: common species to both sites/(common species at both sites + unique species from both sites).

Results

Species Richness

Wetland species diversity varied considerably among the sites (table 10.2, appendix 10.1). Although the largest site—Pocket Mesa—had the highest number of wetland species (both total species and common species),

TABLE 10.2. Wetland species data by site

Location	Total area (m²)	Maximum number of common species	Maximum number of species	Total percent cover[a]				Ratio of maximum: minimum percent cover	Percent cover by type, 1997–2000[b]			
				1997	1998	1999	2000		Forb	Grass	Rush/ sedge	Woody cover
Grand Canyon National Park (GCNP)												
Garden Creek	21	2	3	53	46	55	51	1.1	0	0	75	25
Santa Maria	24	5	7	115	107	89	90	1.3	5	59	36	0
Zion National Park (ZNP)												
Temple Lower	9	8	12	68	3	2	2	43.2	100	0	0	0
Temple Upper	20	8	16	63	33	27	49	2.4	98	0	0	2
Potato Hollow	18	10	14	55	70	91	40	1.8	4	1	95	0
Parunuweap	36	11	12	36	62	64	56	1.8	54	16	10	20
Pocket Mesa	121	15	22	124	90	109	95	1.4	74	9	12	5

[a] Data for total percent cover by year are adjusted for rock cover and upland.
[b] Parunuweap, Temple Upper, and Temple Lower data represent 1998–2000 owing to major impacts during August 1997.

the species-area relationship was not significant for the study as a whole (r_s = 0.429, N = 7, p = 0.337 for richness of common wetland species; r_s = 0.364, N = 7, p = 0.429 for richness of maximum wetland species). Common wetland species comprised 50–100 percent of wetland species present.

Overall, Zion sites, regardless of size, had higher wetland species richness (6–12 species common in at least one year, 12–20 species total), than Grand Canyon sites (2 and 5 species common in at least one year, 3 and 7 species total). However, the Temple and Potato Hollow sites in Zion had variable species richness through time; species richness dropped to levels similar to those in Grand Canyon at some times. Potato Hollow lost species as the site recovered after the burn, ending with two species in 2000. The Temple plots were more erratic through time. No wetland species found during this study was unique to either park. Each park contains spring-associated endemics, but many of these are hanging garden species (see chapter 9).

Community Similarity

There was no overlap in wetland species between the two Grand Canyon sites. Overlap among the Zion sites varied among pairs of sites and over time. For taxa identified at the species level, sites ranged from as little overlap as 1 species in common among 24 present at Parunuweap and Pocket Mesa to 8 species in common among 20 present at Pocket Mesa and Potato Hollow, and 9 species in common among the nearly adjacent Temple Lower and Temple Upper sites, which together had 12 species present.

Between-site similarity for the Zion sites measured with Jaccard's coefficient ranged from 4.2 percent between Pocket Mesa and Parunuweap to 40 percent between Pocket Mesa and Potato Hollow. Temple Lower had a nested subset of 9 of the 12 species present at Temple Upper (similarity of 75 percent). Elevation difference did not have a consistent impact on similarity (r_s = −0.536, p = 0.110; n = 10 pairs of sites). The lowest similarity was between the sites with greatest elevation difference (Pocket Mesa and Parunuweap: 4.2 percent), and the highest between-site similarity was between one of the pairs of sites with the lowest difference in elevation between sites (Pocket Mesa and Potato Hollow). Nevertheless, sites at similar elevations but in different main canyons (e.g., Parunuweap and the Temple sites) had less similarity (10.5 and 12.5 percent) than sites at different elevations in the same main canyon (e.g., Potato Hollow and the Temple sites: 31.3 percent and 33.3 percent). Using only Temple Upper of the two Temple plots, to avoid same-site plots, the relationship between similarity and elevation was still not significant ($r_s = -0.618$, $p = 0.191$, $n = 6$ pairs).

Juncus ensifolius was the most common species encountered, being present at all but the Garden Creek site; an additional 3 species (*Muhlenbergia asperifolia*, *Carex aurea*, and *Agrostis semiverticillata*) occurred at both parks but at only one site in each (appendix 10.1). Of the remaining 32 taxa identified to species, *Juncus bufonius* and *Mimulus floribundus* were present at three sites, 11 species were present at two sites, and 21 species were present at a single site (5 from Grand Canyon, 16 from Zion).

Habitat Affinity

Fourteen species categorized as "facultative" (equally likely to be found in wetland or upland conditions) accounted for at most 3.3 percent of total wetland cover—the remaining cover was contributed by 32 species that ranged from slightly more likely to be found in wetland than in upland conditions (wetland indicator status, FAC+) to obligate wetland species (wetland indicator status, OBL) (appendix 10.1).

Low-gradient wetlands below hanging gardens (Parunuweap, Temple Lower, Temple Upper) contained more hanging garden species (Welsh 1989a, 1989b) than did low-gradient wetlands distant from hanging gardens, despite the obvious edaphic and topographic differences of both these wetlands from hanging gardens. At Parunuweap, Jones' reedgrass (*Calamagrostis scopulorum*), giant helleborine (*Epipactis gigantea*), scarlet lobelia (*Lobelia cardinalis*), asters (*Aster* cf. *welshii*), and golden sedge (*Carex aurea*) were present at the study site, as well as at one or more of the associated hanging gardens.

Below the weeping wall in the Temple of Sinuwava, one or both of the Temple sites contained cardinal monkeyflower (*Mimulus cardinalis*), Zion daisy (*Erigeron sionis*), and columbine (*Aquilegia chrysantha*, *A. formosa*, and hybrids). Welsh (1989a) surveyed the hanging garden at the Temple of Sinuwava. In addition to the species I observed on the sandy flats below the garden, he observed maidenhair and northern maidenhair ferns (*Adiantum capillus-veneris*, *A. pedatum*), Jones' reedgrass, Zion shootingstar (*Dodecatheon pulchellum*) and rock spiraea (*Petrophytum caespitosum*). These species apparently did not find appropriate growing conditions in the sand substrates of study sites below the wall.

Goldenrod (*Solidago* spp.) occurred commonly at Pocket Mesa, Santa Maria Spring, Temple Upper (in 1997), and Parunuweap; only the latter two sites are associated with hanging gardens, and only the Parunuweap hanging garden contained goldenrod. Because of the timing of site visits, identification to species could not be made; however, Welsh (1989b) lists *S. canadensis*, *S. sparsiflora*, and intergrades as components of hanging gardens that also use other habitats. Thus, all sites may contain the same goldenrod species regardless of proximate association with hanging gardens.

Bundle panic (*Panicum acuminatum*) and golden sedge, species that regularly occur in hanging gardens but are not limited to them, were common at Santa Maria Spring, a steep site (more than 100 percent slope in parts) but with well-developed soil. No hanging gardens occurred within several hundred meters of the Santa Maria Spring source.

Cover Changes at Springs from 1997 to 2000

The ratios of maximum wetland plant cover to minimum wetland cover at the study sites ranged from 43.2 to 1.1 (table 10.2). The Temple Upper and Lower sites had the greatest change in percent cover of wetland species. In 1997, these sites below a weeping wall were approximately two-thirds vegetated, discounting areas of rocks. Both sites had substantial cover of cardinal monkeyflower and columbine, as well as the more dry-adapted scouring rush (*Equisetum arvense*). During August 1997, a large storm brought heavy rains to the Zion highlands. The drainage on the plateau above the weeping wall flooded. The waterfall from the plateau rim, usually dry or at most a faint veil of mist in summer, flooded the small amphitheater below. Mature trees on the low scree walls around the study area were stripped of leaves, limbs, and bark; some were topped entirely. Sandstone slabs up to nine cubic meters were repositioned, and substantial sediment, including ash and burned wood from the previous year's fire on the plateau, was deposited on the sandy flat below the wall.

The weeping walls above the Temple sites were apparently not subjected to the force of the flood, and they have continued to support hanging garden species throughout the study. However, most vegetation in the study sites was scoured or buried by the flood. Temple Lower was partly covered by a repositioned boulder and lost approximately half of its previous available area for rooting plants. The upper site has recovered approximately three-quarters of its plant cover (table 10.2), but this is almost entirely scouring rush. The lower site supports many seedlings of a variety of species, but these rarely reach maturity, and cover has not approached 10 percent since the inundation. Neither site now supports the hanging garden species (except one or two Zion daisies) that it previously contained, despite surface-water flow through portions of the sites and water tables high enough to produce damp soil at the surface of other portions.

The Parunuweap site had the third-largest change in percent cover of wetland species; however, its greatest change—removal of substrate—is not reflected in table 10.2. The same flood that drastically rearranged the Temple sites removed a three-square-meter area of substrate at the Parunuweap site, widening a section of the channel of the hanging garden outflow that runs at the edge of the study site wetland. The scouring produced somewhat shallower, wetter substrates and reduced litter cover, probably contributing to the higher wetland plant cover of the following years.

Changes at the Potato Hollow site (table 10.2) also resulted from a single agent of change: a forest fire burned across Horse Pasture Plateau and surrounding drainages in July 1996. In 1997, the six-meter by three-meter study area was at the edge of a part of the valley floor where the fire had burned well into the ground, damaging or destroying the original apparently continuous root mat of Nebraska sedge (*Carex nebrascensis*) that dominated the immediate vicinity. One end of the study plot was an intact sedge clone; the other end was primarily bare ground, with sparse forb cover, particularly various monkeyflowers (*Mimulus* spp.). The following year, vegetation in the area, apparently responding to the fertilizing effect of the fire and having recovered somewhat in the intervening year, was up to 100 percent taller than in 1997. Nebraska sedge averaged 50 centimeters in 1997 and 115 centimeters in 1998. Smooth brome (*Bromus inermis*) in the drier portions of the valley was over 1.5 meters tall. In 1999, Nebraska sedge height was intermediate, at 80–93 centimeters, and sedge cover continued to expand, reducing species diversity (table 10.2). In 2000, a late spring freeze killed back the ends of the young sedge leaves. Sedge growth was shorter and less dense than in any of the previous study years but more uniform because the sedge had regrown in the lower part of the plot.

The Pocket Mesa site showed the most uniform vegetation cover of the Zion sites over the four-year period but still had an approximately 25 percent decrease in vegetation from 1997 to 1998 (table 10.2). An early summer visit to the site in 1998 showed that the small drainage above the meadow had flooded at some point after the 1997 site visit, and sediment and debris had been deposited in lines of drift in the upstream three-quarters of this essentially flat site. Drifts of pine needles and pinecones, as well as soil, were very evident, and "understory" plants such as yarrow (*Achillea millefolium*) were less common during 1998. The deposition observed in 1998 was the greatest of any of the observation years and also may have been attributable to the August 1997 precipitation event that affected the Temple and Parunuweap study areas.

The Grand Canyon sites had the least overall variation in plant cover during the reporting period. Both were grazed and browsed by native ungulates, but grasses and sedges showed no obvious sign of having been eaten. Emory seepwillow (*Baccharis emoryi*) was browsed at Garden Creek by mule deer (*Odocoileus hemionus*), which were commonly observed at the adjacent Indian Garden campground, and once on the study site. Bighorn sheep (*Ovis canadensis*) were observed drinking at Santa Maria spring, and cattail (*Typha* sp.) was browsed in some years.

Discussion

Mechanisms of Change

FLOOD. During this study, location in the path of an ephemeral stream was most likely to lead to major change in vegetative cover. Such a location not only exposes wetland vegetation to inundation and sedimentation but also may be associated with rockfall and large-scale debris flows. Size of the watershed, breadth and slope of the floor, substrate, and vegetation all contribute to the nature of flood impacts for springs located in the path of floodwaters. In the often-abrupt topography of the Intermountain

West, flow paths may include substantial vertical drops, such as the nearly four-hundred-meter fall above the Temple sites.

Potential flood impacts are obviously reduced for springs, such as the two in Grand Canyon, that are situated on the sides of their respective drainages. During this study, those two springs had the most consistent vegetation cover. These springs are also fed by older aquifers that are known to have relatively stable flow, further reducing water-related change (Monroe et al. 2005).

FIRE. Impacts of fire on wetlands are reported from a variety of individual sites and vary with soil moisture, the nature of the surrounding vegetation, and the size of the wetland (e.g., Timmins 1992); however, no summary review of such fires has been assembled. The fire that burned Potato Hollow was caused by a lightning strike during a minor drought. Mature ponderosa pines growing in the valley bottom around the wetlands continued to burn even after falling into the wet meadows and obviously contributed to the destruction of root mats on which they fell. In addition, the fire completely burned large historic waterbars made of ponderosa pine logs 25–100 centimeters in diameter that had been emplaced by the Civilian Conservation Corps in the 1930s to slow erosion. The troughs created in the wetland by their burning held water the following year but had filled in with sediment and vegetation by 2000. I noted no signs of erosion at these depressions.

Recovery time following fire may depend not only on the character of the fire but also on events that occur during the period before vegetation regrowth, when soils are vulnerable to erosion. In Potato Hollow, the August 1997 rainstorm caused some gullying, particularly in drier areas of the valley bottom, but I observed no downcutting deeper than approximately 15 centimeters, and that occurred only in short stretches of a meter or two in length. Root mats persisted over much of the valley floor, and soil loss was minimal. Areas in which wet-meadow root mats had been eliminated were completely reinvaded by neighboring root mat by the summer of 1999. Interestingly, the dense root mats of rhizomatous species that permitted quick recovery after the fire may have supported dense vegetation that provided an abundance of fuel in the dry 1997 summer. *Spartina pectinata*, a rhizomatous cordgrass growing in tallgrass prairie wetlands, responds so positively to annual burning that, although it is a wetland species, researchers have suggested that it may be fire-dependent (Johnson and Knapp 1993). *Carex nebrascensis*, the dominant wetland species at Potato Hollow, grows east into the prairies of Kansas and South Dakota and may also have some degree of adaptation to fire. However, information on such characteristics is unavailable for most wetland species.

The fire-induced diversity in the burned gap in the Potato Hollow study plot was short-lived. The proportion and number of nongraminoid (and also nonrhizomatous) species declined steadily after 1997, the first year postfire. Some forb species were still present in 2001, although as a minor presence in an otherwise near-monoculture of Nebraska sedge. Of the four monkeyflowers, two were present only in 1997, and all were gone by 2000 when the sedge monoculture was clearly reestablished. Graham (see chapter 15) indicates that fire effects in hanging gardens can be far more persistent than those observed here.

HERBIVORY. Most of the study sites discussed here have been protected from livestock grazing since before 1920. The Potato Hollow and Pocket Mesa sites were protected in the 1950s and in 1979, respectively. During this study, herbivory was an agent of only minor change. During the summer of 2001, hornworms (*Manduca* sp.) were present at Pocket Mesa in high densities (one to five per square meter in much of the 11 × 11 m plot) but they focused their feeding primarily on *Epilobium* plants and did not seem likely to greatly defoliate the site. Deer, bighorn sheep, and elk (*Cervus elaphus*) were variously present at the study sites, but despite the lack of hunting pressure, no signs of strong grazing or browsing pressure were present with the possible exception of deer browse on coyote brush at Garden Creek. Whereas mountain lion (*Puma concolor*) impacts on large herbivores were evident at Zion, human presence near Garden Creek campground in Grand Canyon provided both a deterrent for predators and an attraction for mule deer.

The lack of evidence of overgrazing at protected seeps and springs should in no way be taken as evidence that they are difficult to overgraze. Unmack and Minckley (chapter 2) and the Grand Canyon Wildlands Council (2002) have clearly demonstrated that uncontrolled livestock access to springs destroys vegetation and may change water availability.

Most springs are too small or too precipitous to support continuous beaver presence. However, within the general area of this study, beaver herbivory has been observed at springs along the Colorado River and its tributaries in Grand Canyon (e.g., Johnson 1991; L. Stevens, pers. obs.; V. Meretsky, pers. obs.) and at springs around Lake Powell (Spence, pers. comm.). Removal or reduction of woody vegetation can alter light availability, but if trees are not killed by beaver damage, shade may be reestablished within only a year or two (Meretsky, unpubl. data). Thus, although the visible effects of beaver herbivory are striking, the changes in the vegetation community may be of short duration. Damming does not seem to occur, possibly because of the usually short supply of woody debris; however, damming was once an important feature in southwestern desert riparian corridors (Coleman and Dahm 1990).

Methods of Studying Springs

Our understanding of the finer details of vegetation community dynamics at springs is limited by the opportunities for careful experimental study. Although desert springs are not as scarce as the popular literature may suggest, relatively unaltered springs are rare and are generally restricted to steep terrain in parks. Not only are such sites logistically difficult to study experimentally, their fragility often precludes regular close observation. Furthermore, their history and context (physical, biological, and social) vary widely. Sites similar enough to provide good experimental replicates may suffer from pseudoreplication—high spatial autocorrelation that complicates analysis and effectively reduces sample size (Hurlbert 1987). In addition, the patchy nature of some agents of change (e.g., fire) may introduce new variables and alter replication at any point during the observations. Precisely defining the effects and interactions of agents of change is thus dependent on experimental design that may be possible at very few sites. Improved spring inventories with appropriate contextual information and long-term monitoring will improve our understanding of vegetation dynamics and interspring variability in dynamics.

The Implications of Change in Spring Vegetation Communities

Even relatively stable, well-protected springs changed measurably during the course of this study. In the relatively short duration of the observations, some known agents of change had either no opportunity or insufficient opportunity to effect major alterations. Climate change, in particular, undoubtedly has affected western springs during the long wet and dry cycles since the Pleistocene and will continue to cause change as anthropogenic factors increasingly affect the climate signal. Ecologists and managers alike tend to focus concern on more-immediate anthropogenic impacts—water extraction, grazing, mining—because these pose immediate risks to spring ecology or spring existence. But even springs in protected areas are subject to change.

Efforts to restore vegetation communities at springs are an important part of conservation in the West. These restoration efforts cause change, and monitoring efforts are needed to determine the success of restoration. The natural variability of spring communities provides an interesting challenge for researchers seeking to determine when a restored ecosystem has "found its balance." Additional, longer-term research can provide a fuller picture of the natural amplitude and frequency of changes in these variable communities.

ACKNOWLEDGMENTS

I gratefully acknowledge the advice and assistance of Tina Ayers, Eric Brueck, Steve Floray, Don Jones, Laurie Kurth, Denise Louie, Karen Mason, Laird Naylor, Rod Parnell, Carolyn Sandolin, Dave Sharrow, John Spence, Abe Springer, Jim Starling, Larry Stevens, and Debbie Wong. Field assistance from Jeffrey Huntsman, Karin Kettenring, Denise Louie, Margaret Malm, Eric North, Patty West, Pat Whitesell, and Julie Zimmerman was invaluable, as were the plant identification skills of Jason Alexander, Tina Ayers, David Hammond, Stan Welsh, and the staff of the S. L. Welsh herbarium. Portions of this research were supported by the School of Public and Environmental Affairs, Indiana University.

Appendix 10.1

WETLAND SPECIES BY STUDY SITE AND YEAR

Site name/ park/size (m)	Species	Presence value				Wetland indicator status
		97	98	99	00	
Garden Creek	*Baccharis emoryi*	2	2	2	2	FACW
GCNP	*Carex hystricina*	2	2	2	2	OBL
7 × 3	*Celtis reticulata*				I	FAC
Parunuweap	*Adiantum capillus-veneris*				I	FACW
ZNP	*Agrostis semiverticillata*	2	2	2	2	OBL
9 × 4	*Aster* sp.[a]	(2)	2	2	2	OBL
	Baccharis sp.	2	2	2	2	OBL
	Calamagrostis scopulorum	2	2	2	2	OBL
	Carex aurea	2	2	2	2	OBL
	Epipactis gigantea	2	2	2	2	OBL
	Equisetum sp.	2	2	2	2	FACW
	Juncus ensifolius	2	2	2	2	FACW+
	Lobelia cardinalis	2	2	2	2	OBL
	Populus fremontii	2	2	2	2	FACW*
	Typha sp.	2	I			OBL
Pocket Mesa	*Agrostis alba*	2	2	2	2	FACW
ZNP	*Carex athrostachya*	2	2	2	2	FAC
II × II	*Carex microptera*	I				FAC
	Carex nebrascensis	2	2	2	2	OBL
	Carex praegracilis	(I)	I	I	I	FACW
	Cirsium vulgare				I	FAC
	Eleocharis patula	2	I		I	OBL
	Epilobium spp.[b]	2	2	2	2	FAC
	Equisetum arvense	2	2	2	2	FAC+
	Gnaphalium cf. *chilense*	I			I	FAC(W)
	Juncus arcticus	2	2	2	2	FACW
	Juncus bufonius	I	2		2	OBL
	Juncus ensifolius	2	2	2	2	FACW+
	Juncus longistylis	2	I		2	FACW+
	Juncus tenuis	2	2	2	2	FAC
	Mimulus floribundus	I		I	I	OBL
	Rumex crispus	2	I	2	I	FACW
	Rumex salicifolius	I	I	I	I	FAC*
	Salix sp.	2	2	2	2	FAC+
	Veronica peregrina				I	FACW+

Site name/ park/size (m)	Species	Presence value				Wetland indicator status
		97	98	99	00	
Potato Hollow[c]	*Agrostis alba*		2	2		FACW
ZNP	*Carex nebrascensis*	2	2	2	2	OBL
6 × 3	*Cirsium vulgare*		1			FAC
	Epilobium spp.[d]	2	2	2		FAC
	Equisetum sp.	1	2			FACW
	Gnaphalium cf. *chilense*		1			FACW
	Juncus bufonius	2	1			OBL
	Juncus ensifolius		2	2	2	FACW
	Juncus tenuis		1	2		FAC
	Mimulus cardinalis	2				OBL
	Mimulus floribundus	2	1			OBL
	Mimulus guttatus	2	2	2		OBL
	Mimulus rubellus	1				FAC+
	Rumex crispus		1			FACW
Santa Maria	*Agrostis semiverticillata*	2	2	2	2	OBL
ZNP	*Carex aurea*	2	2	2	2	OBL
8 × 3	*Juncus ensifolius*			1		FACW+
	Juncus torreyi	1		1		FACW+
	Muhlenbergia asperifolia	2	2	2	2	FACW+
	Panicum acuminatum	2	2	2	2	FACW
	Typha sp.	2	2	2	2	OBL
Temple Lower	*Acer negundo*	1				FACW
ZNP	*Aquilegia* spp.	2	1			FAC
3 × 3	*Baccharis* sp.	1				FACW
	Carex spp.		1			(FACW)
	Equisetum arvense	2	2	2	2	FAC+
	Juncus bufonius		2	1		OBL
	Juncus ensifolius	2	2		1	FACW+
	Mimulus cardinalis	2	2			OBL
	Mimulus floribundus		2			OBL
	Mimulus guttatus		1	2	2	OBL
	Populus fremontii	2				FACW*
	Tamarix spp.			1		FACW
Temple Upper	*Acer negundo*	1			1	FACW
ZNP	*Aquilegia* spp.	2			1	FAC
5 × 4	*Baccharis* sp.				2	FACW
	Equisetum arvense	2	2	2	2	FAC+
	Juncus bufonius		1			OBL
	Juncus ensifolius	2	1			FACW+

Site name/ park/size (m)	Species	Presence value				Wetland indicator status
		97	98	99	00	
	Mimulus cardinalis	2	I		I	OBL
	Mimulus floribundus	I			I	OBL
	Mimulus guttatus	I	2	2	2	OBL
	Mimulus rubellus	I				FAC+
	Muhlenbergia asperifolia	2				FACW+
	Populus fremontii				I	FACW*
	Salix sp.		I		I	FAC+
	Smilacina stellata	I				FAC
	Tamarix spp.				I	FACW

Note: Wetland species described as facultative to obligate wetland species in the national PLANTS database (U.S. Department of Agriculture NRCS 2001). A presence value of 2 indicates that the plant was common (at least 5 percent of one grid square or present in at least 5 percent of grid squares). A presence value of I indicates a plant that was present at the study site but not common enough to qualify as a 2. Values in parentheses are species probably missed during the first survey. Abbreviations: GCNP = Grand Canyon National Park; ZNP = Zion National Park. Wetland indicator status abbreviations are from the PLANTS database.

[a]Wetland *Aster* species include at least one of *A. hesperius* or *A. welshii*.

[b]Wetland *Epilobium* species include at least *E. ciliatum*, *E. glandulosum*, and *E. saximontanum*.

[c]Only ten squares were relocated and measured in 1998, but the remainder were dominated by *Carex nebrascensis*.

[d]Wetland *Epilobium* species include at least *E. ciliatum*.

Chapter 11

Biodiversity and Productivity at an Undisturbed Spring in Comparison with Adjacent Grazed Riparian and Upland Habitats

BIANCA S. PERLA AND LAWRENCE E. STEVENS

Springs and associated wetland and riparian ecosystems are among the most productive, biologically diverse, and threatened habitats, particularly in the arid southwestern United States (Knopf et al. 1988; Johnson 1991; Williams and Danks 1991; Erman 1992; Ferrington 1995; Noss et al. 1995; Botosaneanu 1998; Glennon 2002). Although Odum's (1957) studies of Silver Springs in Florida laid much of the foundation of ecosystem ecology, patterns of wetland and riparian productivity, biodiversity, and essential ecological processes at springs in arid regions have received scant attention. Stevens and Ayers (2002) reported that springs wetlands make up less than 0.01 percent of the area of Grand Canyon in Arizona yet support nearly 11 percent of the plant species in the overall landscape, as well as many invertebrate and bird species and facultative plant, herpetofaunal, and mammal species. Thus, springs may function as keystone ecosystems that, given their generally small size, exert disproportionately large impacts on the biodiversity and ecological processes of the surrounding landscapes.

Although much progress has been made on evaluations of riparian ecosystem ecology and health (e.g., Vannote et al. 1980; U.S. Department of the Interior 1993, 1998; Pellant et al. 2000; Stevens et al. 2005), there have been too few systematic inventories or assessments of springs in the West to address basic questions about geomorphologic commonalities, basic ecological processes, biota, degree of impairment, assessment protocols, or the landscape-scale importance of springs in comparison with adjacent upland habitats. Inventory and research on the ecology of spring ecosystems are needed, as are studies describing and assessing the impacts of human activities, such as livestock grazing, on such systems. Here we characterize basic ecological properties of an ungrazed spring ecosystem in southern Utah, in comparison with those on adjacent grazed riparian and upland ecosystems. Such baseline knowledge of spring ecosystem processes is essential for informed habitat management and restoration.

Human activities have greatly reduced the ecological integrity of many riparian and spring ecosystems in the West (Erman 1992). Overall estimates of riparian habitat loss range from 40 percent to 90 percent among the southwestern states (Dahl 1990). Although the extent of alteration of spring ecosystems has not been explored, the Grand Canyon Wildlands Council (2002) reported that 80–93 percent of the springs within several large land-management units in Arizona had been severely altered by human activities. Springs and other western riparian habitats are focal points of competing exploitative uses, such as timber harvest, recreation, water diversion, and livestock grazing (Thomas et al. 1979; Johnson et al. 1985). Livestock grazing continues to exert pervasive adverse influences on springs and other riparian habitats because riparian zones provide water, shade, and succulent vegetation (Bauer and Burton 1990; Chaney et al. 1990; Fleischner 1994; Stevens et al. 2005). Although springs and other riparian habitats are highly altered throughout the West, undiverted springs are ecologically resilient and may respond positively to improved management practices. Because of their biological importance, threatened status, and potential resilience, the protection and restoration of spring ecosystems should become a high priority for land-management and conservation agencies.

Productivity is an important ecological variable (Odum 1957; Bonham 1989), not only affecting ecosystem function and dynamics but also influencing biodiversity (Huston 1979; Zervas 1998). Unfortunately, research on riparian recovery has been focused primarily on vegetation structure, with less attention given to primary productivity and invertebrate (food-base) population dynamics. Previous investigations on the effects of livestock grazing impacts on riparian habitats have revealed that grazing reduces the number, total biomass, and condition of shrubs and trees, but such measurements have not been well quantified (Stevens et al. 2005). Similarly, differences in productivity between spring and upland sites have not been quantified in the western United States.

We studied patterns of biodiversity, productivity, and the impacts of grazing in 2002 at the undisturbed Seaman Spring and in adjacent grazed riparian areas and uplands in Grand Staircase–Escalante National Monument, Utah. Our purpose was to determine basic ecological characteristics (biodiversity and productivity) of this spring system and explore the impacts of grazing on those processes. To do this, we gathered information on plant, invertebrate, and vertebrate species diversity, abundance, and composition and quantified differences in organic standing mass and estimated net above-ground annual primary productivity along a springs-to-upland ecological gradient. We provide some of the first comparative productivity estimates for a rheocrene spring gradient in western North America.

Methods

Study Site

Seaman Spring is located in the southwest corner of Grand Stair-case–Escalante National Monument, in Kane County, Utah. The spring emerges on the floor of a small, structurally controlled canyon in the pin-yon pine–juniper zone at an elevation of 1,867 meters. The parent rock is Mesozoic sandstone. The climate is highly variable, with average springtime high temperatures of about 26°C, 34–35°C in summer, and 13–15°C in win-ter, with winter lows well below freezing. Rainfall occurs bimodally, with winter and summer precipitation peaks. Seaman Spring has a flow of about 5.5 liters per minute and is a bicarbonate/magnesium spring with moderate water quality (table 11.1) in relation to data presented in Mundorff (1971) for southwestern Utah.

Study Design

We divided the Seaman Spring site into three reaches of approxi-mately equal length and area and also studied an adjacent upland site with equivalent slope and aspect. These four study sites were designated as fol-lows:

REF: The reference site, the spring source riparian habitat that has received little or no grazing
WG: A wet grazed spring riparian habitat that has received rela-tively high grazing intensity

TABLE 11.1. Water quality data from Seaman Spring, August–September 2002

Water quality variable	Mean
Flow—field	5.5 L/min
Specific conductance—field	724.33 μmho/cm
pH—field	7.83
Temperature—field	19.83°C
Alkalinity (carbonate, as $CaCO_3$)	366.50 mg/L
Bicarbonate	447.00 mg/L
NO_2, NO_3, NH_3	0.00 mg/L
Phosphorus	0.03 mg/L
Dissolved solids	429.00 mg/L
Hardness (Ca, Mg)	377.73 mg/L
Calcium	63.58 mg/L
Magnesium	53.25 mg/L
Potassium	3.43 mg/L
Sodium	24.05 mg/L
Iron	265.75 μg/L
Manganese	107.75 μg/L
Chloride	26.65 mg/L
Arsenic	6.83 μg/L
Barium	259.50 μg/L
Chromium	7.43 μg/L
Selenium	1.23 μg/L

Note: $N = 3$ samples; data provided courtesy of J. Vanderbilt, Grand Staircase–Escalante National Monument.

DG: A dry riparian reach that has received relatively high grazing intensity
UG: A dry upslope shrubland and woodland site that has received relatively high grazing intensity

REF is naturally excluded from grazing by steep hillslopes and a large rock-fall. Livestock grazes freely on the other three experimental sites.

Data Collection

We compared plant and invertebrate biodiversity, standing plant and litter biomass per square meter and net annual productivity (measured in grams of carbon per square meter per year dry weight) with that of grazed wet and dry riparian systems and with the grazed upland habitat. Total dry organic standing mass (TDOSM) and estimated net aboveground annual primary productivity (NAAPP) were measured on plots in a stratified random design. We clipped and sorted all existing vegetation from twelve randomly selected and georeferenced plots (100 cm^2) in each study reach from ground cover (<1 m tall, comprising grasses and herbs) and from shrub and canopy cover (>1 m tall) strata. A 5-meter-tall survey rod was held vertically and used to measure where to clip shrub and canopy cover. The tree trunk or branch material too large to clip was estimated by relating measured wood mass-to-volume relationships from samples of the same species. We air-dried clipped vegetation to a constant mass at low humidity. Because this was the first year of monitoring, we clipped growth from the 2002 growing season, as well as growth from previous seasons. We separated green (2002) growth from woody growth and litter (pre-2002) and adjusted some upland samples (e.g., *Juniperus* sp. and *Pinus* spp. needles) for green growth that persists for more than one year. Vegetation was dried and weighed and productivity calculated, following the standard clipping methods of Bonham (1989) and Brower and colleagues (1998). This method allowed us to distinguish total organic standing biomass from estimated NAAPP. The NAAPP samples represent cumulative estimates of 2002 estimated NAAPP, and we recognize that these values are conservative underestimates because of losses attributable to seasonality and both invertebrate and vertebrate herbivory.

Plant cover, demography, and vigor were measured by visual estimation of percent cover by stratum and demography of the dominant native and nonnative plant species in each study site. These data were recorded on a site vegetation sketch map. The percentage of the total native plants on site that were healthy, in marginal condition (less than 50 percent of an individual plant wilted, burned, or dead), in poor condition (more than 50 percent of an individual plant wilted, burned or dead), or dead was visually estimated by three researchers and averaged.

We inventoried plants and aquatic and terrestrial macroinvertebrates, and we recorded but do not present data on herpetofauna, avifauna, and mammals. We searched intensively for all species of plants on each site. Two to five individuals or diagnostic portions of any unrecognized plants were collected for identification, and all taxa encountered were recorded. We visually estimated the percent cover of each native and non-native species in four strata: ground cover (annual deciduous nonwoody and <2 m tall); shrub cover (woody perennial, 1–4 m tall); midcanopy (woody perennial, 4–10 m tall); and high (woody perennial, >10 m tall) canopy cover classes.

Aquatic macroinvertebrates were inventoried in each reach using quantitative kick net sampling (0.09 m²) and spot sampling, with particular emphasis on aquatic Mollusca, various Coleoptera (especially Dytiscidae, Dryopidae, Hydrophilidae, Haliplidae, and Elmidae), semiaquatic Hemiptera, and Diptera (especially Tipulidae, Simuliidae, and Chironomidae).

Terrestrial macroinvertebrates were inventoried by collecting two to five individuals or diagnostic portions of all species encountered or by recording other taxa observed. Netting and other spot collecting were conducted with particular emphasis on Isopoda, various Coleoptera (especially Carbidae), and semiaquatic Hemiptera and Diptera (especially Chironomidae and Empididae). In addition, fifty sweeps with an aerial sweep net were performed at each study site for a quantitative comparison of terrestrial invertebrate species richness and abundance. Invertebrate specimens were mounted on pins in the field (especially mosquitoes and mirid bugs) or preserved dry (hard-bodied invertebrates) or in 70 percent ethyl alcohol (soft-bodied forms), labeled, and transported to the laboratory for preparation. Host plant and habitat affinities were recorded for all specimens. Invertebrate specimens are housed at the Museum of Northern Arizona in Flagstaff.

Analyses

We compiled the above data and analyzed data using nonparametric statistical tests because data were non-normally distributed. We conducted nonparametric Kruskal-Wallis and Mann-Whitney statistical tests with sequential Bonferroni corrections to compare TDOSM, estimated NAAPP, wood production, and leaf litter between sites. Statistical analyses were conducted using the statistical analysis computer program SPSS Version 6.0 (Norusis 1993).

TABLE 11.2. Plant species composition of different sites at Seaman Spring

Site	Date	Native species (N)	Non-native species (N)	Exotic species (%)	Total species (N)
UG	8/11/02	17	2	11	19
DG	8/12/02	16	0	0	16
WG	8/13/02	19	8	30	27
REF	8/14/02	17	4	19	21

Note: UG = grazed upland; DG = dry grazed riparian; WG = wet grazed riparian; REF = spring source, riparian with no detectable grazing.

Results

Vegetation

Plant species diversity, vegetation cover characteristics, plant vigor, and proportion of exotic species varied greatly between study sites. Overall, the two sites with surface water (REF and WG) had the highest plant species richness, as well as the highest proportion of exotic species (table 11.2). Nearly one-third of the plant species in the wet grazed site at Seaman Spring were non-native. This was 11 percentage points higher than that in the ungrazed reference site, where 19 percent of the flora was non-native (table 11.2).

Vegetation structure also varied considerably by study site. The REF site was the only site with a strong representation of all cover strata (fig. 11.1). The WG site had a noticeable absence of shrub, midcanopy and tall canopy cover; however, it had abundant low ground cover of wetland grasses, sedges, and rushes. The dominant stratum for both the dry UG and DG sites was the shrub layer, and those sites had little to no ground cover, midcanopy, or high canopy cover.

Invertebrate Biodiversity

Similar to the plant diversity data, the wet riparian sites (WG and REF) had far higher abundance and species richness of terrestrial invertebrates than did the dry sites. The WG site had the highest overall abun-

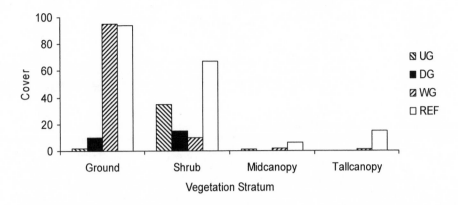

Figure 11.1. Percent cover of each vegetation stratum for each site at Seaman Spring visually estimated by three observers over the total area of the site.

dance of terrestrial invertebrates, but morphotype richness was statistically equivalent to that of the REF site. The higher value on WG was attributable to Diptera associated with cow manure.

The fifty-sweeps data revealed great differences among sites in terrestrial invertebrate abundance, morphotype richness, and order dominance. The wet riparian sites (REF, WG) had a much larger proportion of Diptera than did the dry sites (table 11.3; fig. 11.2). WG also had a higher proportion of Homoptera than did the reference site. REF had a higher proportion of insect predators (Hymenoptera and Arachnida), indicating a more trophically integrated assemblage than that on the WG site. Predatory insects (carabid and cicindelid beetles, spiders, asilid flies, etc.) occupy higher trophic levels and may be more likely to disappear from impaired ecosystems. The dominant insect order in the dry grazed reach was herbivorous, nonpredatory Coleoptera. A high abundance of leaf-feeding Coleoptera may indicate a drought-stress-induced nutrient concentration in heavily browsed riparian shrubs.

Our fifty-sweeps sampling (table 11.4) produced no invertebrates in the UG site. This does not mean there are no insects in the uplands surrounding Seaman Spring, but it does indicate that insect abundance is extremely low in that habitat as compared to the riparian sites. Unfortunately, the absence of ungrazed upland habitats precludes our assessment of the impacts of grazing on invertebrate biodiversity in uplands in this region.

TABLE 11.3. Percent composition of terrestrial invertebrates collected in fifty-sweep samples by site at Seaman Spring

Site	Dominant insect order		
	First	Second	Third
UG	—	—	—
DG	Coleoptera (36)	Hymenoptera (18)	Diptera (18)
WG	Diptera (84)	Homoptera (12)	Coleoptera (2)
REF	Diptera (80)	Hymenoptera (9)	Arachnidae (4)

Note: UG = grazed upland; DG = dry grazed riparian; WG = wet grazed riparian; REF = spring source, riparian with no detectable grazing.

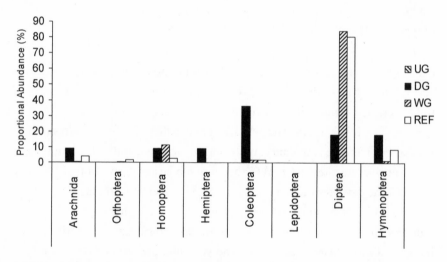

Figure 11.2. Proportional abundance (percentage of total insects collected) of eight invertebrate orders by study site at Seaman Spring. Data were collected from fifty sweeps of each study site with an aerial insect net.

Ecological Processes — Productivity

Large and significant differences existed in TDOSM, estimated NAAPP, wood production, and leaf litter between sites (Kruskal-Wallis $p < .001$, $\chi^2 > 24.89$, df = 3 for each variable). TDOSM was defined as all plant material, including green leaves and ground cover of plants as well as the woody stems of shrubs and trees, in addition to litter. The REF site had up to thirty-four times greater TDOSM than did the dry sites and an order of magnitude higher values than did the WG site (fig. 11.3, table 11.5; Mann-

TABLE 11.4. Morpho-species richness and abundance of terrestrial invertebrates by site at Seaman Spring

Site	Area (m²)	Total richness	Richness/ m²	Abundance/ m²
UG	650	0	0	0
DG	750	9	0.003	0.015
WG	718	58	0.08	1.41
REF	710	56	0.08	0.145

Note: Fifty-sweeps data. Abbreviations: UG = grazed upland; DG = dry grazed riparian; WG = wet grazed riparian; REF = spring source, riparian with no detectable grazing.

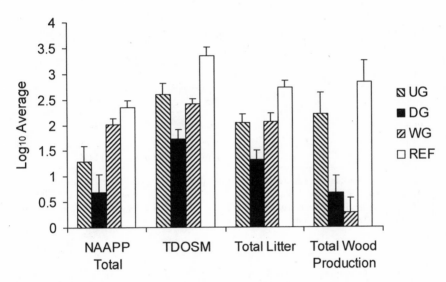

Figure 11.3. Log$_{10}$ average NAAPP, total biomass, leaf litter, and total wood production for each study site at Seaman Spring. Error bars represent one standard error.

Whitney $U < 17.00$, $p < .001$, $n = 12$ per site for all comparisons). TDOSM is a useful indicator of stored, produced, and decomposing carbon in these ecosystems. Ecosystems with higher TDOSM generally have higher levels of net above- and below-ground productivity, more available nitrogen, and higher levels of litter decomposition.

Mean estimated NAAPP differed greatly between sites as well (tables 11.5, 11.6; figs. 11.3, 11.4; Kruskal-Wallis $\chi^2 = 29.79$, $p < 0.0001$, df = 3). Wet

TABLE 11.5. Comparison of mean total dry organic standing biomass among study sites in 2002

Category	DG	UG	WG	REF
Productivity (g C/m²/yr)	14.1	97.9	124.8	281.6
Shrub, canopy stems (g C/m²)	57.1	330.0	15.4	2,501.0
Litter (g C/m²)	31.4	191.2	154.7	687.9
TDOSM (g C/m²)	102.6	619.4	294.9	3,470.5

Note: Twelve samples taken at each site. Statistically significant differences exist ($p \ll .05$) between all sites for each variable except between total productivity in WG and REF, litter UG-WG, shrub and canopy stems DG-WG. Abbreviations: UG = grazed upland; DG = dry grazed riparian; WG = wet grazed riparian; REF = spring source, riparian with no detectable grazing.

TABLE 11.6. Comparison of mean estimated net aboveground annual primary productivity among study reaches in 2002

Category	Reach			
	DG	UG	WG	REF
Ground cover (g C/m²/yr)	3.9	22.7	120.2	122.3
Shrub and canopy leaves (g C/m²/yr)	10.3	75.2	4.5	159.3
Total productivity (g C/m²/yr)	14.1	97.9	124.8	281.6

Note: Twelve samples taken in each reach. NAAPP and ground cover NAAPP were significantly different between all reaches (except WG and REF) at $p \ll .05$. Shrub and canopy cover NAAPP differed significantly among all sites. Abbreviations: UG = grazed upland; DG = dry grazed riparian; WG = wet grazed riparian; REF = spring source, riparian with no detectable grazing.

riparian sites were two to twenty times more productive than upland and dry sites. Estimated NAAPP did not differ significantly between the WG and REF sites; however, estimated NAAPP in WG was due primarily to ground cover, while that on the REF site was more balanced between ground cover and shrub and canopy cover layers (table 11.6; figs. 11.3, 11.4). Multistoried vegetation provides better wildlife habitat, as ground, shrub, and canopy layers provide diverse cover, invertebrate and bird food and habitat, and leaf fall that contributes to stream invertebrate diversity (Vannote et al. 1980).

Measurements of litter and other fallen and decaying vegetative matter revealed significant differences between sites, with the REF site having significantly higher levels of litter than other sites (table 11.5, fig. 11.3; Kruskal-Wallis $\chi^2 = 28.5$, $p < 0.0001$, df = 3). The WG site had no

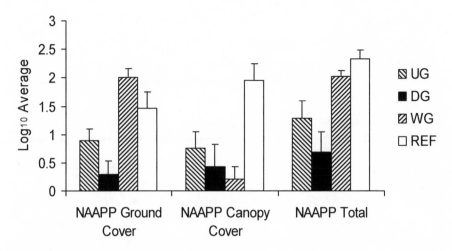

Figure 11.4. Mean estimated $\log_{10}$ of NAAPP by vegetation stratum for each site at Seaman Spring. Error bars represent one standard error.

more litter than did the upland site (Mann-Whitney $U = 70$, $p = 0.908$, $n = 24$). This was because reduction in shrub and canopy layers prevented litter deposition. Fallen canopy leaves and wood made up a large proportion of the litter in the REF site, and WG had comparatively little canopy cover. Cow manure made up a minor but conspicuous component of the litter in WG, and that source of litter was absent in the REF site. Carbon storage in woody stems was highest in the REF site by one to two orders of magnitude over other sites (Kruskal-Wallis $\chi^2 = 25.4$, $p = 0.0001$, df = 3).

Discussion

Our data indicate that spring ecosystems are focal points of biodiversity, non-native plant invasion, productivity, and organic matter accumulation. Wet riparian sites (WG, REF) had higher plant species richness and two to three times higher vegetative cover than did the upland grazed (UG) and grazed dry riparian (DG) sites. Wet riparian sites also were more prone to exotic plant species invasions and had 8–30 percent higher exotic species richness than did dry sites, a finding consistent with those of Stohlgren and colleagues (1999, 2003) and Stevens and Ayers (2002). Grazing disturbance of wet riparian sites may further increase the exotic plant diversity: WG had 30 percent non-native species, whereas the REF site had 19 percent non-native species.

Vegetation structure on sites varied in relation to water availability, as well as grazing disturbance. The ungrazed REF site was the only site that had a strong representation of all structural vegetation levels, including ground, shrub, midcanopy, and high canopy cover. Such factors influence not only microclimate, microhabitat structure, and wildlife habitat and food resource availability, but also ecological processes such as NAAPP, carbon storage, and probably decomposition. The REF site had far higher litter and wood standing mass values because of the extensive cover of shrubs and trees that created canopy cover, woody stem carbon storage, and leaf litter.

Grazing can greatly alter or reduce riparian ecosystem structure. Marcuson (1977) found that differential browsing by livestock reduced the number and total biomass of shrubs and trees in riparian habitats, and Glinski (1977) demonstrated that livestock strongly reduced Fremont cottonwood (*Populus fremontii*) seedling establishment. Knopf and Cannon (1982) similarly demonstrated that grazing significantly altered the size, shape, volume, and quantities of live and dead willow stems. The impacts of grazing on tree production have a critical impact on the riparian ecosystem because of the importance of the woody vegetation to wildlife habitat and its effects on riparian microclimate (Kauffman and Krueger 1984; Stevens et al. 2005).

Structural vegetation differences, some of which were attributable to grazing impacts, affected productivity at Seaman Spring and are likely to influence other basic ecological processes. The REF and WG sites had significantly higher estimated NAAPP levels than did the UG and DG sites. Productivity differed between the WG and REF sites not in the quantity of estimated NAAPP but rather in where carbon was being produced (ground cover versus canopy cover, respectively). Because litter levels differed greatly among sites, decomposition rates are likely to differ considerably as well because the litter layer is where decomposition is likely to be most rapid.

Terrestrial invertebrate assemblages likewise differed between sites. Wet riparian sites had higher abundance and greater morpho-species richness than did upland and dry sites, by more than two orders of magnitude. Some of the differences in the terrestrial invertebrate assemblages between the WG and REF sites were directly attributable to grazing and manure distribution. For example, WG had a greater relative abundance of Diptera, while REF had a greater abundance of predatory insects, suggesting a more complex trophic structure on the ungrazed site. Such patterns indicate that invertebrate production and potential food resources for terrestrial and aquatic vertebrates are greater in riparian and perhaps spring ecosystems and that dewatering spring-fed channels may substantially reduce and change the invertebrate food base for wildlife.

Our studies of Seaman Spring provided considerable insight into fundamental ecological processes at an undisturbed spring ecosystem in comparison with adjacent grazed wet riparian, dry riparian, and upland habitats. However, most springs in the Grand Staircase–Escalante National Monument and throughout the West have been highly modified for livestock use and culinary water supplies. It is increasingly difficult to find undisturbed springs study sites, such as Seaman Spring, at which to conduct basic research on ecological processes. Such sites are important laboratories for research into ecological processes and restoration potential of these unique, highly productive, biologically diverse, and poorly understood ecosystems.

ACKNOWLEDGMENTS

Funding for this project was provided by the Grand Canyon Wildlands Council. We greatly appreciate assistance on this project provided by the staff of Grand Staircase–Escalante National Monument, particularly that of Joni Vanderbilt, Thomas O'Dell, and Laura Welp. Preliminary water quality data were provided by the national monument through the State of Utah Department of Water Quality and the Kane County Water Conservancy District. We thank Margaret Erhart, Ann Hadley, Vicky Meretsky, Eric North, Chad Runyon, Phoebe Stevens, Steven Till, and Huarong Zhang for field and laboratory assistance on this project.

Chapter 12

Plant Diversity Influenced by Indigenous Management of Freshwater Springs

Flora of Quitovac, Sonora, Mexico

GARY PAUL NABHAN

As the freshwater resources potentially available to wildland plants and animals become increasingly usurped for use by humans alone (Postel 1992), it may be worth determining where historic cultural stewardship practices have demonstrated means of managing springs without diminishing the biodiversity associated with them (Nabhan 1982). Just as comparative anatomy and physiology may reveal to botanists the many adaptations that desert plants have evolved to deal with moisture deficits and high heat loads, a comparative inquiry into cultural traditions of springs management may reveal options for managing these habitats, options that conservation biologists might not otherwise recognize. One such comparative approach to springs habitat management has emerged out of a series of studies focused on the spring-fed oases of Quitovac and Quitobaquito in the Sonoran Desert (Nabhan et al. 1982; Mellink 1985; Felger et al. 1992; Reichhardt et al. 1994). They reveal that indigenous as well as immigrant desert residents of the desert have distinctive but always-dynamic relationships with spring-fed habitats, with each of their management practices positively or negatively influencing the biodiversity of these oases in ways not previously recognized in the ecological literature. The purpose of this chapter is to infer which cultural influences on desert springs may be ancient and perhaps co-evolutionary in nature, as opposed to more recent anthropogenic influences on biodiversity that arrived with Euro-American settlers. Given that springs were probably among the first places where desert hunter-gatherers wished to settle—and that the water, plants, and wildlife there were essential to their subsistence (Nabhan et al. 1989)—it should not surprise conservation biologists that many desert springs are not "pristine" but have been managed through indigenous practices for millennia. Given this fact, the dilemma for land managers becomes how to manage springs and their floras to sustain natural and cultural diversity, not simply how to protect them from any and all cultural influences.

Quitovac has been maintained as a traditional O'odham village, natural oasis, and de facto wildlife refuge for millennia. It is currently managed as common lands for the Sonoran O'odham in the municipality of Plutarco Elias Calles, Sonora, Mexico. Located at 112°45′ N longitude and 31°31′ W latitude, and at an elevation of 350 meters, it can be reached by traveling 50 kilometers southeast from the U.S.-Mexico border crossing of Lukeville/ Sonoyta along the highway to Caborca, Sonora. Road signs guide visitors approximately 2 kilometers west on a dirt road to the village; the oasis "lagoon" is 0.4 kilometers westward of the public school buildings.

The mean annual rainfall for Quitovac was 21.9 centimeters for the three years reported in Hastings and Humphrey (1969) or an estimated 19.2 centimeters based on calculations from a regional precipitation database maintained by the U.S. Geological Survey. Nevertheless, the rainfall at the oasis is highly variable, and no measurable precipitation falls at all in some years (Ezcurra and Rodrigues 1986). Like the *pozos*, or hand-dug wells, excavated by coyotes or early human migrants along the coast of the Gran Desierto de Sonora 50 kilometers west of Quitovac, the springs below the Quitovac hills have played an essential role in the survival of humans and wildlife in a region otherwise devoid of surface water resources for many months at a time (Ezcurra et al. 1988).

The oasis of Quitovac was first identified in a non-native language by Coronado, who in the 1540s referred to this O'odham settlement as Bacapa and as Quitobapc, the latter term referring to *tules* (reeds, rushes, or cattails) (Rodriguez-L. et al. 1993). In the 1690s, Padre Kino called the site San Luis de Bacapa ("Marshy Place") while still recognizing another O'odham name for its surrounding landscape, Moicaqui ("Soft-Textured Wash"). Kino observed that the site offered "very good water and good pasturage . . . close to a high peaked mountain at whose foot were some springs of water and some lakes" (Bolton 1984). This landscape has also been richly described by Lumholtz (1912) and others (cited in Rodriguez-L. et al. 1993). The oasis ecology and biodiversity associated with the Quitovac village have been documented in detail elsewhere (Nabhan et al. 1982; Mellink 1985; Reichhardt et al. 1994). Although the flora of the Quitovac Hills were earlier collected by

Blumer in 1911 and by Gentry in the 1940s, I have not relocated these historic specimens. I have only been able to ascertain the historic presence of a few plant and animal species noted by Davis (1921) in his observation of the Vi'igita ceremony that was held in Quitovac after being moved there by the Hia c-ed O'odham of the Pinacate.

Given that there are a few hundred spring-fed oases in hyperarid portions of the U.S. Southwest, the flora and vegetation of Quitovac continue to generate unparalleled scientific interest because of three factors:

1. Even though it is among the few naturally spring-fed wetland oases of more than one hectare in area found anywhere in the Sonoran Desert, it is potentially being threatened by groundwater pumping and cyanide-heap leaching within its aquifer at La Choya mine.
2. It has been inhabited by indigenous people for millennia, with excellent archaeological and historical evidence of occupational history and resource use.
3. It is in many ways ecologically analogous to Quitobaquito, Arizona, in Organ Pipe Cactus National Monument, with the major difference being that O'odham farmers have persisted longer in managing Quitovac, whereas O'odham residence in Quitobaquito was terminated in the 1950s (Nabhan 1982; Felger et al. 1992). As such, the cultural ecological context extant at Quitovac can inform the National Park Service on how to better manage and interpret both the natural and the cultural resources of Quitobaquito, using strategies now endorsed by park service policy initiatives to "nurture living cultures and communities" (National Park Service Advisory Board 2001) even though park resource managers rejected such a vision in most of their management decisions made between 1950 and 1990.

Methods

My associates and I collected voucher specimens from an eight- to ten-hectare area surrounding the oasis pond during the following months: November 1979, May 1980, September 1980, January 1981, April 1981, August 1981, December 1981, January 1982, March 1982, and during December 2001–January 2002. Voucher specimens have been deposited in the herbarium of the University of Arizona (ARIZ), Tucson, unless the collection number listed below is annotated "A.M.R.," indicating deposition in the herbarium of the San Diego Natural History Museum (SDO). Flowering dates, corolla color, and habitat notes for each species collected are derived largely from voucher specimen labels, supplemented by field notes at the site; most of these notes and specimens were collected within the five-hectare area closest to the pond.

Vegetation transects and soil sample data were collected in August 1981, and a water sample from one of the three major springs was collected in September 1980. Chemical analyses were made by the Soils, Water and Plant Tissue Testing Laboratory at the University of Arizona. At least five 20-meter line transects were randomly placed and read for woody perennials in each mappable vegetation association. At least 30 meters of point frame transects were run for coverage by late summer herbaceous annuals and perennials. The mapping units of vegetation associations are based on March 1982 aerial photos taken by Peter Kresan, supplemented by data from hand-drawn, paced maps that I made of vegetation prior to bulldozing.

For many of the voucher specimens, data concerning O'odham knowledge and use of the plants were volunteered by Luciano Noriego (born approximately 1915, died 1995), a long-time resident and elder leader in O'odham ceremonies at Quitovac. In addition, I have learned about some plant uses from Mariano Salcido, Rafael Garcia Valencia, David Manuel Lara, Hector Manuel Velasco, and their immediate families. Whenever I have come upon widespread northern Piman (O'odham) names or usages of plants found locally that Luciano Noriego was unfamiliar with, I have noted this information in the flora (appendix 12.1) in parentheses; however, this should not be construed as documentation of current local use.

Results

In total, 167 species of vascular plants in 134 genera and 52 families are listed for the ten-hectare area (appendix 12.1). Unlike the Quitobaquito flora of fifty to one hundred hectares (Felger et al. 1992), this flora concentrates on the five hectares closest to the pond. The annotations indicate several range extensions, including a few records (such as *Stephanomeria exigua* and *Hordeum arizonicum*) new to Sonora. The flora includes at least

34 species that were probably first introduced by transplanting or intentional sowing as cultivated plants in this locality, as well as 23 additional introduced species that are weedy in nature. Subtracting these 57 exotic species, this local flora remains relatively rich in natives (110 species) compared to Quitobaquito and to localities of this size elsewhere within western North America (McLaughlin and Bowers 1999). The aerial photos and on-ground, paced maps provided data enough to discern six mappable vegetation associations that existed prior to bulldozing. The August 1981 coverage data are presented in table 12.1, which also highlights data from the soil and water analyses that potentially affect plant growth and distribution. Table 12.1 also describes vegetation associations in terms of their areal extent, physiognomy, important soil factors, and (pre-bulldozing) O'odham management.

Discussion

Of the 167 species of vascular plants found at Quitovac, 15 species are found there but not shared with the flora of the Gran Desierto immediately to the west (Felger 2000). Whereas spring-fed habitats cover a minimum of 25 percent of our five-hectare study plot, this habitat type makes up a miniscule proportion of the Gran Desierto overall. In our five-hectare plot at Quitovac, through 1981, 38 more plant species were harbored than at the five-hectare plot at Quitobaquito, where only 93 wild, weedy, and cultivated species persist. If I exclude 32 domesticated species now incapable of persisting without man, the "free-living" Quitovac flora would be reduced to 135 taxa. The spring-fed Quitovac oasis also harbors a large number of "wild" plants—whether native or introduced—that are adapted to ruderal (roadside), agrestal (field-plowed), or flood-disturbed conditions. Many of these species are conspicuously absent from the Gran Desierto, and some are no longer found at Quitobaquito, though they were once recorded there (Felger et al. 1992). As defined here, the Quitovac "native" flora as a whole includes many species translocated to the oasis by the O'odham during prehistoric or historic (post-Kino) times. Some of these are the very species that have been lost from Quitobaquito and from some of the coyote-well pozos formerly managed by the Hia c-ed O'odham of the Gran Desierto.

The vegetation transects in the seven habitat types at Quitovac were dominated by just nineteen species, of which seven genera can be considered spring-fed wetland obligates: *Distichlis*, *Heliotropium*, *Potamogeton*, *Scirpus*, *Typha*, *Wislizenia*, and *Zannichellia*. The areal extent and condition of these taxa and their spring-fed habitats have been extremely dynamic, even prior to the 1981–82 draining of the lagoon. The same is true with field habitats affected by the 1984 bulldozing of the area to the north of the lagoon to open up irrigated agricultural land for which the Mexican government never provided infrastructure. In essence, modern irrigated agriculture has not yet replaced traditional subsistence practices at Quitovac, but it did reduce the amount of area dedicated to seasonal gardening and irrigated pasture production.

Nevertheless, several traditional spring, wetland, and riparian management techniques continued to be practiced there by the contemporary O'odham. This study is the first report of indigenous Sonoran Desert dwellers involved in seasonal burning of dried, choked patches of emergent wetland vegetation to open up surface water for waterfowl use and to stimulate new growth of plants for basketry materials and other uses. In addition, the O'odham are involved in constructing and seasonally cleaning irrigation ditches from springs; plowing and temporarily fallowing gardens, fields, and orchards; planting living fencerows of trees; transplanting medicinal and ornamental perennials from other localities; grazing domestic livestock; firewood cutting; and wild-food gathering. These and other cultural activities have all modified the nature of the spring-fed wetland vegetation and created new patches of emergent plants in microenvironments where they might not grow otherwise. Many of the biotic associations at Quitovac are thus anthropogenic, as they are at other desert oases around the world.

Ethnobotanically and nutritionally, the Quitovac flora is rich in wild green leafy vegetables, herbal medicinal plants, and edible fruits that the O'odham and other southwestern tribes have long utilized (Castetter and Underhill 1935; Castetter and Bell 1942; Curtin 1949; Nabhan 1982; Nabhan et al. 1989). Many of these are rich in antioxidants and soluble fibers historically important in preventing or treating type-2 diabetes and other nutrition-related diseases. The O'odham of Quitovac had few of the native varieties of domesticated food crops remaining by the late 1970s but did cultivate a surprising mix of ornamentals, including *Hymenocallis sonorensis*, beyond its "natural" range. Four multipurpose tree species (*Populus fremontii*, *Sambucus nigra*, *Salix gooddingii*, and *Washingtonia filifera*) appear to have

TABLE 12.1. Habitats at Quitovac, Sonora

Location	Percent of total area	Soil or water characteristics	Dominant plant species (highest cover first)	Life-form mixture	Diversity indices[a]	Land uses
Open water of lagoon and springs	10	Spring water: pH 7.6; soluble salts, 689 ppm; EC × 10³, 1.10; NH_4-N, 0.10 ppm; K, 5.83 ppm	*Potamogeton pulcinatus, Zannichellia palustris*	Submergent macrophytes and floating algae	—/—	Swimming; aquatic bird hunting; use of water for irrigation
Cultivated field of annual crops irrigated from crops and springs	6.5	Sandy loam: pH 8.4; soluble salts, 2,121 ppm; EC × 10³, 3.03; N, 3.75 ppm; P, 1.28 ppm; K, 1.15 meq/L	*Cynodon dactylon, Cucurbita argyrosperma, Citrullus lanatus, Ambrosia confertifolia*	Herbaceous weedy ephemerals and perennials, plus crop annuals	0.947/ 0.856	Tillage, seed sowing, and crop harvest; wild greens harvesting
Tufa mesa rimming the pond, and nearby scrubland (including abandoned fields)	27.5	Sandy loam: pH 7.4; soluble salts, 994 ppm; EC × 10³, 100; N, 32.13 ppm; P, 11.82 ppm; K, 2.27 meq/L	*Suaeda torreyana, Prosopis velutina, Lycium andersoni*	Open, mixed spinescent, drought-deciduous, and evergreen shrubs and trees	0.526/ 0.670	Wild fruit gathering; woodcutting; hunting and trapping
Cultivated orchard of irrigated fruit trees, and adjacent field-side hedge	2.5	Sandy loam: pH 8.4; soluble salts, 2,121 ppm; EC × 10³, 3.03; N, 3.75 ppm; P, 1.28 ppm; K, 1.15 meq/L	*Ambrosia confertifolia, Ficus carica, Sarco-stemma cynanchoides, Cynodon dactylon*	Broadleaf deciduous tree canopy with broadleaf deciduous shrubs, ephemerals, and vines	0.831/ 0.818	Cultivated fruit harvesting; irrigation; wild and cultivated perennial transplanting; hunting

					Diversity indices[a]	
Ephemeral water-course (arroyo) and adjacent uncultivated floodplain	4.5	Loamy sand: pH 7.6; soluble salts, 504 ppm; EC × 10³, 0.72; N, 10.2 ppm; P, 8.47 ppm; K, 1.26 meq/L	*Hymenoclea monogyra, Lycium berlandieri, Ambrosia ambrosioides*	Microphyllous shrubs, cacti, broadleaf shrubs, and few ephemerals	0.514/0.604	Grazing; hunting or trapping; cactus harvesting
Lagoon edge, shallow holding pond and ditches, and spring-to-pond channels	15	Silty loam: pH 8.3; soluble salts, 6,181 ppm; EC × 10³, 8.83; N, 5.35 ppm; P, 8.23 ppm; K, 2.51 meq/L	*Typha domingensis, Scirpus olneyi, Distichlis spicata*	Emergent perennial reeds and grasses	0.616/0.736	Burning; grazing; fiber gathering; medicinal plant gathering from ditches
Meadowlike flats with alkaline seeps	34	Sandy loam: pH 9.1; soluble salts, 70,427 ppm; EC × 10³, 100.6; N, 6.32 ppm; P, 4.78 ppm; K, 31.48 meq/L	*Distichlis spicata, Wislizenia refracta, Heliotropium curassavicum*	Perennial mat-forming grasses, few herbaceous root perennials, and ephemerals	0.089/0.078	Periodic grazing and burning

[a] Diversity indices: Shannon (Weaver)/Simpson (see Nabhan et al. 1982).

been vegetatively propagated and transplanted to Quitovac, as has the rare domesticated agave, Hohokam agave (*Agave murpheyi*), which is known only from Hohokam and O'odham terraces and dooryard gardens.

For centuries, the springs and hand-dug ponds at Quitovac provided sources of drinking and irrigation water, as well as habitat for a diversity of useful plants. These habitats functioned as a magnet for wildlife, since there is no permanent surface water for at least ten kilometers in any direction. Although there have been "improvements" in recent years in the main pond banks using heavy machinery, artesian-fed irrigation has remained a process of directing surface water via earthen ditches to the fields without mechanical pumping.

In the winter of 1981–82, the long O'odham tradition of gravity-fed surface-water irrigation was temporarily suspended. To make the area accessible to bulldozers, the pond was drained; only five centimeters of water remained in the pond by December 1981. By March 1982, the pond depth had been increased and bulldozing was nearly completed. In January 1984, most of the bulldozed land was covered with the invasive tumbleweed (*Salsola tragus*), except for a 0.25-hectare jojoba (*Simmondsia chinensis*) nursery, which has since failed. The following flora (appendix 12.1) therefore includes both wild and cultivated species found from 1981 to 2001; but one must recognize that the Quitovac oasis is now facing new threats from gold mining, groundwater extraction, population growth, and woodcutting on an unprecedented scale.

ACKNOWLEDGMENTS

An ethnobotany is possible only with the collaboration of local residents conversant with and expert in the traditional ecological knowledge of their indigenous community. This chapter could not have been written without the cultural memory of one such local expert, Luciano Noriego, who, unfortunately, died before this work was published. He graciously worked with me, Amadeo M. Rea (A.M.R.), Eric Mellink, and others on ethnoecological studies at Quitovac over a five-year period. I thank the O'odham community of Quitovac for their always-generous cooperation. In addition to the Noriego family, I thank Hector Manuel, David Manuel, and Rafael

Garcia of the O'odham Tribe in Sonora. Other O'odham from north of the border—including Culver Cassa, Philip Salcido, and Delores Lewis—helped us appreciate and gain perspective on Quitovac, its people, and their language. For assistance with plant identifications and taxonomic problems, I thank, in particular, Richard Felger (R.S.F.), Reid Moran, and Becky Van Devender. Others who participated in fieldwork at the oasis include Larry Toolin, Janice Bowers (J.B.), Alan Zimmerman, Bryan Brown (B.B.), Peter Warshall (P.W.), Julian Hayden, John Somner, Peter Kresan, Barney Burns, and Takashi Ijichi (T.I.). Other individuals cited in the flora include Delores Lewis (D.L.), Steve McLaughlin (S.M.), and Karen Reichhardt (K.R.). For comparisons with Quitobaquito Oasis in Organ Pipe Cactus National Monument, Arizona, I thank National Park Service personnel, particularly Roy Johnson and Keith Anderson. I inspected specimens from Howard Scott Gentry and Juan Arquelles and from Larry Toolin and Amadeo Rea, held at the University of Arizona herbarium. This work was funded by the Consortium for the Study of Man's Relationship with the Global Environment, the Man and Biosphere program, the Tinker Field Foundation, the Pew Scholars Program, and Agnese Haury. It is dedicated to Ofelia Zepeda, the MacArthur Award–winning O'odham linguist and poet, whose family emigrated from Quitovac to southern Arizona, in gratitude for her guidance and inspiration, and to the late Luciano Noriego.

Appendix 12.1

ANNOTATED SPECIES ACCOUNTS

Species are listed by family. Within each entry, the binomial is followed by the common name in English, Spanish, and/or O'odham (following Nabhan et al. 1989 and Felger et al. 1992). The use of an asterisk at the beginning of an entry signifies the species' historical introduction into North America. Life-form categories include winter ephemeral, summer ephemeral, nonseasonal ephemeral, herbaceous perennial, shrub (less than two meters high), tree (more than two meters high), succulent (including cacti), and epiphyte (including several aboveground parasites). The distribution categories refer to status within the ten-hectare surrounding oasis pond: *widely distributed* indicates numerous individuals found in more than one vegetation association; *localized* indicates at least several individuals, usually restricted to one or two associations; *rare* indicates species for which a careful search had to be made to encounter more than one or two individuals. Initials of plant collectors are noted with each voucher and are explained in the acknowledgments.

Acanthaceae

Justicia californica (Bentham) D. N. Gibson. Desert honeysuckle, chuparosa, vipismal. Shrub; flowers red-orange; mesa rimming pond, wash; widely distributed. *G.N. & J.B. 172, G.N., A.M.R. & T.I. 249.*

Agavaceae

Agave deserti Engelmann subsp. *simplex* Gentry. Mescal, a'ud. Succulent; flowers pale; in hills to the southeast; "cabezas" (heads) pit-roasted and eaten, leaves also possibly used for fiber. *Gentry & Arguelles 21203.*

Agave murpheyi Gibson. Bulbil-producing century plant, mescal, [a'ud] nonhakam. Succulent; bulbils usually develop instead of flowers; grown in yard outside study area; "cabezas" pit-roasted and eaten like *panocha*; a Hohokam and O'odham cultigen not reported in the Gran Desierto.

Aizoaceae

Trianthema portulacastrum L. Horse purslane, verdolaga de cochi, verdolaga blanca, ka:svañ. Summer ephemeral; field, ditch, yard, mesa rimming pond; widely distributed; eaten as cooked green. *G.N., A.M.R. & D.L. 112, G.N., A.M.R. & D.L. 156, A.M.R. 339, 360.*

Amaranthaceae

Amaranthus palmeri S. Watson. Carelessweed, pigweed, quelite de las aguas, cuhukkia. Summer ephemeral; field; eaten with beef bones. *G.N., K.R. & A.M.R. 225.*

Tidestromia lanuginosa (Nuttall) Standley. Woolly tidestromia, hierba ceniza, espanta vaquera. Summer ephemeral; scrubland and probably elsewhere. *G.N., A.M.R. & D.L. 118.*

Amaryllidaceae

Hymenocallis sonorensis Salisbury. Spider lily. Perennial herb; flowers white, August and September; planted in orchard; ditch; rare; ornamental transplanted/cultivated, native elsewhere in Sonora. Not reported native to Gran Desierto. *G.N., A.M.R. & D.L. 117.*

Apiaceae

Bowlesia incana Ruiz & Pavón. Hairy bowlesia. Winter-spring ephemeral; in wash. *G.N. 210.*

*Coriandrum sativum L. Coriander, cumin, comino. Cultivated annual, eaten; flowers white, flowering almost year-round. No specimen.

*Daucus carota L. Carrot, zanahoria. Cultivated annual, eaten. No specimen.

Asclepiadaceae

Sarcostemma cynanchoides Decaisne subsp. *hartwegii* (Vail) R. W. Holm. Climbing milkweed, guirote, vi'ibgam. Perennial vine; flowering and fruiting at various seasons including April and September, flowers are white and purplish; orchard, wash, scrubland, widely distributed; for chewing like gum (*chicles*). *G.N., A.M.R. & D.L. 157, G.N. 207, A.M.R. 148.*

Asteraceae

Ambrosia ambrosioides (Cavanilles) W. W. Payne. Canyon ragweed, chicúra, ñuñuwi je:j. Shrub, wash, and fencerow; medicinal. No specimen.
Ambrosia confertiflora D.C. Slimleaf bursage, estafiate. Perennial herb; fruiting in September; field, mesa rimming pond, yard. *G.N., A.M.R. & D.L. 113.*
Ambrosia deltoidea (Torrey) Payne. Triangle-leaf bursage, chamizo forrajero, tadṣad, va: gita. Small shrub; fruiting in April and May; in wash; localized. *G.N. 215, G.N., A.M.R. & E.M.B. 287.*
*Artemisia dracunculus L. Wormwood. Perennial herb; yard; cultivated. Not reported in Gran Desierto. *G.N. & P.W. 95.*
Baccharis sarothroides A. Gray. Desert broom, romerillo. Broomlike shrub; flowers white; scrubland. *A.M.R. 232.*
Bebbia juncea (Bentham) Greene var. *aspera* Greene. Chuckwalla delight, sweetbush, hauk 'u'us. Suffrutescent bushy perennial; flowers yellow; field, fencerow. *G.N., A.M.R. & D.L. 158, G.N., A.M.R. & E.M.B. 296.*
Brickellia coulteri A. Gray var. *coulteri*. Brickellbush. Small shrub; in wash. No specimen.
*Calendula officinalis L. Pot marigold. Annual herb; cultivated. Not reported in Gran Desierto. No specimen.
Eclipta prostrata (L.) L. False daisy, hierba de tajo, chile de agua. Annual herb; emergent; flowers white; at spring. *G.N. 100.*
Encelia farinosa A. Gray var. *phenicodonta* (Blake) I. M. Johnston. Brittlebush, incienso, hierba del vaso, tohaves. Suffrutescent shrub; flowers yellow; nonseasonal; mesa rimming pond, hedgerow; localized. *G.N. 213.*
Geraea canescens Torrey & A. Gray. Desert gold, desert sunflower. Winter-spring ephemeral; flowers yellow; mesa rimming pond; localized; ceremonial. *G.N. 212.*
Hymenoclea monogyra Torrey & A. Gray. Slender burrobush, jécota, i:watoḍ. Shrub; fruiting in December; in wash; the brush used for ramadas. *G.N. 133.*
Hymenoclea salsola Torrey & A. Gray. White burrobush, jecota, (i:watoḍ). Willowy shrub; in wash. *G.N. 132.*
Hymenothrix wislizenii A. Gray. Annual or perennial herb; fruiting in September; flowers yellow; field. *G.N., A.M.R. & D.L. 131, G.N., K.R. & E.M.B. 243.*
Isocoma acradenius (Greene) Greene var. *acradenia*. Alkali goldenbush. Small shrub; flowers yellow in late summer; salt flats, mesa rimming pond, houseyard, ceremonial(?). *G.N. & P.W. 93, G.N., D.R. & A.M.R. 261, A.M.R. 242.*
*Lactuca serriola L. Prickly lettuce, compass plant, lechuga, (i:wakï). Spring-summer annual or ephemeral; flowers pale yellow; field; cultivated. No specimen.

Machaeranthera carnosa (A. Gray) G. L. Nesom var. *carnosa* [=*Aster intricatus* (A. Gray) S. F. Blake]. Alkali aster. Perennial herb; flowers yellow, in November, etc.; salt flats, mesa rimming pond. *G.N. & P.W. 163*.

Machaeranthera coulteri (A. Gray) B. L. Turner & D. B. Horne var. *arida* B. L. Turner & D. B. Horne [=*Machaeranthera arizonica* Jackson & Johnson]. Annual to perennial herb; flowering response nonseasonal, flowers yellow and purple; field, mesa rimming pond, yard, roadside; widely distributed. *G.N. & P.W. 107, G.N. 110, G.N. 138, A.M.R. 197*.

Machaeranthera tagetina Greene. Spring-summer annual; flowering in April. *G.N. 271*.

Palafoxia arida B. L. Turner & Morris. Spanish needles. Aseasonal herbaceous ephemeral; flowers lavender. *G.N. 139*.

Pectis papposa Harvey A. & A. Gray. var *papposa*. Desert chinchweed, manzanilla del coyote, ban mansani:ya. Summer ephemeral; flowers yellow; mesa rimming pond. *G.N. 202, A.M.R. 235*.

Pluchea sericea (Nuttall) Coville. Arrowweed, cachanilla, komagɪ 'u'us. Shrub; flowers pink; stems used for arrows and house building. *G.N. & B.B. 86, G.N., A.M.R. & D.L. 102*.

*Sonchus oleraceus L. Sow-thistle, chinita, (ha:wĭ hehewo). Ephemeral/annual herb, mostly winter-spring; flowers pale yellow; ditch, mesa rimming pond; widely distributed. *G.N. & P.W. 90, G.N., A.M.R. & D.L. 103*.

Stephanomeria exigua Nuttall var. *pentachaeta* (D. C. Eaton) H. M. Hall. Wire lettuce, annual mitra. Winter-spring (or nonseasonal?) ephemeral; flowers pink; field; localized. Not reported in Gran Desierto. *G.N., A.M.R. & D.L. 130*.

Stephanomeria pauciflora (Torrey) A. Nelson var. *pauciflora*. Desert straw. Bushy perennial herb; field; flowers pink; localized. *G.N., A.M.R. & E.M.B. 279, G.N., A.M.R. & D.L. 130*.

Thymophylla concinna (A. Gray) Strother. Dogweed, manzanilla del coyote, ban mansanita. Winter-spring ephemeral; flowers white and yellow; roadside; localized. *G.N. 274*.

*Xanthium strumarium L. Cocklebur, cardillo, abrojo, vivul. Annual herb; fruiting in November; field, ditch. *G.N. & P.W. 79*.

Boraginaceae

Amsinckia tessellata A. Gray. Checker fiddleneck, cedkam. Winter-spring ephemeral; flowers yellow; mesa rimming pond; localized. *G.N. 210*.

Cryptantha barbigera (A. Gray) Greene. Bearded cryptantha. Winter-spring ephemeral; flowers white; field, wash. *G.N. 211, ?.T., A.M.R. 281*.

Cryptantha maritima (Greene) Greene. White-haired cryptantha. Winter-spring ephemeral; flowers white; field, wash. *?.T., A.M.R. 1745*.

Cryptantha micrantha (Torrey) I. M. Johnston subsp. *micrantha*. Winged-nut cryptantha. Winter-spring ephemeral; flowers white; in wash. No specimen.

Heliotropium curassavicum L. Alkali heliotrope, hierba del sapo, ba:bad 'i:wakĭ, kakaicu e'es, kakaicu sisivoda, (kakaicu 'iwa:gi). Succulent perennial; in ditch, at spring, salt flats, and marsh; medicinal for coughs and sore throat. *G.N. 88, G.N. 99, G.N., A.M.R. & K.R. 219*.

Brassicaceae

Brassica oleracea L. Kale. Winter-spring annual; cultivated for food. Introduced vegetable. No specimen.

Brassica tournefortii Gouan. Sahara mustard, wild turnip, mostaza, mo:stas. Winter-spring ephemeral; flowers pale yellow; ditch and other habitats; eaten as greens. *G.N. 204.*

Lepidium lasiocarpum Nuttall ex Torrey & A. Gray. Sand peppergrass, (ka:kowani). Winter-spring ephemeral, flowers white; field, orchard; localized. *G.N., A.M.R. & E.M.B. 292.*

Lyrocarpa coulteri Hooker & Harvey ex Harvey var. *coulteri*. Lyrepod, ban censañïg. Perennial herb; in wash. *G.N., A.M.R. & E.M.B. 291.*

Sisymbrium irio L. London rocket, pamita, ban censañig. Winter-spring ephemeral; flowers yellow; field. *G.N. 201.*

Cactaceae

Carnegiea gigantea (Engelmann ex Emory) Britton & Rose. Saguaro, sahuaro, ha:sañ, bahidaj. Flowers white; in wash; fruit (pulp) for ceremonial wine and food, the seeds for food and chicken feed, the dry stem ribs for building and ceremony. No specimen.

Cylindropuntia arbuscula (Engelmann) F. M. Kunth. Pencil cholla, siviri, tasajo, vipinoi. Succulent, fruiting in August; in wash; localized; buds eaten after boiling. *G.N., K.R. & A.M.R. 218.*

Cylindropuntia fulgida (Engelmann) F. M. Kunth var. *fulgida*. Jumping cholla, chain-fruit cholla, ha:nam. Succulent; fruiting in August; in wash; widely distributed; wild/transplanted; flower buds pit-roasted or boiled, used in hedgerow. *G.N., K.R. & A.M.R. 222* [not at ARIZ], *A.M.R. 240.*

Cylindropuntia leptocaulis (D.C.) F. M. Kunth. Desert Christmas cactus, tasajillo, 'acĭ vipinoi, ce'ecem vipinoi. Succulent; fruiting in August; in wash; localized; ripe fruit eaten, root a medicine for sore throats, bellyache. *G.N., K.R. & E.M.B. 231, G.N., K.R. & A.M.R. 264.*

Echinocereus engelmannii (Parry) Lemaire var. *acicularis* L. D. Benson. Hedgehog cactus, i: swig. Succulent; mesa rimming pond, yard; wild/transplanted; fruit eaten. Albino-flowered form locally frequent. These plants may be somewhat intermediate with *E. fasciculatus* (Engelmann) L. D. Benson, which replaces *E. engelmannii* var. *acicularis* eastward in northern Sonora. *G.N., A.M.R. & T.I. 251.*

Ferocactus emoryi (Engelmann) Orcutt [=*F. covillei* Britton & Rose]. Barrel cactus, bisnaga, jiawulu. Succulent; flowers reddish; fruiting in December; mesa rimming pond, yard, wash; localized; fruit eaten raw. *G.N., K.R. & E.M.B. 230.*

Mammillaria thornberi Orcutt. Thornber's fishhook cactus, cabeza del viejo, baban hamauppa. Succulent; flowers white and pink; fruiting in mid-December; mesa rimming pond, yard; fruit eaten. *G.N., K.R. & E.M.B. 298.*

Opuntia engelmannii Salm-Dyck ex Engelmann var. *engelmannii*. Prickly-pear, tuna, nopal, i: bhai, naw. Succulent; flowering in May, flowers yellow; orchard, mesa rimming pond, wash; widely distributed; wild/transplanted; young pads (stems or cladodes) eaten with chiles in May and June, fruit eaten later in summer. *G.N., E.M.B. & A.M.R. 273.*

Opuntia ficus-indica (L.). Prickly-pear, tuna, nopal, na:wi:bhai. Perennial arborescent succulent; flowers yellow; ornamental, cultivated. Fruits and tender young pads eaten. No specimen.

Opuntia gosseliniana F.A.C. Weber. Purple prickly-pear, nopal, tuna, gisokĭ. Succulent; in wash; localized; buds eaten. Not reported in Gran Desierto. *G.N., A.M. & E.M.B. 275, A.M.R. 190.*

Pachycereus schottii (Engelmann) D. R. Hunt var. *schottii.* Senita, sina, ce:mĭ. Columnar cactus, succulent; flowers pink; in wash; fruit eaten. No specimen.

Stenocereus thurberi (Engelmann) F. Buxbaum. Organpipe cactus, pitaya dulce, pitahaya, cucuvis. Columnar cactus; in wash; wild; skeleton used as crossbeams in house, fruit made into sweet preserve as local cottage industry, with each village member harvesting an average of 50 kg of raw fruit for jam making each summer through the 1990s. No specimen.

Cannaceae

*Canna × generalis L. H. Bailey. Common garden canna. Perennial herb; flowering in September; orchard, ditch, spring, localized; cultivated. Not reported in Gran Desierto. *G.N. & A.M.R. 165.*

Capparaceae

Wislizenia refracta Engelmann subsp. *refracta.* Jackass-clover, spectacle pod. Nonseasonal annual; flowering in May and September; field, spring, mesa rimming pond, scrubland; widely distributed. Probably the yellow flower mentioned by Davis (1921) as being used in the Vi'igita ceremony; used for snakebites among the Seri and the Cachanillas. *G.N. & B.B. 87, G.N., A.M.R. & D.L. 104, G.N., A.M.R. & D.L. 107.*

Caprifoliaceae

Sambucus nigra L. subsp. *cerulea* (Presl) Boll. Mexican elder, tapiro (tree), sauco (flowers), tahapidam. Tree; fruiting in May, August, and September; marsh, mesa rimming pond, roadside; localized, probably transplanted/cultivated; eaten, medicine. Not native in Gran Desierto flora. *G.N. 83, G.N., A.M.R. & D.L. 114, G.N., A.M.R. & K.R. 222.*

Caryophyllaceae

Spergularia salina J. & C. Presl. Saltmarsh sand-spurry. Spring annual; fruiting in May; spring, salt flats; localized. *G.N. & B.B. 162.*

Chenopodiaceae

Atriplex elegans (Moquin) D. Dietrich. Wheelscale saltbush, chamizo cenizo, orach, 'oñk i: wakĭ. Annual herb; fruiting in September; mesa rimming pond; eaten as greens. *G.N. 257.*

Atriplex polycarpa (Torrey) S. Watson. Allscale, cattle spinach, chamizo cenizo, 'oñk i:wakĭ. Shrub; scrubland; localized. *G.N., K.R., S.M. & J.B., G.N., A.M.R. & D.L. 152.*

Atriplex wrightii S. Watson. Wright's annual saltbush, 'oñk i:wakĭ. Annual herb; flowering in August; field; wild; eaten as cooked greens. *G.N., K.R. & A.M.R. 270.*

Beta vulgaris L. Swiss chard. Annual herb; yard outside study area; localized. No specimen.

Chenopodium murale L. Nettleleaf goosefoot, quelite de puerco, 'oñk i:wakĭ. Winter-spring annual herb; marsh, mesa rimming pond, roadside; edible greens. *G.N. 91, G.N. 140.*

Nitrophila occidentalis (Moquin) S. Watson. Alkaliweed. Emergent perennial herb, succulent; flowering in January; ditch, marsh. *G.N., K.R., J.B. & S.M. 265.*

Salsola tragus L. (=*S. australis* [R. Brown]). Russian-thistle, chamizo volador, hejel 'e'esadam, (wo'opo sa'i). Annual herb; flowering in September; mesa rimming pond. *G.N., A.M.R. & D.L. 109.*

Spinacia oleracea L. Spinach, espinaga. Annual herb; yard outside study area; cultivated. No specimen.

Suaeda moquinii (Torrey) Greene. Desert seepweed, sosa, quelite salado, s-cuk oñk. Suffrutescent shrub; fruiting in January and September; salt flats, mesa rimming pond, scrubland; widely distributed. *G.N., A.M.R. & D.L. 119, G.N., K.R., J.B. & S.M. 169, G.N., K.R., J.B. & S.M. 255, A.M.R. 235.*

Convolvulaceae

Cuscuta indecora Choisy. Dodder, cuscuta. Parasite, warm-weather ephemeral vine; flowers white; field; widely distributed. *G.N. & A.M.R. 167.*

Ipomoea carnea Jacq. subsp. *fistulosa* (Mart.) D. Austin. Campanilla. Shrub; winter-dormant; flowers pink; yard; cultivated. *G.N., A.M.R. & K.R. 295.*

Merremia dissecta (Jacquin) H. G. Hallier. Bi:biak, bi:bhiag. Vine; flowering in August; orchard, mesa rimming pond, yard; cultivated by house or in shade of trees. *G.N., A.M.R. & K.R. 259.*

Cucurbitaceae

Citrullus lanatus (Thumb.) Mansf. Watermelon, sandia, gepĭ. Field, eaten. No specimen.

Cucumis melo L. Melon, melón, milon. Field, eaten. No specimen.

Cucurbita argyrosperma Merrick. Silver-seeded cushaw, calabasa, ha:l. Annual vine; flowering in August; field. *G.N., K.R. & A.M.R. 226.*

Cucurbita digitata A. Gray. Coyote gourd, calabacilla, chichi coyota, 'adavĭ, ad. Vining root perennial; flowers yellow; in wash; medicinal. No specimen.

Lagenaria siceraria Standley. Bottlegourd, bule, guaje, vapko. Vine; flowers white; field. No specimen.

Cyperaceae

Cyperus laevigatus L. Flat sedge. Perennial herb; orchard, ditch, spring; localized. *G.N., A.M.R. & D.L. 148.*

Cyperus niger R. & P. Perennial herb; orchard, ditch, spring; localized. Not reported in Gran Desierto. *G.N., K.R. & E.M.B. 235.*
Scirpus (Schoenoplectus) americanus Persoon. Bulrush, tule, vak. Perennial emergent herb; fruiting in September; ditch, spring, salt flats, marsh; ceremonial; part of origin myth of oasis. *G.N. 97, G.N. 127, G.N., A.M.R. & D.L. 144.*

Euphorbiaceae

Euphorbia eriantha Bentham. Desert beetle spurge, golondrina. Nonseasonal ephemeral; salt flats. *G.N., A.M.R. & D.R. 268.*
Euphorbia polycarpa Bentham. Small-seeded sand-mat, golondrina. Nonseasonal ephemeral and herbaceous perennial; field; localized. *A.M.R. 230.*
Jatropha cinerea (Ortega) Muller Arg. Ashy limberbush, sangregrado, va:s. Shrub; in wash; basketry. No specimen.
*Nerium oleander L. Oleander, laurel. Shrub; scrubland, yard; cultivated; ornamental providing shade. *G.N., A.M.R. & D.L. 117.*
Sebastiana bilocularis S. Watson. Mexican jumping bean, hierba de la flecha, hierba mala, 'ina hita'. Shrub; fruiting in December; in wash; rare; medicine. Name refers to rattling sound made by *Rothschildia* moth cocoons collected from this host plant by Pascola dancers as ornament. *G.N. & K.R. 137.*

Fabaceae

Acacia greggii A. Gray. Catclaw, gatuña, u:paḍ. Tree; flowering in spring and also in August and fruiting in December; flowers cream-white; wash, scrubland; localized; not now used here. *G.N. 135, G.N., K.R. & A.M.R. 221.*
*Melilotus indica (L.) Allioni. Yellow sweetclover, alfalfilla, trebol agrio, puwl. Winter-spring ephemeral; ditch, marsh; cultivated; widely distributed. *G.N. 253.*
Olneya tesota A. Gray. Desert ironwood, palo fierro, ho'idkam. Tree; flowers pale lavender; in wash; the wood used for *batea* troughs, fence posts, and crafts, the seeds formerly eaten, the flowers and tea for kidney ailments. Overexploited by charcoal makers in nearby regions. *G.N., A.M.R. & E.M.B. 286.*
Parkinsonia florida (Bentham ex A. Gray) S. Watson. Blue palo verde, palo verde, ko'okmadk, kalisp. Tree; flowering in spring and fruiting in May and also in September; flowers yellow; in wash, hedgerow; used in fencerows and seeds or pods eaten green, second choice to *P. microphylla*. *G.N., A.M.R. & D.L. 121, G.N. & A.M.R. 272.*
Parkinsonia microphylla Torrey. Littleleaf palo verde, palo verde, kek cehedagĭ. Tree; flowering in April; flowers pale yellow; roadside; localized. *G.N. & A.M.R. 276.*
*Phaseolus acutifolius A. Gray. White tepary, tepari blanco, s-toa bavĭ. Warm-weather ephemeral; flowers pink; cultivated. No specimen.
*Phaseolus vulgaris L. Common bean, frijol comun, mu:ñ. Warm-weather ephemeral; cultivated. No specimen.
*Pisum sativum L. O'odham pea, chicharro, vihol. Winter-spring ephemeral; cultivated. Not reported from Gran Desierto. No specimen.

Prosopis pubescens Bentham. Screwbean mesquite, tornillo, kujil, (kuwidculis). Tree; in wash; seeds ground for food, drunk with cold water. *G.N., A.M.R. & T.I. 245, G.N. 295.*
Prosopis velutina Wooton. Velvet mesquite, mesquite, kui. Tree; flowering in May, fruiting in September; flowers yellow; mesa rimming pond, wash, scrubland, hedgerow; widely distributed; pods harvested, ground into flour for food, the gum used as medicine, the wood used in utensils, crafts, and as fuelwood, the living trees as hedgerows. Possible introgressive hybridization with nearby populations of *P. glandulosa. G.N. & B.B. 87, G.N., A.M.R. & D.L. 123.*
*Trifolium repens L. White clover, trebol blanco, pu:hl. Annual herb; flowering in December; ditch, marsh; localized; cultivated/feral; eaten as greens. Not reported in Gran Desierto. *G.N., K.R. & E.M.B. 233.*

Fouquieriaceae

Fouquieria splendens Engelm. Ocotillo, melhog. Perennial shrub; flowers red-orange; wild and cultivated; used for fences, ramadas, and lath. No specimen.

Geraniaceae

*Erodium cicutarium (L.) L'Héritier. Filaree, storksbill, alfilerillo, hohoi 'ipad. Winter-spring ephemeral; flowers pink; field. *G.N. 179, G.N. 197.*

Hydrophyllaceae

Phacelia distans Bentham. Wild heliotrope, caterpillar phacelia. Winter-spring ephemeral; in wash; localized. *G.N. 199.*

Liliaceae

*Allium ampeloprasum L. Lee. Shallot. Perennial herb; field, orchard; rare; cultivated; eaten. *G.N., A.M.R. & E.M.B. 290.*
*Allium cepa L. Onion, cebolla, ciwol. Perennial herb; field; cultivated. No specimen.

Loasaceae

Petalonyx thurberi A. Gray subsp. *thurberi.* Sandpaper plant, hadsadtdam. Suffrutescent shrub; flowering nonseasonally, including May, flowers white; in wash; localized. *G.N., A.M.R. & E.M.B. 277.*

Malvaceae

Abutilon palmeri A. Gray. Indian mallow. Suffrutescent shrub; flowers orange-yellow; mesa rimming pond. *G.N., K.R., J.B. & S.M. 168.*

Malva parviflora L. Little mallow, cheeseweed, malva, quesito, ñadum cuikam. Winter-spring ephemeral; ditch, marsh. *G.N. 198.*
Sphaeralcea coulteri (S. Watson) A. Gray. Coulter globe-mallow, mal de ojo, hadam tatk, ñiatum. Winter-spring ephemeral; scrubland. *G.N. 205, G.N., A.M.R. & E.M.B. 297.*

Martyniaceae

Proboscidea parviflora (Wooton) Wooton & Standley subsp. *parviflora*. Devil's claw, uña de gato, torito, gatuña, i'hug. Summer ephemeral; flowering and fruiting in September; flowers pink-purple and white; field. Not intentionally cultivated or used for basketry here, indicating that devil's claw domestication in the nineteenth century probably began to the north, with the Akimel O'odham or eastern Tohono O'odham. *G.N., A.M.R. & D.L. 105.*

Moraceae

Ficus carica L. Fig, higo, su:na. Fruiting in June and July; orchard; fruit eaten fresh or dried, and used in jellies. *G.N., A.M.R. & E.M.B. 289.*
Morus alba L. Mulberry, mora, gohi. Tree; orchard; localized and cultivated; fruits eaten fresh and made into jellies. No specimen.

Nyctaginaceae

Boerhaavia coccinea Miller. Red spiderling, juaninipili, mochi. Perennial herb; flowers red-purple; orchard; localized. This widespread weedy species is probably not native in this region. Not reported in Gran Desierto. *G.N. 151, A.M.R. 237.*
Boerhaavia wrightii A. Gray. Spiderling, ma:kkum ha-jewed. Large-bracted spiderling. Hot-weather ephemeral; flowers pink; field. *G.N., K.R. & A.M.R. 267.*

Onagraceae

Gaura parviflora Douglas ex Hooker. Lizardtail. Nonseasonal ephemeral; flowers pink; field, scrubland; widely distributed. *G.N. 206, G.N., E.M.B. & A.M.R. 284, G.N., E.M.B. & A.M.R. 206.*

Palmaceae

Phoenix dactylifera L. Date palm, datil, cukuḏ sosa. Tree; mesa rimming pond; cultivated; fruits eaten, the leaves used for ramada roofing and shade, the trunks used for corrals. No specimen.
Washingtonia filifera (Linden) H. Wendland. California fan palm, palma, pa:lma. Tree; fruiting in late fall and winter; field, orchard, yard, mesa rimming pond; localized; cultivated/seminaturalized; trunks for building material for corral, leaves used in weaving baskets and hats and as roofs for ramadas. Possibly in area as early as the 1690s, if Kino's use of native palms on Palm Sunday in the Pinacate is valid (Bolton 1936). *G.N. & A.M.R. 143.*

Plantaginaceae

Plantago ovata Forsskal. Indian wheat, woolly plantain, pastora, mumṣa. Winter-spring ephemeral; field; forage. A major winter seed food of the Hia c-eḍ O'odham, who could collect an abundance of seeds in late winter between Quitovac and the Pinacate. Also an important food for the Sonoran pronghorn herds of the area. *G.N., A.M.R. & E.M.B. 284.*

Poaceae (Gramineae)

*Avena fatua L. Wild oat, avena silvestre, ('agsi mudatkam, kosam). Winter-spring ephemeral; marsh. *G.N. & B.B. 94.*

Bouteloua aristidoides (Kunth) Grisebach. Six-weeks needle grama, navajita, aceitilla. Summer ephemeral; mesa rimming pond, yard, roadside; widely distributed. *G.N., A.M.R. & D.L. 142.*

Bouteloua barbata Lagascea. Six-weeks grama, zacate liebrero, navajita, (cuk mudatkam). Summer ephemeral; field, orchard, scrubland. *G.N., A.M.R. & D.L. 128.*

*Cynodon dactylon (L.) Persoon. Bermuda grass, zacate ingles, pata de gallo, (ki:weco wasai, 'a'ai hihimidam wasai). Perennial; field, orchard, ditch, salt flats, marsh, mesa rimming pond, roadside; widely distributed. *G.N., A.M.R. & D.L. 147.*

Distichlis spicata (L.) Greene. Desert saltgrass, zacate salado, 'onk vasi ('salt grass'). Perennial; orchard, ditch, salt flats, roadside; widely distributed; forage. *G.N., A.M.R. & D.L. 145.*

*Echinochloa colonum (L.) Link. Jungle rice; zacate pinto, zacate rayado, (so'oi wasai). Summer annual; fruiting in August; field, orchard, ditch; localized. *G.N., A.M.R. & K.R. 220.*

Elymus multisetus (J. G. Smith) Jones. Wild rye. Winter-spring ephemeral; field; localized. Not previously reported in Gran Desierto. *G.N., K.R. & A.M.R. 266.*

*Eragrostis cilianensis (Allioni) Vignolo-Lutati ex Janchen. Stinking love grass, zacate apestoso, zacate de aguas. Summer ephemeral; field, orchard. *G.N., K.R. & A.M.R. 224.*

Eragrostis pectinacea (Michaux) Nees var. *pectinacea.* Lovegrass. Summer ephemeral; ditch; localized. *G.N., A.M.R. & K.R. 256.*

Hordeum arizonicum Covas. Arizona barley. Winter-spring ephemeral; field, orchard, scrubland; probably localized. This is the only confirmed record for this species in Sonora; not otherwise reported from the Gran Desierto or Sonora. *G.N., A.M.R. & E.M.B. 283.*

*Hordeum murinum L. subsp. *glaucum* (Steudel) Tzvelev. Arizona barley. Winter-spring ephemeral; field, orchard, marsh, mesa rimming pond; widespread. *G.N. 203.*

Leptochloa fusca (L.) Kunth. subsp. *uninervia* (J. Presl) N. Snow. Mexican sprangletop, zacate salado. Nonseasonal annual; flowers in August and September; orchard, ditch, spring, salt flats. *G.N. 161, G.N., K.R. & A.M.R. 263, A.M.R. 233.*

*Pennisetum ciliare (L.) Link. Buffelgrass, zacate buffel. Perennial; field, orchard; localized. *G.N., A.M.R. & E.M.B. 280.*

*Phalaris caroliniana Walter. Carolina canary grass, alpiste, (ba:bkam). Winter-spring ephemeral; marsh; localized. *G.N. 196.*

*Polypogon monspeliensis (L.) Desfontaines. Rabbitfoot grass, zacate cola de zorra. Nonseasonal ephemeral; orchard, ditch, salt flats, marsh. *G.N. 98, G.N. 160, G.N. 293, A.M.R. 241.*

Schismus barbatus (L.) Thelling. Mediterranean grass. Winter-spring ephemeral; field, orchard, wash; localized. *G.N. 269.*

Sporobolus pyramidatus (Lamarck) Hitchcock. Whorled dropseed, zacate piramide. Summer ephemeral; marsh, mesa rimming pond; localized. *G.N., A.M.R. & K.R. 162b, G.N., D.R., S.M. & J.B. s.n.*

*Triticum aestivum L. Sonora wheat, trigo, 'olas pilkañ. Winter-spring annual crop; field, orchard; localized; cultivated; ground for tortillas. *G.N. 195.*

*Zea mays L. subsp. *mays.* Corn, maiz, hu:ñ. Summer annual crop; field; cultivated; ceremonial, food. No specimen.

Polygonaceae

Eriogonum deflexum Torrey. Skeleton weed. Spring-summer ephemeral or annual; fruiting in December and September; field, yard, wash. *G.N., A.M.R. & D.L. 129, G.N. & A.M.R. 166, G.N., K.R. & E.M.B. 237.*

*Rumex crispus L. Curly dock, canaigre, vakwandan. Perennial herb/emergent; fruiting in May; ditch, marsh, mesa rimming pond, roadside; widely distributed. *G.N. 84, G.N. 258.*

Portulacaceae

*Portulaca oleracea L. Common purslane, verdolaga, ku'ukpalk. Warm-weather ephemeral; field, orchard, ditch; localized; used as cooked greens. *G.N. & B.B. 84, A.M.R. 238.*

Potamogetonaceae

Potamogeton pectinatus L. Slender pondweed. Dense masses of slender, filmy stems submerged in ponds and ditches, more obvious in warm months. *G.N. & R.S.F., s.n., July 10 1982.*

Punicaceae

*Punica granatum L. Pomegranate, granada, galnayu. Fruiting in September; flowers red; orchard, scrubland; localized; cultivated; fruit eaten. *G.N., A.M.R. & D.L. 120.*

Rhamnaceae

Condalia globosa I. M. Johnston. Bitter condalia, crucillo, kauk kuavul. Shrub; fruiting in January; flowers yellow; ditch, scrubland. *G.N., K.R., S.M. & J.B. 170, G.N., K.R., S.M. & J.B. 269.*

Ziziphus obtusifolia (Torrey & A. Gray) A. Gray var. *canescens* (A. Gray) M. C. Johnston. Graythorn, abrojo, 'u:spad. Large shrub; fruiting in December; scrubland; localized; fruit eaten, seeds spit out. *G.N., K.R. & E.M.B. 232, G.N., K.R. & A.M.R. 262.*

Salicaceae

Populus fremontii S. Watson subsp. *fremontii*. Frémont cottonwood; alamo, 'auppa. Large winter-deciduous tree, possibly transplanted from elsewhere; mesa rimming pond, scrubland; rare; possibly used as a wood source. No specimen.

Salix gooddingii Ball. Goodding willow, sauce, sauz, ce:'ul. Winter-deciduous tree; orchard, mesa rimming pond, roadside, fencerow; localized; wild/transplanted; planted in hedgerow, and leafy twigs or branches made into crosses for blessing houses on Dia de Santa Cruz. *G.N. 92, G.N. 214.*

Saururaceae

Anemopsis californica (Nuttall) Hooker & Arnott. Hierba-mansa, hierba del mansa, hierba del manso, va:visa. Perennial herb; flowering most of the year, winter-dormant; flowers white; orchard, ditch, marsh, mesa rimming pond; medicine, the root boiled, for flu, colds, and impetigo or to break a fever, formerly collected at the Colorado River delta, now grown by Hia c-ed O'odham elders in their gardens. *G.N. & P.W. 78, G.N. 81.*

Simmondsiaceae

Simmondsia chinensis (Link) C. K. Schneider. Jojoba, hohovai (the O'odham origin of the Spanish and English term). Woody perennial shrub, flowering in late winter; usually on rocky slopes, transplanted here onto the sandy soil of the paleo-floodplain. Used as a salve for burns and sores on humans and livestock. Introduced into fields as a Mexican government promotion of jojoba as a new industrial crop for the shampoo business. No specimen.

Solanaceae

Datura discolor Bernhardi. Desert thornapple, toloache, kotadopĭ. Nonseasonal ephemeral; flowers white; field, scrubland; wild; medicine/ceremonial herb, chewed rarely, hallucinations resulting, relationship with hawkmoths (*Hylea lineata* and *Manduca quincemaculata*) and their larvae referred to in curing songs. *G.N., A.M.R. & D.L. 106.*

Lycium andersonii A. Gray. Desert wolfberry, manzanita, salicieso, s-toa kuavul. Shrub; flowering and fruiting in December and at other seasons; flowers pale lavender; mesa rimming pond, scrubland, hedgerow; fruits eaten [the edible fruit should be from *L. fremontii*]. *G.N., A.M.R. & D.L. 116, G.N. 141.*

Lycium berlandieri Dunal. Barchata, hosó, kuavul. Shrub; flowering in January; flowers white; in wash; localized. Not reported in Gran Desierto. *G.N., A.M.R. & T.I. 244.*

Lycium exsertum A. Gray. Wolfberry, tomatillo, kuavul. Shrub; fruiting in January; flowers white. *G.N., K.R., J.B. & S.M. 171.*

Lycium fremontii A. Gray. var. *fremontii*. Frémont thornbush; tomatillo, kuavul. Shrub; fruiting in May; flowers lavender. *G.N. 96.*

Lycium parishii A. Gray. var. *parishii*. Parish thornbush, kuavul. Shrub; flowering in January; flowers lavender; in wash; rare. *G.N. & A.M.R. 246.*

Nicotiana obtusifolia M. Martens & Galeotii. Desert tobacco, tabaquillo de coyote, ban vi:v, ban ha-vivga. Perennial herb; in wash; localized; used ceremonially in curing. *G.N., A.M.R. & T.I. 250.*

**Solanum americanum* Mill. Black nightshade, chichiquelite, cuvĭ wu:puĭ. Flowering in September; orchard, ditch. *G.N., A.M.R. & D.L. 159.*

**Solanum lycopersicon* L. Tomato, jitonate, toma:di. Annual herb; orchard; localized; cultivated in pots. No specimen.

Tamaricaceae

*Tamarix aphylla (L.) H. Karsten. Athel, salt-cedar, 'onk 'u'us. Large tree; orchard, mesa rimming pond; cultivated; trunks used for ramadas. No specimen.

**Tamarix ramosissima* × *chinensis* Lour. Salt-cedar, tamarisk. Shrub; flowers pink; marsh, mesa rimming pond. *G.N. 89, G.N., A.M.R. & D.L. 115.*

Tropaeolaceae

**Tropaeolium majus* L. Perennial ornamental; marsh edge; cultivated. No specimen.

Typhaceae

Typha domingensis L. Southern cattail, tule, uduvad. Perennial herb/emergent; fruiting in May and November; marsh and holding pond; fiber used for tall baskets, the tender stalks are eaten raw by children. *G.N. & P.W. 80, G.N. & B.B. 82.*

Ulmaceae

Celtis pallida Torrey. Desert hackberry, garambullo, granjeno, cumbro, kuawul. Shrub; fruiting in fall; mesa rimming pond, scrubland. *G.N., A.M.R. & D.L. 122.*

Urticaceae

Parietaria floridana Nutt. Desert pellitory. Winter ephemeral, spring flowering. *G.N. 217.*

Viscaceae

Phoradendron californicum Nuttall. Desert mistletoe, toji, hokowad, to:ki. Epiphytic parasite on *Olneya* and *Prosopis velutina*; fruiting in December and at other seasons; in wash; widely distributed. *G.N. 134, G.N., K.R. & E.M.B. 227.*

Vitaceae

Vitis vinifera L. European grape, uva, vid, uduwig. Perennial woody vine; cultivated for table grapes and homemade wine. No specimen.

Zannichelliaceae

Zannichellia palustris L. Horned pondweed. In open water in lagoon, rarely in ditches and springs. No specimen.

Zygophyllaceae

Larrea divaricata Canavilles subsp. *tridentata* (Sesse & Mocino ex D.C.) Felger & Lowe. Creosote, gobernadora, hediondilla, segai. Shrub; flowering nearly year-round; used medicinally as tea for sore throats and diabetes, as a salve for burns, or for funerals. No specimen.

Chapter 13

Historic and Prehistoric Ethnobiology of Desert Springs

AMADEO M. REA

I want to sketch out very briefly, and roughly, three major topics related to the ethnobiology of desert springs: (1) a mythology of springs as sacred places; (2) a concrete example of biodiversity at two spring-fed oases; and (3) some thoughts on indigenous management and biodiversity. The first and the third are essentially ignored in scientific discussions, while only the second might be considered truly scientific, that is, quantified and published (preferably in English).

The Southwest Mythology of Springs

Old World herbivores have caused profound change in the Southwest, both environmentally and culturally. At least some O'odham (that is, Pimans) understood the ecological consequence of cattle raising quite early in the postcontact period. The Pima of the *ranchería* of Remedios on the upper San Miguel River in what is now Sonora, Mexico, opposed Fr. Kino in 1687 "and presented some reasoned arguments against having a mission. They said that the Spaniards hanged Indians, that they required too much work on mission lands to the detriment of the Indians' own farming, that too many cattle were pastured around the missions so that the water supply was seriously reduced, that people were killed by the holy oils of the missionaries, and that the missionaries were not able to keep their promises that they would protect the Indians from forced labor by the Spaniards" (Spicer 1962:121–123; see also Bolton 1936:259 [Bolton breaks up his translation with commentary]). Another translation of this passage says, "Their springs were drying up." Although each of these points voices a legitimate concern of the indigenous people, the association between cattle and water loss, particularly the drying of springs, where Spicer paraphrases the term as "water supply," Bolton translates it as "watering places," is of special interest to the ethnologist.

For these desert people, springs were sacred places. Springs were mentioned in creation epics and in memorized orations, and they figured in O'odham contemporary ceremonies. In the desert, the place where water emerged had to be sacred. O'odham toponyms naming places where water emerged are still scattered across the geography of southern Arizona and northern Sonora, an area called Pimería Alta (see appendix 13.1). Some places represent the spot where whole clans reportedly emerged from the mythic underworld into the presently inhabited world.

Throughout the Southwest (as well as in many other parts of the world), caves, mountaintops, and springs are power places or sacred sites. They are mysterious places that elicit awe. Some aspect of that primal wonderment may be what motivates scientists to study springs today.

From Hopi and Zuni country southward throughout Mesoamerica, a Water Serpent (the Horned or Winged Water Snake) is associated with springs (Fewkes 1900; Parsons 1939). Among the Mayans, he is Kukulcan, the Feathered Serpent, associated with the famous Sacred Cenote at Chichén Itzá (Devereux 2000). This being is perhaps best known as Quetzalcoatl, the Plumed Serpent of the Nahua or Aztecs. He is associated not only with springs but also with clouds, rainstorms, and irrigation water. The large Plumed Serpent is seen in the kivas, particularly during ceremonies in March. He is painted on kachina masks. He is a collective being who lives in all springs. Some ceremonial dances begin at these springs, such as those at Ojo Caliente south of Zuni Pueblo. From there, the masked dancers march for many miles back to the village to perform.

The Plumed Water Serpent is also associated with natural disasters such as floods, earthquakes, and landslides. Local mythology records that he sometimes has been appeased with human sacrifice. For instance, among the Zuni and Hopi, a boy and girl were sacrificed into a flood to abate the water caused by the serpent. The diagnostic morphological characteristics of the Plumed Water Serpent, found in no current field guide, are given in appendix 13.2.

A somewhat similar mythic event gave rise to the Shrine of the Four

Children on the Tohono O'odham reservation. In this case, it is a dry spring associated with Wind. Here a man chased a badger, a tabooed animal, into its burrow, intending to kill it. But the badger dug deep, and the waters of the ocean threatened to come up through the hole and flood the country. A boy and girl from each of the two moieties, the Buzzard and the Coyote, were sacrificed; then the hole was covered with a pile of stones and enclosed in a fence of ocotillo, another Piman rain metaphor. Periodically the enclosure is renewed, the ceremonies conducted, and disaster again averted. A condition of stasis is thus reestablished between human culture and the powerful forces of nature.

Farther south in Tohono O'odham country, a giant serpent called Ñeebig lived at the oasis called Quitovac (Davis 1920; Saxton and Saxton 1973; see appendix 13.2). It could draw people in with its breath and swallow them, a common Mesoamerican theme. In this case, so many people were being eaten that the People (O'odham) sought the help of culture hero I'itoi, living in his cave in the Baboquivari Mountains. Swept into the creature's mouth, I'itoi propped its mouth and gullet open with sticks and cut out its heart with an obsidian knife. The thrashing monster knocked much of the water out of the lake (call it Holocene drying, if you will), and died, leaving its flesh as the white deposits that are still seen about the springs. Its bones and huge teeth are still to be dug up. Western culture assigns these remains to Pleistocene megafauna.

Every summer, following the first full moon of July, a composite ceremony at Quitovac combines a rain-making wine feast, a Navichu kachina dance, a deer dance, and a renewal of offerings to the stone heart of the Ñeebig water monster (Davis 1920). The ceremony's purpose, I was told, is to renew the world and bring blessings to humans.

Griffith (1992) has traced the Water Serpent broadly throughout the borderlands of Pimería Alta. Here among the mestizo population it is called *corúa*. The folk belief persists perhaps more strongly in Sonora than in the U.S. Southwest. The Water Serpent is large, shaped like a boa constrictor, inclined to be gentle, and definitely beneficial. It is the guardian of a spring. If you kill one at a water hole, the water will dry up. This is the overriding belief system of both indigenous and mestizo peoples to explain and help maintain a fragile relationship between humans and water sources in arid regions. Some biological ramifications of this long-term relationship are to be seen in a particular example that has been quantitatively studied in Pimería Alta.

The Story of Quitovac and Quitobaquito

There are two spring-fed oases in the Southwest that have been home to native peoples for millennia and to the O'odham for three centuries or more: Quitobaquito on the international border at Organ Pipe Cactus National Monument near Lukeville/Sonoita in Arizona, and Quitovac some 25 miles (41 kilometers) south in Sonora. The National Park Service manages the first. The O'odham who once farmed and ranched there have been removed, as has most evidence of their former settlement. A few remnants of their orchards persist in inconspicuous places that do not mar the perception of this "virgin" desert oasis.

In our earlier survey of the two oases, we had found more than 65 species of birds at the farmed oasis, Quitovac, and fewer than 32 species at the U.S. National Park Service–managed wildlife area (Nabhan 1982:91). These preliminary data suggested that the agricultural habitat diversity maintained by the native culture of Quitovac was more successful in attracting wildlife than the "nonmanagement" approach exemplified at Quitobaquito, which is part of the Man and the Biosphere Reserve system.

The difference in bird species richness in these two localities can be explained partially by the vegetation composition in the immediate area of the aquatic habitats. Bowers's (1980) flora of the vicinity of Quitobaquito spring yielded 54 species of plants. Nabhan identified 167 plant species in five hectares at Quitovac, including many seed-bearing annuals and shrubs that are not found at Quitobaquito (see chapter 12). The anthropogenic plant community at Quitovac includes intentionally planted living fence-rows, orchards, and palm groves, modifications that encourage the presence of other species.

In subsequent studies, we set out to quantify with greater precision what was happening to biodiversity at these two oases as well as at similar nearby farmed and unfarmed areas that served as controls. The team consisted of Karen Reichhardt, data keeper, plant ecologist, and statistician;

Gary Paul Nabhan, project mastermind, botanist, and mischief-maker; Eric Mellink, mammalogist and sanity manager, and myself as ornithologist. The results have been published in a number of journals (Nabhan et al. 1982; Mellink 1985; Rea et al. 1983; Reichhardt et al. 1994; Nabhan, this vol.). What I want to touch on very briefly are the diversity results from the two oases.

The greater plant diversity at traditionally farmed Quitovac versus park-managed Quitobaquito may be attributed to the protected status of the park oasis. At Quitobaquito, annuals were found only at the parking lot or along paths, where some soil disturbance created a favorable habitat for weedy growth. At Quitovac, however, clearing, irrigating, plowing, and other disturbances still occur on a sporadic basis, opening microhabitats for colonization by plants excluded from more stable sites.

The bird diversity and especially the sheer numbers of birds visiting the O'odham-managed oasis were far greater than at any of the four control sites (see Reichhardt et al. 1994). The higher diversity and species richness values at Quitovac than at Quitobaquito appear to be directly related to differences in habitat heterogeneity and succession dynamics. Unfarmed sites may offer a combination of habitats required for attracting diverse birds, but in general, traditional agricultural sites offer more microhabitats and seral stages in vegetation development. Thirty species were found to be unique to agricultural conditions. Of the three farms (including control sites), Quitovac attracted the most bird species (103), Emiliano Chico's *ak(i) chiñ* or arroyo dry-farmed field the second-largest number, and Buckelew Farm (using Anglo farming technology) the fewest.

The rodent diversity was not what might have been expected, no doubt due to the fact that about the time of our study the Mexican government initiated a program at Quitovac to "improve" the oasis. Despite the intensity of agricultural disturbance at Quitovac, the site lacked populations of the characteristic colonizing rodents adjacent to the fields. At the time we studied the oasis, the habitats adjacent to the fields had just been cleared and drained of water. Irrigation of crops was being accomplished by spring-fed irrigation from spring waters confined in ditches; consequently, flooding rarely reached the land surface. Small rodents were probably the quickest to respond to the ecological disturbance and habitat clearing under way and to succumb to raptors.

We found that small-scale farming in arid zones can substantially increase local habitat heterogeneity and associated biodiversity. Even when the twenty to thirty alien, anthropogenically favored plants and three obligate anthropogenic birds (the Inca dove, *Columbina inca*; great-tailed grackle, *Quiscalus mexicanus*; and house or English sparrow, *Passer domesticus*) were deleted from the inventories, the trends stayed the same. Despite the fact that Quitovac had been bulldozed, drained, and converted by recent government development efforts, it continued to harbor biological diversity comparable to or greater than that found in the Quitobaquito oasis.

The habitat management techniques that promoted this enhanced diversity in the Quitovac oasis included the intentional burning of plant communities of the side pond, grazing and browsing by livestock, the disturbing of soils through plowing and the creation of trails, the planting and maintaining of living fence rows, the making of fences by piling up dead branches, the transplanting and tending of domesticated perennials, and irrigation of selected areas.

Quitovac is not unique. I conducted a long-term study of birds and habitats in Gila River Pima country, a linear Sonoran Desert oasis of the middle Gila, whose waters were once supplemented by numerous local springs and desert washes, such as Santa Cruz, Greene's, and Santa Rosa washes (Rea 1983; see also Rea 1997:29–66). Here I found that traditionally managed Pima agricultural fields harbored thirty-six breeding bird species, while modern mechanized fields nearby (Anglo-style "clean" farming) provided nesting habitat for only seven (whether modern mechanized farming is being done by Native Americans, Mexicans, or Anglos is ecologically irrelevant[1]).

Indigenous Management and Biodiversity

It is perhaps advisable at this point to look at the broader implications raised by indigenous management of desert springs and oases. This is the issue of the relationship of the Fourth World hunter-gatherer-subsistence agriculturalists to the preservation of biodiversity. (I use the term "Fourth World" as do some Native American authors and anthropologists to indicate peoples with communal resource ownership who do not aspire to a peasant economic system.) There are now many case studies that examine perspectives and values of the peoples themselves (see, for instance, Posey 1983,

1985; Anderson and Posey 1989; Balée and Gély 1989; Hecht and Posey 1989; Salick 1989; Nazarea 1998; Atran 1999; Arabagali 2001; Cafaro and Verma 2001; Diegues 2001; Parajuli 2001; Sarkar 2001).

In looking at this question of indigenous management, we might be aided by an essay by Alcorn (1994:7) in which she asks, "Is biodiversity conserved by indigenous peoples?" She hastens to add that it is not a very good question, at its best being asked from the moral high ground by those seeking to preserve biodiversity as a global heritage and at its worst echoing colonial-era debate over Europeans' right to ignore indigenous people's human rights and their preexisting tenurial rights over their resources. She notes,

> Northern biological preservationists wish to keep biodiversity in untouched natural settings free of any human habitation. They see people who live and work there as threats to be educated and removed if at all possible. To achieve their desires, they trust strong governments to wield the stick to protect biodiversity. They believe in a "noble State."
>
> On the other hand, Northern cultural preservationists wish to see exotic peoples preserved as idealized, superior cultures which live in "harmony with nature," untainted by the market economy. They believe in "noble savages." Where is the reality? (Alcorn 1994:7–8)

We should know, of course, that both of these positions are faulty, and that both are Northern Hemisphere—that is, First World—projections. What should make us trust the "noble State" and its claim that it will change its past behavior and in the future act to conserve biodiversity? Look at the uphill (and always tenuous) battles to protect endangered species in the United States, the most powerful (and, one might presume, most enlightened) nation on earth. Free-market environmentalism is alive and well in the United States, and it has not been particularly effective at preserving biodiversity.

However, indigenous peoples have rarely had the power and the position to uphold their cultural values in the colonial and postcolonial world (Arabagali 2001). Alcorn makes this point admirably in stating that "indigenous people conserve biodiversity—particularly those indigenous peoples with de facto tenurial security, cohesive cultures and strong social institutions that function to maintain the community while at the same time enabling communities to adapt to new stresses on the basis of their own cultural values" (Alcorn 1994:9).

Let me interject here that such conservation of biodiversity is predicated on an intact system of indigenous ecological knowledge and folk taxonomy, which was more likely to have been found in parts of North America seventy-five or a hundred years ago. After several generations of overt assimilationist policies by the U.S. and Canadian governments, such traditional ecological knowledge has eroded and is largely relictual (Anderson and Medina Tzuc 2005). Today, reservations and reserves for the most part function in a market economy. However, self-sustaining peoples are still to be found in much of the Orinoco, Amazonia, New Guinea, and Indonesia, as well as in parts of Mexico and Central America. In such places, indigenous peoples with traditional knowledge and authority (tenurial rights, cultural sanctions) can best manage local fisheries, game, timber, grasslands, and fields. Indigenous communities need the political power to defend protected areas against such outside economic interests as gold mining, oil extraction, timber cutting, cattle raising, and so forth.

Redman's *Human Impact on Ancient Environments* (1999) seems to me to be a calmly written, coolly reasoned, and nonsimplistic analysis of the culture/nature problem. Free of New Age rhetoric, the work examines the various values and attitudes that different cultures hold toward their environments. Redman (1999:4–5) points out, "The archaeological record encodes hundreds of situations in which societies were able to develop long-term sustainable relationships with their environments, and thousands of situations in which the relationships were short-lived and mutually destructive. The archaeological record is 'strewn with the wrecks' of communities that obviously had not learned to cope with their environment in a sustainable manner or had found a sustainable path, but veered from it only to face self-destruction."

But there is a record that is perhaps even more impressive than that revealed in the archaeological record. It is this: our heritage of high biodiversity that has survived anywhere on the earth today is to be found in areas that have for countless millennia been under the management of indigenous, almost always tribal, peoples of the world. Their technology may be labeled "primitive" and their belief system "mythological," but the practical consequences of their lifestyles are embodied in the surviving local biotas.

Whether in the rain forests of South America or the hinterlands of Australia, this legacy is really the only rich biodiversity we know today.

Perhaps we who live in First World societies, whether in deserts or elsewhere, need to ask: Will *we* kill *our* Water Serpents?

Preserving Biodiversity

Springs in the desert were, and in many places still are, sacred sites for native peoples. These locations were often associated with a benign Water Serpent, or corúa, which served as their protector. Additionally, many specific springs are associated with mythic events from indigenous origin narratives or creation stories.

Desert springs and oases were managed throughout the Greater Southwest for agriculture as well as for their natural products. In the case study above (and associated citations), we quantified species abundance and diversity at two oases in Tohono O'odham (Papago) country, one managed by the U.S. National Park Service as a "pristine" wildlife reserve, the other the center of an indigenous Papago village in Sonora, Mexico. The indigenous-managed oasis ranked higher in most parameters (plants, birds, small mammals). The latter springs ecosystem has survived in coexistence with human cultures with value systems different from our own.

It is the consensus of many researchers who have carefully examined people-biota relationships that indigenous communities with actual tenurial rights, intact cultures, and strong social institutions can maintain rich biodiversity and live in stasis with the environment as long as they are protected against exploitative economic interests outside their culture. In actuality, all terrestrial areas of high biodiversity that have survived into the modern world have been culturally managed by indigenous peoples practicing a hunting, gathering, and often subsistence agriculture mode of life for millennia.

Appendix 13.1

TOPONYMS NAMING SPRINGS OR WELLS IN O'ODHAM (PIMAN)

Some background in historical linguistics is offered to provide the keys to interpret some spring and well place names still currently used in Pimería. Dialects of Tepiman languages were once spoken from the Gila River in Arizona southward to the Río Santiago in northern Jalisco, with a single break, the Taracahitan bridge, from Yaqui-Mayo country eastward into Tarahumara (Rarámuri) country. Colonial Spaniards named the southern Tepimans "Tepehuanes" and northern Tepimans "Pimas." The self-designation of most of these people is "O'odham," or some variant of that term. Colonial powers, both Hispanic and Anglo, further segregated, for juridical purposes, the northern O'odham into Sobaipuri, Pima, and Papago. Ecclesiastical jurisdiction broke the northern Tepimans into Pima Bajo, who were missionized first, and Pima Alto, who were missionized later. However, all the lowland groups spoke mutually intelligible dialects, and dialect group was (and to a large extent still is) the really defining ethnic identity within O'odham communities. Variations between dialect groups are, for the most part, extremely subtle, partly at the level of word morphology and partly in grammar and syntax. The most obvious divergence in sounds occurred historically. The strong palatalization of certain sounds (for instance *s* to *sh*, *n* to *ñ*, *d* to *j*, *t* to *ch*) preceding certain vowel environments occurred among the northernmost Piman speakers. Because numerous place-names (toponyms) in the Southwest were recorded throughout colonial times, we know that palatalization began sometime after the Jesuit expulsion of 1767 and was thoroughly in place within less than one hundred years. In contrast, Lowland (Desert) Pima Bajo and Mountain Pima retained the original nonpalatalized sounds in their dialects.

The O'odham (Piman) word *son* means the base or beginning of something, such as the trunk of a tree, foot of a mountain, bank of a wash, or spring of water. With palatalization, all *son* words became *shon* in modern spoken Pima and Papago, but those that had been taken into Spanish as toponyms remained *son*, as linguistic fossils. There also are derivative words. *Shonag* means "there is a spring (somewhere)." *Shongam* is more frequently heard. According to River Pima Sylvester Matthias, *shongam* means "a spring that turns into a creek." River Pima Joseph Giff said, "A *shon*, 'spring,' has a starting point, *shongam*."

A related word *vavhia* (pl. *vaipia*, *vaiphia*) means "well," often with the meaning of a spring that has been modified (Mathiot 1973). In Spanish orthography, the *v* was often written as *b*, an interchangeable sound, although this is incorrect in O'odham, where the sounds are not equivalent. The root word *va-* signifies water or something associated with it, meaning "aquatic." Many place-names survive in Spanish or English preserving the O'odham names for spring or well.

Here are a few of the numerous spring-related toponyms from Pimería Alta:

ari sonac, ali shonac, Arizona, 'place of the little spring'; but see Gavate (1999).
ari tuk son, ali chuk shon, Little Tucson, 'little black base [or] spring'
ari vaiphia, Arivaca, Aravaipa, A'al Waiphia, 'little wells'
ban vavhia, 'Coyote Well'
kaav vavhia, Cobabi, 'Badger Well'
koom vavhia, Comobabi, 'Hackberry Well'
nod vavhia, 'Sacaton Grass Well'
siv vavhia, 'Bitter Well'
s-toha vavhia, 'White Well'
koson vavhia, 'Packrat Well'
tukud vavhia, chukud vavhia, Tucurubabia, 'Great Horned Owl Well'

Appendix 13.2

DIAGNOSTIC CHARACTERISTICS OF THE PLUMED WATER SERPENT
(NORTHERN SUBSPECIES)

EXTERNAL MORPHOLOGY: Short, thick-bodied snakes. Two morphs: one all black dorsally, the other brown. Both are white ventrally with a red lateral streak dividing dark from light. Black morph with symbols. Protuberant goggle-like eyes; slender red tongue, forked, generally extended. Red teeth but no fangs. Head, most diagnostic feature, anteriorly bears medial horn, curving forward, sometimes jointed (as in arthropods), sometimes solid (as in bovids). Proximal to horn, a medial crest of plumes projecting vertically, fanned laterally. Wears necklace of feathered strings.

INTERNAL ANATOMY: Backbone of wood; ribs hoops of wood; heart of crystal, sack of all seeds (all colors of corn, cotton, melons, squash, etc.), and black prayer stick, these three all suspended from the backbone in the neck region; otherwise hollow.

HABITAT: At springs in arid country and in kivas.

BEHAVIOR: Generally shy and retiring but given to loud roaring in public; non-venomous, not dangerous to humans; sensitive to environmental disturbance, particularly modifications of watersheds. Perhaps parthenogenic, as only male populations are known.

DISTRIBUTION: In kiva environments, appears during the March moon.

SYNONYMY: Also known by the names of Horned Water Serpent, Winged Water Serpent, Quetzalcoatl, Corúa. Various other names among different pueblos. May be conspecific (at least congeneric) with Ñeebig of northwestern Sonora, currently known only from fossil remains.

NOTE: Ñeebig has been variously referred to as a mythological monster, a whale, a serpent, and even a dinosaur (see Rea 1998:262, Mathiot n.d.:114). While it is true that some O'odham speakers will use the term today polysemously to label whale, in an original Piman narrative context the reference is to "a big snake (*ge'e wamad*) called a 'neebig', where there was a spring" (Saxton and Saxton 1973:316). The term *vamad* refers specifically to nonpoisonous snake, in contrast to *ko'oi*, poisonous snake. Its association is with a specific spring (*shongam*) called Gi'ito Vak.

NOTE

1. In the four decades since my study began, traditional River Pima farming has almost totally vanished. The so-called Gila River Tribal Farm is an economic enterprise. No attempt is made to preserve and plant traditional Piman land races of crops or to emulate the time-tested farming methods. Foods that might help buffer the Pima and Maricopa population against the epidemic onslaught of adult onset diabetes are ignored. Its goal is revenue generation. Because of multiple ownership of allotted lands, most other acreage is now leased and farmed as agribusiness—that is, modern mechanized farming. However, Ramona Farms, a Pima Indian owned and operated farming enterprise located on the Gila River Indian Reservation, still supplies certain indigenous local crops on a commercial basis.

Chapter 14

Vegetation Dynamics of Great Basin Springs

Potential Effects of Groundwater Withdrawal

DUNCAN T. PATTEN, LEIGH ROUSE, AND
JULIET STROMBERG

In 1989, the Las Vegas Valley Water District filed 146 applications to pump 800,000 acre-feet per year of groundwater from twenty-six basins in east-central and southern Nevada. The number of proposed basins was later reduced to seventeen, which would result in the pumping of 180,800 acre-feet per year. Recent plans include the pumping of 17,000 acre-feet per year, but this is only the beginning of an extended and expanding groundwater withdrawal schedule to serve Las Vegas (Jenkins 2005; Nash 2005). Additional groundwater withdrawal is planned to serve other rapidly expanding communities in southern Nevada.

The withdrawal of groundwater may alter flow rates and water table levels at many springs in the hydrological basins of central, eastern, and southern Nevada and is considered the single greatest threat to these ecosystems (Brussard et al. 1999; Tiner 2003). Results from various studies indicate that springs discharge can be affected by regional (Schaefer and Harrill 1995) and local (Dudley and Larson 1976) pumping. If a spring's discharge decreases or the shallow water table drops, there may be significant changes to the spring's riparian vegetation, a common phenomenon in arid regions where groundwater is affected by anthropogenic activities (Stromberg et al. 1992, 1996; Scott et al. 1999). The extent of change depends on the cumulative effects of withdrawal activities (Winter 1988).

Springs are important for maintaining regional biodiversity. Although many springs have been altered, they provide habitat for many aquatic and terrestrial organisms, some highly endemic (Hendrickson and Minckley 1985; Heino et al. 2003; Sada et al. 2005). The objective of this study was to evaluate the effects of potentially changing water table levels on riparian vegetation at springs in the Great Basin and adjacent deserts of southern Nevada.

Methods

Study Locations

Springs were selected from areas in east-central and southern Nevada that may be impacted by groundwater withdrawal by Las Vegas. The study areas in southern Nevada were Ash Meadows National Wildlife Refuge (managed by the U.S. Fish and Wildlife Service) and Lake Mead National Recreation Area (managed by the National Park Service); in northern Nevada, the study areas were Spring Valley and Railroad Valley (east and west of Ely, respectively), both administered by the Bureau of Land Management (fig. 14.1). At each location, two primary springs were selected for long-term study, and two secondary springs for short-term comparison studies. Each pair included a large and a small spring.

Eastern Nevada is part of the Basin and Range province. Some groundwater flow systems here are contained in individual basins, with the water flowing through basin-fill deposits (Fenelon and Moreo 2002). Others are connected hydraulically between basins (Plume 1996; U.S. Geological Survey 1997). In eastern Nevada, for example, a deep Paleozoic carbonate aquifer underlies the basins and acts to hydraulically connect groundwater under mountains between basins (Dettinger et al. 1995; Plume 1996). These aquifers, especially the deep carbonate aquifer extending from northeastern to southwestern Nevada, are what produce springs in Ash Meadows National Wildlife Refuge (Ash Meadows NWR) and in Spring and Railroad valleys (Dudley and Larson 1976; Dettinger et al. 1995; Prudic et al. 1995). Most of the water that recharges the basin-fill and deep carbonate aquifers originates in the mountains. Although little recharge normally comes from direct precipitation on the basin-fill deposits, the aquifers that support the springs of Lake Mead National Recreation Area are recharged from local watersheds (Pohlmann et al. 1998). Another water source for these Lake Mead NRA springs may be the Virgin River subflow system in southeastern Nevada and southwestern Utah, a subsystem of the Colorado River regional flow system (Prudic et al. 1995).

Figure 14.1. Map of Nevada, showing the location of the study areas: Spring and Railroad valleys in the north and Lake Mead National Recreation Area and Ash Meadows National Wildlife Refuge in the south.

Field Data Collection

At each primary spring site, vegetation was sampled in plots located along permanent transects. Transects were perpendicular to the spring or associated outflow stream and extended from the wetland zone to the upland zone. The spring's discharge was measured using a flow meter and/or permanent or temporary weirs. Monitoring wells were installed along the transects at the primary spring sites to determine water table depth. Soil matrix water potential was determined using soil moisture blocks installed at a depth of fifty centimeters. Soil samples were collected at each plot. Soils were analyzed for texture, pH, and electrical conductivity. Electrical conductivity of surface water and groundwater was measured. Ordination analysis (DCA) and regression analyses were used to determine relationships between vegetation and environmental variables.

Results and Discussion

A Description of Vegetation

Plant community composition at the springs varied along gradients of water availability (water table depth), water quality (electrical conductivity and pH), and soil texture, as determined by ordination analysis. Within this continuum, we identified four plant communities, each associated with particular conditions of water quality and quantity resulting from different hydrologic conditions: (1) wetland (hydroriparian) communities, (2) mesoriparian (wetland/upland transition) communities, (3) phreatophytic-upland communities, and (4) upland communities. Plant community composition differed between the southern and northern springs. Ash Meadows was the only study location with endemic species. These were *Ivesia kingii* var. *eremica*, *Centarium namophilum*, and *Grindelia fraxinopratensis*. The latter two are considered relict populations from the pluvial period of the Pleistocene and are listed as federally threatened species.

WETLAND COMMUNITIES. Wetlands had standing water or saturated soil and salinities similar to that of the spring water. Vegetatively, they were dominated by obligate and facultative wetland herbs (Sabine 1994). At Ash Meadows and Lake Mead (the southern springs), typical wetland species were *Eleocharis rostellata*, *Scirpus americanus*, and *Juncus mexicanus*, with *Phragmites australis* often occurring near the spring orifice. Woody species occasionally occurring in or adjacent to the wetland zone at these springs were *Salix exigua* and *S. gooddingii*, *Fraxinus velutina* (only at Ash Meadows), and *Prosopis* species. At Spring and Railroad valleys (the northern springs), wetland communities were composed of *Carex*, *Juncus*, and *Eleocharis* species. *Nasturtium officinale* was common at the spring orifice area. Certain plant species, including *Berula erecta*, occurred in wetland areas with very low (<1 dS/m) soil electrical conductivity, while other species, including *Scirpus americanus*, were associated with intermediate EC values (up to 3 dS/m).

MESORIPARIAN COMMUNITIES. The mesoriparian zone had wet to moist soils, a shallow water table, and less-saline soils than those of the uplands. Facultative and facultative wetland (*sensu* Sabine 1994) woody plants and herbs dominated this zone. *Anemopsis californica* and *Iva acerosa* were common in the mesoriparian zone of the southern springs, shifting to *Sporobolus airoides* and *Distichlis spicata* in slightly drier areas. Woody species found in the southern riparian communities included *Prosopis glandulosa*, *P. pubescens*, *Baccharis emoryi*, and *Atriplex lentiformis*. The mesoriparian zone at northern spring sites was dominated by woody species including *Artemisia tridentata* and *Rosa woodsii* and occasionally *Juniperus scopulorum*. The transition from the riparian zone to the uplands was abrupt at many of these northern spring sites.

PHREATOPHYTIC UPLAND COMMUNITIES. This zone was beyond the zone of direct influence of the surface stream and had moderately deep water tables and high soil salinity. It was vegetated by salt-tolerant phreatophytes. *Distichlis spicata* was the most common herb; *Sporobolus airoides* occurred occasionally at the southern sites. Halophytic, phreatophytic woody plants included *Atriplex confertifolia* and *Suaeda moquinii* near the southern springs and *Sarcobatus vermiculatus* and *Ericameria* (*Chrysothamnus*) species near the northern springs.

UPLAND COMMUNITIES. This zone had dry, often alkaline, soils influenced only by precipitation and was characterized by xerophytic shrubs. *Artemesia tridentata* was part of the upland community at the northern springs, as was *Larrea tridentata* at the southern springs.

Springs Discharge and the Water Table

The discharge (Q) of the study springs ranged from about 0.12 cubic meters per second (cms) to 0.0001 cms. Length of the spring outflow streams was relative to the discharge. Large springs produced a perennial stream that flowed into a larger river, a lake, or a terminal basin. The length of outflow for small springs appeared to depend on factors in addition to the discharge from the spring, including underlying clay layers, cattle activity, and depth to bedrock.

Monitoring wells showed that the water table near the spring orifice and along perennial outflow streams had little seasonal fluctuation. In contrast, the water table declined during the summer growing season at sites distant from the spring orifice and outflow streams and near the terminus of outflow streams at small springs.

Electrical conductivity (EC) of surface discharge and outflow water varied considerably among the study springs. The lowest surface-water EC levels were at the Spring Valley and Ash Meadows springs (<1 dS/m), while the highest surface-water EC levels were at the Lake Mead springs (>3 dS/m). This supports the finding that the springs at Lake Mead have aquifer sources different from those of the other two areas (Prudic et al. 1995; Pohlmann et al. 1998). Groundwater EC levels were generally similar to those of surface waters.

Soil Characteristics

Soil electrical conductivity (EC) at the spring sites ranged from below 1 dS/m to more than 4 dS/m. The pH of most soils was around 8. A few soils in well-leached areas had pH values below 7, and some in upland areas had pH values greater than 9.

Soil texture among springs was highly variable and corresponded to the geographic and topographic location of the spring. Soils sampled at springs near the mountains often had high percentages of gravels (i.e., material larger than 2 millimeters in diameter), while soils from springs in mid or lower areas of alluvial fans often had little or no percentage of gravels. Some locations, including Ash Meadows NWR springs, had high clay and low sand contents, while others, including Lake Mead NRA, were low in clay and quite high in sand content.

Soil Moisture

Soil moisture at the spring sites typically was high (>-0.1 MPa), indicating low plant moisture stress. The few study plots that did show low moisture were located far from outflow streams. However, soils at many other sites located far from outflow streams remained moist most of the year, probably because of capillary rise of water from the shallow alluvial water table. Some moisture blocks placed below an underlying clay layer had low moisture readings (e.g., <-2.5 MPa), indicating little percolation of surface water into this zone.

Plant-Hydrology Relationships

One objective of this study was to develop quantitative models relating vegetation to hydrologic variables and to use these models as tools for predicting biotic response to changes that may occur in response to groundwater withdrawals. One plant community trait analyzed was the wetland indicator score, which was calculated for each study plot by weighting the relative abundance of plants in different wetland classes by their respective wetland scores (scores range from 1 for obligate wetland to 5 for upland).

Wetland indicator scores for herbaceous vegetation varied significantly with mean depth to water (as indicated by regression analysis), for individual springs, and for data pooled among springs. This relationship reflects the different tolerance ranges of plants in each of the wetland classes. Obligate and facultative wetland plant groups, for example, both had very high cover values at sites where water was within one meter (above or below) of the ground surface. Many obligate wetland herbs were found in standing surface water more than one meter deep in ponds at spring orifices. Obligate wetland plants did not occur at sites where the water table depths dropped below two meters, whereas plants in other categories tolerated deeper water tables.

Regression analyses also showed a significant relationship between weighted-average wetland indicator scores for woody communities and water table depths, with lower (i.e., "wetter") wetland indicator scores corresponding to shallower water tables. This relationship of woody species to water table depths indicates, for example, that with a decline in the water table of approximately one meter, the woody plant community would change from facultative wetland to facultative. Wetland indicator scores for woody plots were also significantly related to seasonal changes in water table depth, with lower wetland scores being closely related to less seasonal fluctuation.

Water Table Decline and Alteration of Spring-Associated Vegetation

Groundwater modeling studies suggest that water tables in the basin-fill aquifer in eastern and southern Nevada may decline by 0.3 meters to 3.0 meters in response to pumping of the deep carbonate aquifers (Schaefer and Harrill 1995). Testimony by hydrogeologists at 2006 groundwater withdrawal hearings for Spring Valley using the Schaefer and Harrill model suggests that water table decline following long-term pumping might be as great as 60 meters. This groundwater withdrawal may reduce springs discharge, thereby truncating the outflow stream and causing water tables in the riparian zone to decline or undergo greater seasonal fluctuation (fig. 14.2). This, in turn, will cause the riparian area to contract, as portions of the hydroriparian vegetation area are replaced by mesoriparian vegetation and the mesoriparian vegetation is replaced by upland vegetation. The groundwater withdrawal may terminate the flow of small springs, resulting in complete succession to upland vegetation.

Outflow streams presently maintain localized elevated water tables allowing wetland plants to survive through the hot, dry summers. As these water table levels decline, the plant communities dependent on this water source will be altered. Obligate and facultative wetland herbs presently dominate much of the spring areas, and these plants would be lost entirely if the reduced hydraulic head of the deep aquifer no longer could maintain a shallow basin-fill water table or a spring flow. Plants in these groups would be greatly reduced if the water table decline exceeded one meter. A one-meter drop in the water table might change the composition of woody riparian species along the outflow stream by one "wetland unit," with, for example, facultative wetland plants being replaced by plants in the facultative group.

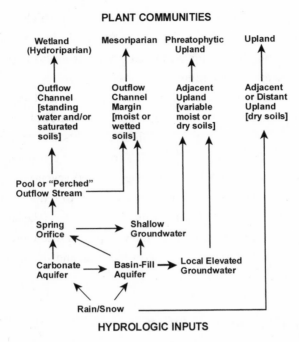

Figure 14.2. A model of the dynamics of the southern Great Basin and adjacent desert springs, showing hydrological inputs, connections among several flow paths, and resulting vegetation types (see text for descriptions of the vegetation types). The diagram illustrates the potential impact of groundwater withdrawal from the deep carbonate layer. Connections (or flows) between it and the basin-fill aquifer and springs orifice may decline or terminate, depending on the magnitude of the groundwater withdrawal. Most vegetation types that are dependent on water from the deep aquifer would change to upland-type communities.

Groundwater withdrawal from the deep carbonate aquifer also would cause the water table to decline in the upland areas surrounding the spring orifice and outflow stream. The halophytic phreatophytes in this zone would be affected as the water table declines below their root zones (Cooper et al. 2006).

Groundwater withdrawals also may affect vegetation by altering soil salinities. Electrical conductivity may increase near the springs, causing a decrease in the percent cover of salt-intolerant wetland species. A declining water table will reduce the amounts of salts and water wicked to the surface by capillary action in the phreatophytic upland zone, which concomitantly may alter plant composition. A large water table decline may cause areas presently vegetated by halophytic phreatophytes to shift to nonhalophytic, nonphreatophytic plant communities.

The extent of change in riparian vegetation will vary among the spring areas because of different degrees of dependency of the springs hydrology on local versus regional groundwater sources. For example, a small water table decline at the Lake Mead NRA area, although difficult to project, will probably have little effect on spring-associated vegetation. In contrast, the projected groundwater withdrawals are likely to reduce the southerly flow of the deep regional carbonate aquifer that recharges the basin-fill aquifer and supports springs in Ash Meadows; springs discharge in this area thus may be reduced and the associated vegetation and species composition altered.

Summary and Conclusion

The orifice pool, outflow streams, and shallow water table associated with springs discharge are critical features creating a diverse vegetation mosaic at springs in the southern Great Basin and adjacent deserts. The springs create riparian habitat and increase diversity within a landscape mosaic comprised of upland shrublands (some of which are phreatophytic and some of which are rainfall dependent). The distance from the orifice to the stream terminus is directly associated with amount of springs discharge; the more discharge, the greater the distance. The outflow streams are often perched on a semi-impervious or impervious layer that influences the distance of outflow. The outflow stream often creates wetland conditions, with standing water or saturated soils, and tends to maintain a localized, elevated water table immediately below or adjacent to the outflow channel. It may also maintain, through percolation or capillary action, wetted or moist soil along the margin of the outflow channel, providing a critical water source during the warm, dry summers. This lateral percolation may leach out salts, thus permitting saline-intolerant species to establish in these areas.

A decline of several meters in the level of regional water tables in response to pumping, as suggested by groundwater modeling studies, will reduce springs discharge and reduce the water table that supports riparian vegetation. Using models of water table decline in concert with information on vegetation-hydrology relationships, we suggest that a water table decline of one to three meters or more in the Spring Valley area will cause a substantial reduction in diversity, cover, and productivity of riparian vegetation at springs, paralleling patterns observed at other sites subject to water withdrawal (Sorenson et al. 1991; Schultz 2001; Naumburg et al. 2005).

Chapter 15

The Knowles Canyon Hanging Garden, Glen Canyon National Recreation Area, Eight Years after Burning

Changes in Vegetation and Soil Biota

TIM B. GRAHAM

On 4 July 1989, a hanging garden in Knowles Canyon (KCHG), in Glen Canyon National Recreation Area, Utah, burned after visitors ignited fireworks in the area. Aboveground vegetation on the upper terrace was completely destroyed, although the rootstocks of some species survived; portions of the lower garden also burned. The fire was extremely hot, causing rock slabs to spall from the alcove ceiling, and most of the soil organic material was oxidized. A few patches of dense litter, mostly leaves and seed pods of redbud (*Cercis occidentalis*) and packrat middens, did not burn completely. The fire, while unfortunate, provided the National Park Service (NPS) with an opportunity to study postfire change in a little-studied habitat type. The objectives of the present research were to document the postfire vegetation and soil changes over an eight-year period in this hanging garden. Understanding the recovery processes may allow the National Park Service to improve management of this and other springs habitats and to more effectively restore anthropogenically altered hanging gardens.

Hanging gardens are wetland and mesic plant communities that grow where groundwater seeps laterally out to a cliff surface from perched aquifers (May et al. 1995). Community composition of hanging gardens ranges from small monospecific patches of vegetation on vertical cliffs to complex assemblages in large alcoves and can include common riparian species, disjunct populations, and endemic species restricted to the hanging garden environment (Welsh and Toft 1981; Welsh 1989a; Fowler et al. 1995).

Welsh and Toft (1981) defined types of hanging gardens based on development of alcove and foot slopes and presence/absence of plunge-pool basins. In their parlance, KCHG was classified as a Type IV (plunge basin) garden (Welsh 1984). Communities in Type III and IV gardens are structured by environmental factors, such as exposure, temperature, water availability, and accessibility, with elevation and latitude influencing the impact of the above factors (Welsh and Toft 1981). Nebeker and colleagues (1977) reported that large-seeded, bird-dispersed plant species were more numerous in hanging gardens in Arches and Canyonlands national parks but that each hanging garden plant assemblage they studied was a random subset of the species pool. Malanson (1980) reported that hanging garden community composition along the Virgin River in Zion National Park was not correlated with any physical environmental variables he examined. He similarly concluded that hanging garden plant assemblages resulted from random establishment (Malanson 1982); however, Malanson and Kay (1980) found that spore-dispersing species dominated frequently flooded hanging gardens in Zion National Park. They suggested that hanging garden assemblages were strongly influenced by the dispersal ability of the associated species and by the frequency of disturbance.

Natural disturbance in Type III and IV hanging gardens is dominated by rockfalls that crush vegetation, scouring it and colluvium off the footwall. Under these circumstances, succession is typically a cyclic, secondary process, with the crushed vegetation generally replaced by the same species (Welsh and Toft 1981). Rockfall is essentially a physical process in those systems. In contrast, severe fire disturbance involves changes in physical, chemical, and biological characteristics that generally do not occur after rockfalls. Fire reduces microbial biomass and respiration, with the extent of reduction related to the heat and duration of the fire (Raison and McGarity 1980; Fritze et al. 1993; Pietikäinen and Fritze 1993; Fritze et al. 1994). Soil biota and vegetation recovery rates also depend on the amount of ash deposited and degree of vaporization of soil carbon (Raison 1979; Fritze et al. 1994). Therefore, postfire ecosystem changes at KCHG may differ substantially from succession after rockfalls.

Soil community structure influences the direction of plant succession in many systems (Ingham et al. 1985; Insam and Haselwandter 1989; Setälä et al. 1991), and knowledge of soil-plant interactions may enable land managers to protect or restore hanging gardens more effectively. However, little is known about soil communities in hanging gardens or how changes in soil community structure affect hanging garden plant succession.

Methods

The Study Site

KCHG emerges from the Mesozoic Navajo Sandstone off the main-stem of Knowles Canyon, at an elevation of 1,160 meters. KCHG is two-tiered: an upper terrace consisting of a wide, steeply sloping ledge in a shallow alcove, and a lower section consisting of vegetation on the cliff face and around the plunge basin in the canyon floor. My study was limited to the upper terrace, which burned completely, because access to the lower garden was too difficult. The study area was defined as the colluvial slope within the alcove, between the overhanging cliff and the steep lower slope and cliff (fig. 15.1).

The burned upper terrace was divided into four distinct habitats based on location and degree of revegetation (fig. 15.1). Different conditions for plant establishment exist in each habitat. The backwall (BW) habitat is a narrow ledge at the top of the detrital slope and the cliff above it. Water does not run down the cliff uniformly; as a result, soil moisture along the BW ledge varies from dry to saturated. The central seep (CS) habitat is a small area in the middle of the alcove. It extends from the top to the bottom of the detrital slope and is defined by wet soils and thick vegetation where water drips off the cliff above. The east dry (ED) habitat is southeast of the CS and covers the entire colluvial slope from the BW strip to the lower cliff. It was largely unvegetated in 1993 but had redbud growing along the upper and lower boundaries of the dry slope and a patch of common reed (*Phragmites australis*) along the eastern edge. The west dry (WD) habitat is northwest of the CS and encompasses the detrital slope from the BW strip to the lower cliff. Like the ED, this area was completely unvegetated in 1993, although redbud and long-leaf brickellbush (*Brickellia longifolia*) grew along the BW boundary above the WD and redbud, Gambel's oak (*Quercus gambelii*), and Baltic rush (*Juncus arcticus* var. *balticus*) were growing just below the WD.

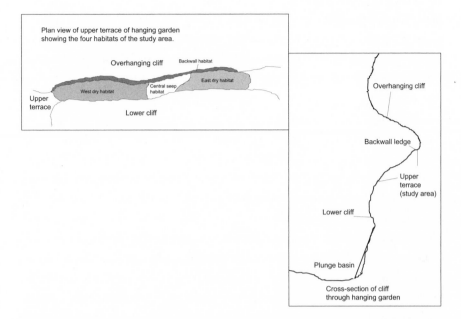

Figure 15.1. A diagram of KCHG. A plan view of the upper terrace, showing the four habitats found after the fire. The cross section shows the relation of the upper terrace (the subject of this study) to the rest of the hanging gardens surveyed by S. Welsh.

The ED receives less insolation than the WD, but otherwise the two habitats appear similar.

Field Data Collection

A list of all plant species seen in the burned areas was compiled each year. This was used as an estimate of the maximum number of species in each habitat to evaluate the effectiveness of the sampling protocol. Species on the cliff below the terrace, on the overhanging cliff, or found outside the alcove on the terrace were excluded.

To facilitate locating random points in the dry areas, I established a permanent 50-square-meter (5 m × 10 m) plot in each dry habitat, overlaid a grid on the plot, and located twenty permanent quadrats randomly on each grid. Quadrat locations and transect end points were chosen based on the presence of large boulders that provide more permanent markers so that the quadrats could be relocated even if the stakes were lost.

Because of the difficulties of sampling an overhanging cliff, only the narrow ledge of the BW was sampled. Quadrats were located from random numbers along a 30-meter transect in the middle of the BW and were permanently marked. Only herbaceous plants were recorded along the BW. Canopy cover of shrubs was not recorded, because of the difficulty in trying to estimate cover above the quadrats in the confined space of the BW ledge.

Quadrats were established along four transects across the CS portion of KCHG (two running downslope from the BW and two traversing the garden roughly along contours). Only the ends of the transects were permanently marked. Quadrats were located from random numbers along a tape measure stretched between the transect endpoints. Species accumulation curves from the 1993 data indicated that more quadrats should be sampled in the BW and in the two dry habitats. In 1994, the BW transect was extended to 60 meters, adding twenty more quadrats, and the size of the sampling plot in each of the two dry habitats was expanded to 100 square meters (5 m × 20 m), adding twenty more quadrats in each dry zone. The CS was small enough that twenty quadrats adequately sampled the vegetation in that area.

Percent cover for rock, bare ground, ash, charcoal, litter, and each plant species were visually estimated in each quadrat in each habitat as the percentage of the quadrat frame covered by a species, and I recorded cover class according to a modified Daubenmire (1959) system: $1 = <1$ percent, $2 = 1–5$ percent, $3 = 5–25$ percent, $4 = 25–50$ percent, $5 = 50–75$ percent, $6 = 75–95$ percent, and $7 = 95–100$ percent. Cover was estimated for all plant species occupying space in or above the quadrat, whether the plant was rooted in the quadrat or not. Different-sized quadrats were sampled in the four habitats because plant density and total area were different (Daubenmire 1959). The dry habitats were sampled with 0.25-square-meter quadrats (a total of 5 square meters). The central area was sampled with 0.1-square-meter quadrats (a total of 2 square meters). Twenty 0.05-square-meter quadrats (1 square meter in total) were sampled in the BW habitat.

Percent absolute cover for each species was estimated by multiplying the midpoint of each cover class by the number of quadrats assigned to that cover class, summing over all cover classes and dividing by the number of quadrats. Percent cover was calculated for each species and for rock, litter, and bare ground in each area. Percent cover calculations for all species were summed to obtain a total plant cover value for each area. Other vegetation metrics were measured but are not reported here; these include species area curves and canopy diameter changes.

In 1993, twenty soil samples were collected at random points: four each from the BW and CS habitats and six from each dry habitat. From 1994 to 1997, six soil samples were collected from each of the four habitats. In 1995, one BW and two ED samples were lost in transit. In 1994, eight soil samples were collected from a nearby undisturbed hanging garden: three were from under redbud and hackberry canopies thought to be similar to the burned garden dry areas, and five were from under herbaceous hanging garden vegetation that resembled the BW and CS areas of the burned hanging garden.

Each soil sample consisted of a core measuring 5 centimeters in diameter and 15 centimeters deep, except in the BW habitat, where soil depth ranged from 3 centimeters to over 15 centimeters. Depths of BW samples were to bedrock or to 15 centimeters. Samples were collected, placed in plastic zippered storage bags, and kept on ice until shipped to Soil Food Webs (SFW) in Corvallis, Oregon, for analysis. Soil samples were analyzed as described by Coleman and colleagues (1990) and Ingham and colleagues (1989). Soil moisture was measured in each sample by subtracting the weight of dry soil from wet soil weight; it is reported as the weight of water (in grams) that would be associated with one gram of dry soil. Active bacterial biomass and active and total fungal biomass (ABB, AFB, and TFB, respectively) are reported as micrograms of microbe per gram of dry soil. Total nematode abundance (TNA) and trophic composition of the nematode assemblage were determined and are reported as the number of nematodes per gram of dry soil.

Results

Plant Composition Change

Welsh recorded the following KCHG plant species in 1983 (Welsh 1984):

PLUNGE BASIN

Artemisia ludoviciana
Astragalus lentiginosus
Baccharis salicina
Brickellia longifolia
Bromus rubens
Bromus tectorum
Cercis occidentalis
Cirsium rydbergii
Erigeron bellidiastrum
Rhus aromatica
Sporobolus reflexa
Stephanomeria pauciflora

FACE

Adiantum capillus-veneris
Castilleja linariifolia
Cirsium rydbergii
Epipactis giganteus
Lactuca sp.
Mimulus eastwoodiae
Petrophytum caespitosum
Primula specuicola
Solidago sparsiflora

DETRITAL SLOPE

Calamagrostis scopulorum
Cercis occidentalis
Phragmites australis
Quercus gambelii

Only ten (43 percent) plant species of those listed by Welsh were detected at KCHG in 1993 (table 15.1). Welsh probably surveyed the entire site and perhaps concentrated on the lower slope, cliff, and plunge pool, whereas this study was limited to the alcove, accounting for some of the difference between the two studies. The comparison of the two studies did not improve over the five years of this study: only the same ten species found in 1997 were common to both studies.

The number of plant species occurring in each burned area changed little over the five years; however, species composition varied to some extent (table 15.1). Changes in species composition were typically one of three types: (1) ruderal species (*sensu* Grime 1979) colonized habitats early and were absent in subsequent years; (2) species typically found in more arid environments, such as slender wirelettuce (*Stephanomeria tenuifolia*) and Zion milkvetch (*Astragalus* cf. *zionis*) were present in early surveys but were absent by the fifth year; and (3) species generally found in undisturbed hanging gardens or other riparian environments, such as gray sagewort (*Artemisia ludoviciana*) and Rydberg's thistle (*Cirsium rydbergii*), became established.

The mesic BW and CS habitats showed less flux in the number of species present each year, but species composition varied over time (fig. 15.2, table 15.1). Species lost from these habitats were either more-arid-adapted species (e.g., wirelettuce) or ruderal species that may have been outcompeted. KCHG burned in 1989, but my first quantitative sampling did not take place until 1993. Two species found in either the BW or the CS area in 1993 were not seen again during the five-year study; a third species was absent after 1994 (table 15.1). Total noncanopy vegetation cover in the CS remained relatively constant across the five years of study.

Dry habitats showed the greatest change in numbers of species (table 15.1) and in species composition as well (fig. 15.2). Despite these changes, after eight years the dry habitats remained largely unvegetated. Species appearances and disappearances all occurred along the upper and lower boundaries of the dry areas, where moisture was available at the surface at least part of the year. The upper perimeter received some moisture from the BW, and the lower edge received some precipitation that was not blocked by the overhanging cliff. Shrub canopy cover increased dramatically over the study period, with concomitant decreases in open habitat, the preferred habitat of

TABLE 15.1. Species observed in the burned habitats in KCHG, 1993–1997

Species	Backwall 93	94	95	96	97	Central Seep 93	94	95	96	97	East Dry 93	94	95	96	97	West Dry 93	94	95	96	97
Adiantum capillus-veneris	x	x	x	x	x	x		x	x	x										x
Artemisia ludoviciana				x	x			x	x	x	x	x	x	x	x					
Astragalus cf. *zionis*																x				
Brickellia longifolia	x	x	x	x	x	x	x	x	x	x		x	x	x	x	x	x	x	x	x
Bromus rubens	x	x	x	x	x	x	x	x	x	x	x	x	x	x	x	x	x	x	x	x
Carex aurea					x	x	x	x	x	x										
Cercis occidentalis	x	x	x	x	x	x	x	x	x	x	x	x	x	x	x	x	x	x	x	x
Cirsium rydbergii	x	x	x	x	x				x	x	x	x	x	x	x					
Conyza canadensis	x	x	x	x		x		x	x		x	x	x	x					x	
Eriogonum cernuum																x				
Juncus arcticus balticus	x	x	x	x	x	x	x	x	x	x						x	x	x	x	x
Lobelia cardinalis						x	x	x	x	x										
Mentzelia albicaulis	x	x	x	x	x						x	x	x	x						
Mimulus eastwoodiae				x	x															
Muhlenbergia andina						x														
Panicum acuminatum	x	x	x	x	x	x	x	x	x	x	x	x	x	x	x	x	x	x		
Phragmites australis											x	x	x	x	x					
Polypogon monspeliensis	x	x				x	x													
Polypogon semiverticillata	x	x	x	x	x	x	x	x	x	x	x	x	x	x	x					
Pseudognaphalium stramineum	x	x	x	x	x	x	x	x	x	x	x	x	x	x	x					
Quercus gambelii	x	x	x	x	x	x	x	x	x	x	x	x	x	x	x	x	x	x	x	x
Solidago sparsiflora	x	x	x	x	x	x	x	x	x	x		x	x	x	x			x	x	
Sonchus asper		x	x	x	x	x	x	x	x	x		x	x	x	x			x		
Stephanomeria tenuifolia	x										x							x		
Tamarix sp.	x	x	x	x	x	x	x	x	x	x	x	x	x	x	x	x	x	x	x	x
Unknown dicot seedling	x	x	x			x	x		x	x										
Total number of species	15	16	15	14	15	17	16	15	17	16	8	13	13	13	10	9	7	9	9	7

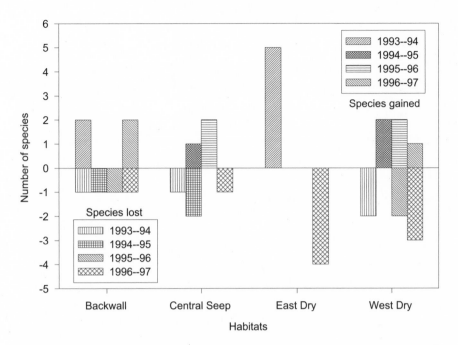

Figure 15.2. Changes in plant species composition in the burned habitats in KCHG, 1993–97.

ruderals such as horseweed (*Conyza canadensis*), cotton cudweed (*Pseudognaphalium stramineum*), and whitestem blazingstar (*Mentzelia albicaulis*).

Total vegetation cover increased fairly steadily across most of KCHG (fig. 15.3); this was true whether shrub canopy (consisting primarily of redbud, brickellbush, and tamarisk) was included or not. Increased cover was derived partly from the establishment of additional species but primarily from the expansion of previously established individuals that had survived the fire. Horseweed and cudweed were absent from the burned areas by 1997 (table 15.1). Horseweed was present in at least one burned area every year from 1993 to 1996; its cover increased from 1993 to 1995, then declined by 1997, from 0.48 percent to 0.20 percent in 1995, dropping to 0 by 1997 (total for all four zones). Cotton cudweed was observed in the CS and BW all five years but did not occur in BW quadrats in 1993 or in CS quadrats in 1996 or 1997. Estimated cover for cudweed showed a pattern similar to that of horseweed, increasing from 0.18 percent in 1993 to 0.54 percent in 1995,

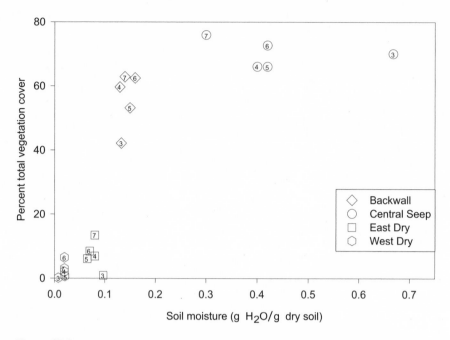

Figure 15.3. Average soil moisture versus total vegetation cover for the four burned habitats at KCHG, 1993–97. Numbers refer to the last digit of the year in which data were collected. The numbers pertaining to each habitat are given in a distinctive font.

then declining to 0.09 percent in 1997. Cudweed was present in the upper ED by the time sampling occurred in 1994 but was not detected in the WD (upper perimeter) until 1996. In 1997, cudweed was absent from both dry habitats owing to unknown physical or biological changes.

Shrub cover increased dramatically over the five years across KCHG (fig. 15.4). The reemergence of shrubs after the fire was almost exclusively from stump resprouting, indicating that KCHG was shrub-dominated prior to the fire, with roughly the same dispersion pattern. Whether much of the dry areas had been vegetated primarily by shrubs or whether herbaceous hanging garden species had occupied some or all of that area is not known. The large chunks of alcove roof that spalled off during the fire covered any shrub root crowns that might have existed in those areas, preventing re-sprouting even if they had survived the fire.

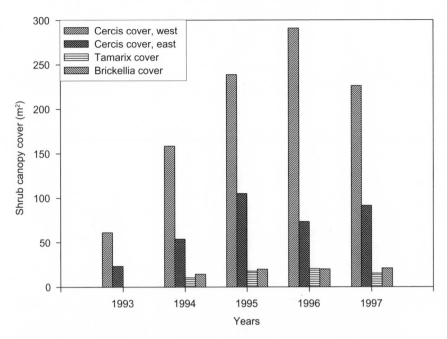

Figure 15.4. The canopy cover of the three major shrubs—*Cercis occidentalis*, *Brickellia longifolia*, and *Tamarix* sp.—across the entire burned upper terrace at KCHG, 1993–97. The amount of cover of *C. occidentalis* was compiled separately for the east and west halves of the garden.

Soil Changes

No consistent relationship was detected between soil moisture and changes in vegetation cover in any habitat, although, as expected, higher soil moisture was related to greater total vegetation cover (fig. 15.3). Soil moisture estimates remained constant in three of the habitats, yet total vegetation cover increased fairly steadily over time (fig. 15.3). In contrast, CS soil moisture content varied interannually while vegetation cover remained fairly constant. Some dry habitat samples were moist at a depth of about 10–15 centimeters owing to subsurface seepage from the wet BW. This moisture was apparently unavailable for plant germination, as these areas remained almost completely barren for eight years. Encroachment of vegetation into these dry habitats consisted of tiller and rhizome expansion, probably exploiting the subsurface water.

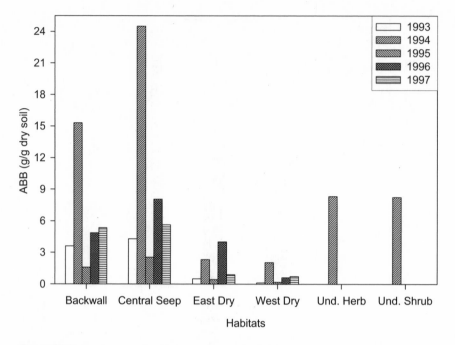

Figure 15.5. The average active biomass of bacteria in soil samples collected from each habitat at KCHG, 1993–97.

Soil Biota

Biomass of soil fungi, bacteria, and nematodes differed significantly among habitats and years. Some but not all components varied from year to year, but not all variables changed in the same ways. For all soil biota measured, the variation between samples within a habitat was large, with one standard deviation often matching or exceeding the mean (see appendix 15.1, figs. 15.5 and 15.6).

Active fungal biomass (AFB) oscillated in all habitats between years (see appendix 15.1). The magnitude of increases and decreases and the absolute mass of fungi differed greatly between habitats, but the pattern was consistent throughout KCHG: CS soils had high levels of AFB, while the other three habitats had little, except the BW in 1994 (see appendix 15.1). Average AFB estimates from the undisturbed site were similar for the herb and shrub communities. Undisturbed soil AFB was much greater than most samples from the burned areas, except on the CS and in 1994 on the BW.

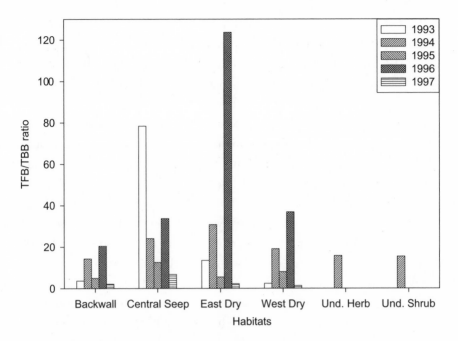

Figure 15.6. The average ratios of total fungal biomass to total bacterial biomass in soil samples from the four burned habitats in KCHG, 1993–97.

Total fungal biomass (TFB) varied in roughly the same way as AFB across the four habitats over the five years of study (see appendix 15.1). There was little change in the mesic samples between 1995 and 1996; both showed slight decreases, as did the WD that year. The ED samples indicated a significant increase in TFB that year. Samples from the undisturbed site in 1994 had TFB levels similar to those of the BW and CS.

Active bacterial biomass (ABB) varied greatly among the four habitats over the study (fig. 15.5). ABB increased from 1993 to 1994, decreased from 1994 to 1995, and increased again the following year in all habitats. From 1996 to 1997, there was little change in BW and WD soils, but CS and ED soils showed marked decreases. Average ABB values in the undisturbed areas were much higher than average ABB values for any burned habitat in any year, except in 1994 on the BW and CS.

Total bacterial biomass (TBB) changed consistently from year to year in the four habitats (see appendix 15.1). In the first four years, TBB oscillated from year to year, increasing and decreasing in all habitats. Large (at least threefold) increases in TBB occurred in all burned habitats from 1996 to 1997, much larger changes than between other years (see appendix 15.1). ABB and TBB thus changed in the same manner during the first four years but diverged significantly in 1996 and 1997. TBB in the undisturbed site was within the range of values found in burned habitats.

The ratio of active fungal biomass to active bacterial biomass (AFB: ABB) varied dramatically among habitats during the five years (see appendix 15.1). The BW went from a highly bacteria-dominated system in 1993 to roughly equal fungal and bacterial biomass over the next three years. In 1997, active bacteria again dominated the soil biota (see appendix 15.1). CS soils were dominated by AFB all five years, although there was considerable variation among years (see appendix 15.1). In the ED soils, AFB:ABB declined over the five-year period, with ratios dropping from 4.19 in 1993 to 0.122 in 1997, although AFB increased dramatically from 1995 to 1996. The WD area showed a pattern similar to that of the ED, but the ratios were all less than 1.0, indicating dominance by bacteria during all five years. The ratios of fungal to bacterial biomass from control sites were higher than those of most of the burned site samples, suggesting greater fungal dominance in undisturbed settings. Except for the 1993 ED ratio, only the CS ratios were greater than undisturbed habitat ratios.

All four habitats (except the CS from 1993 to 1994) showed the same pattern of change in TFB:TBB from year to year (see appendix 15.1). In the CS, this ratio declined from 1993 to 1994 but increased in the other habitats. Ratios oscillated over the next four years. Other than the CS 1993 ratio and the 1996 ED ratio, the magnitude of all TFB:TBB ratios remained similar for each of the years. The ratios of total fungal to bacterial biomass from the undisturbed sites in 1994 were in the same range as most of the burned habitats' ratios over the course of the study (fig. 15.6).

Total nematode abundance (TNA) was generally low, except in the BW and CS in 1993 and the BW in 1997 (see appendix 15.1). Dry habitats in particular had low TNA all five years. However, relative to 1996, the BW and WD showed large increases in 1997, and TNA almost doubled in the CS. TNA decreased significantly in the ED from 1996 to 1997.

Discussion

Vegetation Changes

Postfire vegetation patterns of this hanging garden appeared to be driven by two factors in 1992: surviving rootstocks and soil moisture. Vegetation was limited to herbaceous species growing in the wettest areas, sprouting rootstocks of shrubs (primarily redbud and Gambel's oak), and graminoids, such as common reed and Baltic rush. Herbaceous species covered approximately 10 percent of the alcove, but herb cover reached 100 percent in some patches. Desert annual plants occurred around the periphery of KCHG, including nodding buckwheat (*Eriogonum cernuum*), foxtail brome (*Bromus rubens*), and cheatgrass (*Bromus tectorum*). Shrub cover increased steadily from 1993 to 1996 (fig. 15.4), especially for redbud on the west side of KCHG, then declined in 1997. Shrubs regrew vigorously: some were over two meters tall after eight years. Dry, rocky soil, ash, and boulders from the ceiling covered the rest of the alcove.

Establishment of individual plant species and the development of vegetation structure have been driven by access to water at the soil surface and the survival of rootstocks after the 1989 fire. All aboveground plant material was burned by the fire (NPS files–1989 video), yet by March 1990, redbud, Gambel's oak, brickellbush, and tamarisk had begun sprouting from rootstocks and the CS had been colonized by herbaceous plants (NPS files). A few plants were also growing on the BW by 1990. Vegetation in the dry zones was limited to surviving rootstocks of the shrubs, Baltic rush in the WD and common reed in the ED. Postfire shrub abundance, especially of redbud, indicates that the prefire plant assemblage in the alcove consisted primarily of shrubs, with little representation of typical hanging garden species. However, the pattern of recruitment and establishment in the BW and CS habitats indicates that some hanging garden species were present at least in some of those areas prior to the fire. Those species quickly recolonized areas with high surface moisture availability, and little root-sprouting occurred in those portions of KCHG.

Changes in species number and composition in the dry areas were probably related primarily to fluctuations in the amount of water flowing out of the rock and precipitation patterns from year to year. Species-to-species interactions may have influenced the vegetation structure of some parts of KCHG, even prior to commencement of sampling in 1993. Species compo- sition changes, either displacement or colonization, were detected during this study. Nonetheless, interactions among plant species apparently have been of minor importance in the postfire community structure. Foxtail muhly (*Muhlenbergia andina*), a perennial grass, was present in only one small area in 1993, in the midst of a dense stand of common maidenhair fern (*Adiantum capillus-veneris*) and golden sedge (*Carex aurea*). By 1994, it had apparently been crowded out. The introduced annual rabbitfoot grass (*Polypogon mon- speliensis*) occurred in the sparser parts of the CS and BW in 1993 and 1994. This is a common, non-native wetland species on the Colorado Plateau. It is an early colonizer but appears to be a poor competitor, found primarily in the annual flooding zone where potential competitors may be precluded. By 1995, rabbitfoot grass was completely absent; however, bentgrass (*Polypogon semiverticillata*), a common non-native perennial stoloniferous species, ac- counted for most of the grass cover in the CS and BW from 1995 to 1997. In the ED, a single whitestem blazingstar (*Mentzelia albicaulis*) occurred at the edge of the canopy of a tamarisk from 1993 to 1996; it was absent in 1997. Over the five-year study, tamarisk canopy increased, encompassing the blaz- ingstar habitat, and apparently shaded it out.

Throughout KCHG, the majority of changes in colonizing vascular plant species involved ruderal species (Grime 1979). Some ruderal species, notably horseweed and cudweed, decreased or disappeared during the study. The decline in abundance of horseweed and cudweed in the mesic envi- ronments may be attributed to competition with other species, either for light (shading) or for soil water and nutrients. Some wetland and riparian species colonized KCHG and may contribute to the future recovered hang- ing garden composition. Species such as Rydberg's thistle, gray sagewort, and maidenhair fern colonized and expanded in percent cover during the five years of study. However, the greatest and most persistent cover increases came from shrubs. Total shrub cover of redbud, brickellbush, and tama- risk increased from 0 percent in July 1989 to about 50 percent (410 square meters) in 1996. The trunks of some redbud plants were larger than 10 centi- meters in diameter by 1997 (T. B. Graham, unpubl. data), having grown at a rate of more than 1.0 centimeters per year.

Soil Recovery

Lower microbial biomass and respiration are typical of soils after a fire (Raison and McGarity 1980; Fritze et al. 1993; Pietikäinen and Fritze 1993; Fritze et al. 1994). However, characteristics of an individual fire—such as how hot it burned, the amount of ash it produced, and the degree of soil carbon vaporization—also drive postfire soil recovery (Raison 1979; Fritze et al. 1994). Postfire monitoring at KCHG during this five-year study revealed limited soil biota recovery. There are no comparable data from other, undisturbed hanging gardens, except the 1994 control (unburned) data presented here. Because intersite hanging garden plant community structure is highly variable (Nebeker et al. 1977; Welsh and Toft 1981), even this soil comparison may not provide a clear indication of soil biota conditions or recovery.

Soil Biota Dynamics

Fungi and bacteria play numerous, interacting, but poorly understood roles in root-zone dynamics in southwestern ecosystems. For example, the ratio of both total and active fungal to bacterial biomass may indicate the successional status of the ecosystem (Sakamoto and Oba 1994). The roles of fungi and bacteria in hanging gardens are virtually unstudied and initially should be considered in relation to those in surrounding landscapes. Belnap (1995) reported AFB that ranged from 0 to 0.59 ± 0.67 micrograms per gram of dry soil for undisturbed and disturbed blackbrush (*Coleogyne ramosissima*) and pinyon pine-juniper (*Pinus-Juniperus* spp.) soils in the region. She also reported that average ABB ranged from 0.12 ± 0.14 (1 sd) to 2.3 ± 1.0 micrograms per gram of dry soil, and 3.4 ± 0.6 to 14.2 ± 4.8 micrograms per gram of dry soil in undisturbed and disturbed blackbrush and pinyon pine-juniper communities, respectively (Belnap 1995). Active and total fungal to bacterial biomass ratios of 0.13 and 4.0 respectively were recorded for an undisturbed blackbrush community, and 0.33 and 2.5 for AFB:ABB and TFB:TBB respectively in disturbed blackbrush in Arches National Park, Utah (Belnap 1995). Undisturbed pinyon pine-juniper stands at Natural Bridges National Monument, Utah, had AFB:ABB of 0.74, while disturbed stand ratios approached 0.0. TFB:TBB in those same communities were 7.5 and 3.1, respectively (Belnap 1995). Soil samples from perennial grasslands on the Colorado Plateau had AFB:ABB ranging from 0.2 to 0.54, and TFB: TBB ranged from 0.1 to 0.28 (J. Belnap, unpubl. data). In riparian forest and pasture soils in Oregon, fungal biomass was greater than bacterial biomass throughout the year (Griffiths et al. 1997).

In addition, fungi may influence soil ion balance, including calcium retention (Killham 1995; Tate 1995; Paul and Clark 1996). Calcium retention may not be as critical in Colorado Plateau hanging gardens on sandstone cemented with calcium carbonate ($CaCO_3$), which often receive carbonate-rich water. Levels of Ca^{+2} from three hanging gardens emerging from the Navajo Sandstone in Canyonlands National Park ranged from 26 milligrams per liter to 93 milligrams per liter (Long and Smith 1996).

Adequate protection from plant diseases may accrue if AFB and TFB exceed 25 micrograms per gram of dry soil (E. R. Ingham, pers. comm.); however, the role of mycorrhizal associations among springs plants remains poorly known. In CS soils, AFB exceeded that threshold for three of five years, while in BW and undisturbed garden soils AFB approached 25 micrograms per gram of dry soil (see appendix 15.1). If hanging gardens are similar to other plant communities, and if the undisturbed garden soils were typical of those on Navajo Sandstone elsewhere, then soil fungi may poorly protect hanging garden plants, and plant populations able to recolonize the burned habitats may be at risk from root diseases. In dry zones, AFB did not exceed 7 micrograms per gram of dry soil and in fact rarely exceeded 1 microgram per gram of dry soil (see appendix 15.1).

During most years, all burned habitats exceeded the 25 microgram per gram of dry soil threshold for TFB (see appendix 15.1), although the dry areas fluctuated around that value. TFB in the undisturbed soils was greater than that in the burned dry areas or among BW samples, except for the 1994 BW and ED soils (see appendix 15.1). The soils of the CS had much higher TFB values than did those of either undisturbed site. In the BW, as in the CS zone, TFB was high in 1994 and dropped in 1995, remaining low through 1997. Therefore, vascular plants may be fairly well protected in the burned areas if TFB better reflects the quantity of fungi in the soils.

Bacteria abundance and composition in soils are critical determinants of nutrient dynamics, such as nitrate, ammonium, phosphate, and sulfate ions (Blair et al. 1990; Tate 1995; Paul and Clark 1996). In soils derived from quartz sandstone and cemented with $CaCO_3$ (e.g., the Navajo Sandstone), nutrient availability may be low, and bacterial influences on nutrients may be critical to sustaining a viable plant community. TBB exceeding 100 micrograms per gram of dry soil is generally considered to be a threshold for healthy soil in other systems (E. R. Ingham, pers. comm.). At those levels of bacterial biomass, nutrient retention is effective and bacterial suppression of plant diseases is strong. With a TBB between 50 and 100 micrograms per gram of dry soil, soils are considered to be healthy and support vascular plant growth; below 50 micrograms per gram of dry soil, nutrient retention is lost and plant health may decline. However, almost nothing is known about the nutrient dynamics of hanging garden soils or of nutrient use by hanging garden plant species. These questions lay beyond the scope of this study, but additional research is warranted.

No soil samples from the burned or undisturbed hanging gardens approached the TBB threshold value of 100 micrograms per gram of dry soil. Only in 1994 did the CS samples have TBB exceeding 50 micrograms per gram of dry soil, considered to be the minimal value for nutrient retention. Hanging garden vegetation may operate under different conditions and possibly under different "rules" than other ecosystems.

In the BW, TNA was relatively high, indicating some recovery in parts of the system following the fire. Root-feeding nematodes were rare in all samples in all years except in the CS from 1993 to 1995. Vegetation cover did not change appreciably in the CS from 1995 to 1997, but root-feeding nematodes declined in the CS after 1995. Although the abundance of predatory nematodes varied interannually, it was fairly high in CS soils in 1993 and 1997. Predatory nematodes are typically sensitive to environmental perturbations to soils and may be among the first taxa to be lost from a soil food web (Blair et al. 1990; Beare et al. 1992; Todd 1996). Other patterns between nematode guilds and potential food organisms were not clear in this study. Fungivore nematode abundance was never high in BW soils, but it increased fivefold from 1996 to 1997. No relationship was apparent between predator and prey in that case, although a lagged response may have existed. Bacteriovores in the BW apparently tracked TBB, but their numbers did not exhibit the lag shown in CS soils. The burned dry areas had low TNA, often accompanied by low bacterial and fungal biomass. These data constitute some of the first collected from hanging gardens, and causal relationships between nematode abundance and microbial biomass changes remain uncertain.

Soil Biota–Vegetation Interactions

Soil biota may influence vascular plant community structure and succession, and plants can quickly respond to changes in soil food webs (Gibson et al. 1993; Tate 1995; Paul and Clark 1996; Blair et al. 1998). Fungi can protect plant roots from disease and help retain calcium (Tate 1995). Bacteria contain nutrients, such as nitrate, ammonium, phosphate, and sulfur, and also may suppress diseases (DeAngelis 1992; Paul and Clark 1996). The ratio of fungal to bacterial biomass also influences the success of various plant functional types (Hunt et al. 1988; Coleman et al. 1990; DeAngelis 1992; Garcia 1992; Garcia and Rice 1994; Paul and Clark 1996; Edwards 2000). TFB:TBB values below 0.5 are typical of ruderal communities, and many annual plant species grow well under those conditions (E. R. Ingham, pers. comm.). As fungi become more abundant and TFB:TBB approaches 1.0, some perennial grasses, especially tillering species such as Bermuda grass (*Cynodon dactylon*), are favored. Ratios of 1 to 2 support fast-growing, taller grasses such as fescues (*Festuca* spp.) and other bunch grasses; corn (*Zea mays*) also does well under such conditions. Longer-lived perennial species including shrubs do well when ratios exceed 2. TFB:TBB values exceeding 10 tend to support deciduous shrub communities.

Changes in fungal and bacterial biomass that may influence plant nutrition or disease resistance have been shown to be mediated by nematode grazing (Blair et al. 1990; Beare et al. 1992; Paul and Clark 1996). Separating nematodes into feeding guilds provided some additional insight into these interactions. TFB was quite high in the CS in 1993 and 1994 but in 1995 declined to roughly half those values, where it stayed through 1997. Fungal-feeding nematodes increased dramatically in 1994 compared to 1993 in CS soils; they increased again in 1995, then dropped to much lower levels in 1996 and 1997. These data may represent a numerical response of fungivores to the increased fungal biomass present in CS soils in 1993 and 1994, after which point fungivores declined in numbers as their food base was consumed. The same pattern was apparent with CS total bacterial biomass and bacteria-feeding nematode numbers, except that bacteria increased again after dropping to low levels in 1995, reaching their highest levels in 1997. Bacteriovores declined in 1996, increasing somewhat again in 1997, probably in response to increased bacterial biomass again that year. Bacteria may respond more quickly to a reduction in predation pressure, multiplying rapidly as predation decreases and nutrients are made available in the form of excretory products from the bacteriovores (Douce and Webb 1978; Clarholm 1981; Ingham and Horton 1987; Coleman et al. 1990; Beare et al. 1992; Tate 1995). So little is known about hanging garden soil biota that placing these results in context relative to other sites is difficult. However, it is instructive to compare these data to soil biota data from other plant communities, especially those that have been studied both before and after fires.

The results presented here indicate that Colorado Plateau hanging garden soils may be more dominated by fungi than are upland soils, and in this way they appear to be relatively similar to riparian soils in the Pacific Northwest. Fungal biomass generally increased in KCHG from 1993 to 1996 but decreased dramatically by 1997. Shrub cover, both that of individual species and total shrub canopy cover, was much lower in 1997 than in 1996. Shrubs may respond to stresses associated with the decline in fungal biomass present in soils of the hanging garden. The highest TFB:TBB values occurred in 1996 in all zones except the CS (which had almost no shrubs), and ratios dropped significantly in all four zones in 1997. These differences may indicate that the soil biota and at least the shrub component of the burned hanging garden vegetation were relinking, as might be expected in ecosystems recovering from disturbance. However, other than the correlation between shrub cover and TFB:TBB values, there is little indication that the soil biota has influenced the colonization or establishment of vascular plants in the burned KCHG.

Conclusions

Hanging Garden Recovery from Fire

The low similarity between Welsh's (1984) and my plant species lists, the slow and erratic recovery of KCHG vegetation, and the lack of colonization in much of the dry habitat suggest that fire is probably not a natural form of disturbance in hanging gardens. Fire may strongly and anomalously affect a hanging garden for a long period, and the trajectory of recovery from such a novel, severe disturbance is not necessarily predictable. Because the fire burned so hotly that rock slabs spalled off the alcove ceiling, the fire may have altered the water flow through the rock, affecting the quantity of water flowing along the BW or dripping off the ceiling. Thus, KCHG may have been ecologically restructured by the fire and as a result may now represent a different habitat, with different species better able to colonize and occupy this area than previously. The generalization of the observations at Knowles Canyon to fire disturbance at other hanging gardens requires additional comparative study.

Restoration and Management

Other forms of disturbances may lead to the total denudation of sections of hanging gardens (e.g., trampling by visitors trying to view archaeological sites). The available substrate for reestablishment may not be too different in these cases if the developed soil layers erode. Consequently, revegetation methods are needed for recovery of burned hanging garden soils and vegetation, particularly in dry habitats. Soil and soil community development may be necessary for plant establishment and succession to occur. Addition of soil amendments, such as compost or compost tea, and inoculation with soil from nearby healthy hanging gardens may increase the development of the soil fungal component and aid in recolonization. Applying compost tea also may increase surface moisture and foster germination. Because there is moisture at 10–15 centimeters below the soil surface, at least in some parts of the KCHG, the transplanting of rhizomatous or stoloniferous species to expand cover and stabilize the dry areas may be possible.

Recreational activity at Glen Canyon National Recreation Area is dispersed along the vast and highly convoluted shoreline of Lake Powell. It is impossible to monitor visitors or enforce regulations everywhere, and many illegal activities such as shooting fireworks commonly occur. Wildfires caused by visitors are a continuous threat to the natural and cultural resources of Glen Canyon National Recreation Area. Other visitor activities (e.g., trampling) can also seriously impact hanging gardens. The Grand Canyon Wildlands Council (2002) has recommended that land managers establish trails at all springs that are regularly used for recreation or for administrative purposes, as these habitats are highly susceptible to erosion and degradation caused by administrative or recreational visitation.

Monitoring

The KCHG fire provided an opportunity to learn more about hanging garden vegetation and soil biota, as well as how associated plant species respond to this severe, and probably novel, form of disturbance; however, more questions have been generated than answered with this study, a reflection of how little is known about hanging garden ecosystems. Periodic monitoring of KCHG should be continued to track longer-term changes in species composition and to document colonization and expansion. Collecting and analyzing soil samples at regular intervals at the site and from nearby undisturbed hanging gardens will help determine the recovery of hanging garden soils and biota, whether long-term responses to fire occur, and relations to other ecosystems (Ingham et al. 1986, 1989; Garcia 1992; Garcia and Rice 1994; Rice et al. 1998). Obviously, much remains to be learned about hanging garden ecosystems on the Colorado Plateau and elsewhere.

ACKNOWLEDGMENTS

I thank John Spence for his assistance with plant identification and data collection and for his comments on the manuscript. A number of the resource management personnel of Glen Canyon National Recreation Área helped with fieldwork over the years, as did Ken Brown while working on his book. Elaine Ingham was invaluable in helping me understand what the soil biota data meant. Larry Stevens, Jayne Belnap, Elaine Ingham, and three anonymous reviewers commented on, and improved, the manuscript.

Appendix 15.1

SUMMARY OF SOIL COMMUNITY PARAMETERS IN EACH HABITAT

Habitat	Soil H_2O	Active fungal biomass	Total fungal biomass	Active bacterial biomass
Backwall 1993	0.13 ± 0.07 (0.06–0.24)	1.47 ± 2.033 (0–0.67)	31.62 ± 11.19 (17.86–46.54)	3.58 ± 1.87 (1.84–6.42)
Central Seep 1993	0.67 ± .21 (0.48–1.02)	31.66 ± 20.34 (12.32–64.86)	523.6 ± 622.15 (132.4–1592)	4.26 ± 2.15 (1.91–6.52)
East Dry 1993	0.07 ± 0.05 (0.02–0.16)	3.21 ± 5.33 (0–14.61)	30.42 ± 43.74 (0–112.2)	0.50 ± 0.89 (0–2.47)
West Dry 1993	0.03 ± 0.03 (0–0.09)	0.15 ± 0.28 (0–0.76)	1.56 ± 1.15 (0–2.63)	0.13 ± 0.16 (0–0.39)
Backwall 1994	0.13 ± 0.05 (0.06–0.19)	25.45 ± 20.13 (0.36–45.4)	316.9 ± 232.89 (71.44–685.9)	15.26 ± 5.75 (5.45–19.96)
Central Seep 1994	0.4 ± 0.16 (0.3–0.72)	85.78 ± 64.54 (1.2–198.25)	762.0 ± 184.87 (574.4–1080)	24.49 ± 12.71 (7.91–39.61)
East Dry 1994	0.08 ± 0.09 (0–0.28)	3.92 ± 4.19 (0–12.51)	186.78 ± 173.04 (83.1–564.8)	2.32 ± 1.3 (0.85–4.36)
West Dry 1994	0.02 ± 0.01 (0.01–0.04)	0.92 ± 0.96 (0–2.81)	116.18 ± 53.46 (64.15–230.6)	2.05 ± 2.3 (0–7.08)
Undisturbed herb 1994	0.21 ± 0.04 (0.15–0.27)	23.98 ± 13.38 (0–37.24)	206.95 ± 104.22 (39.39–325.1)	8.33 ± 4.96 (0–15.36)
Undisturbed shrub 1994	0.07 ± 0.08 (0.01–0.18)	23.81 ± 16.82 (5.03–45.84)	171.88 ± 65.12 (115.1–263.1)	8.25 ± 1.08 (6.75–9.22)
Backwall 1995	0.15 ± 0.07 (0.08–0.28)	2.58 ± 2.58 (0.32–7.51)	142.4 ± 149.93 (45.2–439.9)	1.59 ± 0.61 (1.09–2.73)
Central Seep 1995	0.42 ± 0.23 (0.12–0.82)	19.36 ± 28.12 (0.17–81.03)	336.0 ± 320.51 (79.34–919.9)	2.54 ± 1.69 (0.4–5.32)
East Dry 1995	0.07 ± 0.07 (0.01–0.18)	0.15 ± 0.26 (0–0.59)	57.503 ± 11.1 (44.88–74.68)	0.45 ± 0.67 (0–1.6)
West Dry 1995	0.02 ± 0.01 (0.01–0.04)	0 (0)	62.38 ± 20.22 (43.73–95.56)	0.19 ± 0.43 (0–1.16)
Backwall 1996	0.17 ± 0.11 (0.08–0.37)	5.88 ± 3.32 (0–9.52)	139.38 ± 175.18 (23.4–526.3)	4.87 ± 2.45 (2–8.4)
Central Seep 1996	0.42 ± 0.17 (0.11–0.63)	88.94 ± 91.45 (1.85–265.24)	330.6 ± 140.05 (102.6–500.8)	8.05 ± 6.04 (2.3–19.6)
East Dry 1996	0.07 ± 0.11 (0.01–0.07)	6.57 ± 7.12 (0–18.09)	86.66 ± 69.3 (19.57–230.2)	4.0 ± 5.11 (0.1–14.3)
West Dry 1996	0.02 ± 0.02 (0–0.07)	0.50 ± 0.62 (0–1.53)	52.35 ± 21.07 (18.34–78.23)	0.62 ± 1.04 (0–2.9)

Total bacterial biomass	Active fungal: bacterial ratio	Total fungal: bacterial ratio	Total # nematodes
11.91 ± 8.34	0.36 ± 0.49	3.64 ± 1.92	12.38 ± 9.29
(4.16–25.83)	(0–1.21)	(2.34–6.45)	(0–22.4)
6.86 ± 0.52	7.45 ± 2.33	78.38 ± 93.07	34.2 ± 53.22
(6.47–7.75)	(4.95–9.95)	(9.58–237.97)	(0–126.3)
1.82 ± 0.92	4.19 ± 5.01	13.48 ± 20.32	2.17 ± 2.87
(0.23–3.16)	(0–13.9)	(0.57–55.32)	(0–7.7)
0.72 ± 0.22	0.46 ± 0.87	2.36 ± 1.95	0.33 ± 0.75
(0.45–1.08)	(0–2.38)	(0–5.72)	(0–2)
21.84 ± 7.66	1.36 ± 1.02	14.29 ± 9.88	2.10 ± 0.41
(9.25–29.16)	(0.07–2.44)	(5.997–30.9)	(1.6–2.63)
33.03 ± 8.44	3.14 ± 1.64	24.11 ± 5.89	7.00 ± 7.6
(21.17–47.6)	(0.15–5.0)	(12.32–29.07)	(0–19.62)
7.0 ± 2.25	1.51 ± 1.15	30.8 ± 32.57	0.15 ± 0.18
(5.17–11.37)	(0–2.87)	(7.49–101.8)	(0–0.51)
8.3 ± 3.91	0.45 ± 0.42	19.03 ± 12.56	0.22 ± 0.45
(2.73–14.16)	(0–1.31)	(4.53–39.14)	(0–1.21)
13.88 ± 3.53	2.60 ± 1.79	15.75 ± 10.33	0.91 ± 0.73
(9.43–19.43)	(0–5.15)	(3.6–15.63)	(0.04–1.7)
11.48 ± 0.92	3.23 ±2.62	15.36 ± 6.79	1.04 ± 0.45
(10.54–12.72)	(0.55–2.75)	(10.3–24.96)	(0.56–1.64)
30.35 ± 7.8	1.34 ± 0.87	4.93 ± 5.15	2.10 ± 1.79
(24.65–45.58)	(0.3–2.75)	(1.23–15.096)	(0.52–5.24)
25.37 ± 7.1	5.34 ± 5.36	12.5 ± 11.14	6.04 ± 4.71
(18.75–38.91)	(0.23–8.89)	(2.54–31.61)	(0.29–12.18)
12.53 ± 4.25	0.09 ± 0.16	5.44 ± 2.58	0.27 ± 0.44
(7.92–17.63)	(0–0.37)	(2.55–8.53)	(0–1.03)
9.17 ± 2.33	—[a]	8.02 ± 4.98	0.18 ± 0.22
(5.96–11.92)	—[a]	(4.38–15.77)	(0–0.65)
8.32 ± 2.45	1.47 ± 1.39	20.44 ± 28.11	1.43 ± 1.41
(5.5–12)	(0–4.3)	(3.44–82.23)	(0.1–3.5)
13.28 ± 11.47	8.72 ± 5.53	33.75 ± 18.61	1.38 ± 1.14
(5.4–38.3)	(0.82–13.7)	(13.08–66.71)	(0.1–3.3)
5.08 ± 5.15	1.13 ± 1.32	123.67 ± 163.1	0.55 ± 0.66
(0.2–15)	(0–2.99)	(4.09–383.7)	(0–1.7)
2.8 ± 2.28	0.73 ± 1.09	36.9 ± 28.7	0.02 ± 0.04
(0.4–7.2)	(0–3.06)	(4.36–94.03)	(0–0.1)

Habitat	Soil H$_2$O	Active fungal biomass	Total fungal biomass	Active bacterial biomass
Backwall 1997	0.13 ± 0.04	3.3 ± 2.61	70.3 ± 19.25	5.4 ± 2.51
	(0.06–0.21)	(0.97–8.31)	(40.41–103.3)	(2.9–9.5)
Central Seep 1997	0.30 ± 0.12	23.5 ± 17.68	270.3 ± 101.4	5.6 ± 2.78
	(0.19–0.54)	(0–44.56)	(178.7–451.5)	(2.3–11.2)
East Dry 1997	0.08 ± 0.15	0.13 ± 0.29	37.6 ± 18.23	0.9 ± 0.28
	(0–0.41)	(0–0.79)	(12.34–59.75)	(0.6–1.4)
West Dry 1997	0.02 ± .02	0.1 ± 0.22	24.4 ± 14.06	0.73 ± 0.34
	(0–0.06)	(0–0.6)	(12.07–51.96)	(0.3–1.2)

Note: Values given as mean ± 1 sd; range below in parentheses.
[a]Only one WD sample in 1995 had active bacteria, thus fungal:bacterial biomass ratios could not be calculated.

Total bacterial biomass	Active fungal: bacterial ratio	Total fungal: bacterial ratio	Total # nematodes
37.1 ± 10.61	0.59 ± 0.33	2.08 ± 0.89	11.8 ± 13.56
(26.6–51.7)	(0.23–1.05)	(1.15–3.42)	(1.0–41.3)
63.6 ± 40.87	3.93 ± 2.53	6.73 ± 5.04	2.6 ± 2.24
(22.9–130.6)	(1.09–8.51)	(1.87–16.3)	(0.3–6.4)
20.5 ± 9.04	0.12 ± 0.27	2.23 ± 1.54	0.18 ± 0.32
(11.0–32.9)	(0–0.73)	(0.54–5.27)	(0–0.9)
18.2 ± 5.73	0.09 ± 0.21	1.36 ± 0.61	1.08 ± 1.55
(12.5–25.5)	(0–0.55)	(0.57–1.88)	(0–3.9)

Chapter 16

Between the Cracks

Water Law and Springs Conservation in Arizona

NANCY NELSON

Neither state nor federal laws regulating the allocation, permissible uses, or quality of water provide real protection for maintaining the ecological health of desert springs. Some protection is afforded to springs located on federal land that has been dedicated to scientific or habitat conservation purposes, particularly those springs designated as critical habitat for species listed as endangered pursuant to the 1973 Endangered Species Act (ESA), as amended. Whether federal or state law applies to the waters of a particular desert spring is determined by whether the spring is located on private or public land. Generally, state law applies to springs located on private or state land, and federal law applies to those located on federal public lands. However, the ESA applies to all lands and intracontinental waters, both public and private; therefore, if particular springs or their associated habitats are designated as critical habitat pursuant to the ESA, that designation will likely place limits on the spring owner's use of the resource. Additionally, the legal classification of springs as either surface water or groundwater largely determines ownership of the waters and the limitations or restrictions on purported rights to use them. For many other examples of groundwater management inconsistencies, the reader is directed to Glennon (2002).

State Law

An Overview

Western law has historically divided bodies of water into surface water and groundwater and has applied different legal principles and rights to each category. Unfortunately, some western water law developed legal principles based on incorrect hydrological science theories (Leshy 1991).

The now-discredited science on which some of our systems of water law is premised assumes a hydrological "bicycle," in which surface waters and groundwater are not connected and removal of water from one source does not affect the other (Leshy 1991). Although Arizona state courts have considered overturning standing legal precedents based on this incorrect science to conform the law to sound hydrology, they have largely refused the opportunity (*Collier v. Arizona Department of Water Resources* 1986). Nonetheless, other states have established systems that analyze groundwater and surface water resources as parts of the same resource, and some apply the same legal doctrines to both surface water and groundwater.

Surface Water

In the western United States, the legal classification of surface waters includes rivers, riverbeds, and sometimes springs that are the headwaters of western rivers. Surface waters in the western United States are generally governed by the legal doctrine of prior appropriation. The doctrine of prior appropriation is most easily described by the adage "First in time, first in right."

Prior appropriation dictates that all surface water belongs to the public until an individual or group of individuals diverts the water from its natural channel for a beneficial use. The diversion of available surface water is an appropriation, which is a vested right belonging to the diverter, and the right to appropriate public water is a right included in all western states' constitutions. Those who first divert previously public water from a particular source have an appropriation right that is senior to the rights of those who later divert water from the same stream. Holders of senior rights are entitled to their appropriation of water in times of drought before the junior users may take their appropriation from the stream, even if the junior user is upstream of the senior user.

The limit and measure of each appropriation right is the amount of water the diverter has put to beneficial use. Water from a river is fully appropriated when beneficial users hold the rights to all of the water in the stream. What constitutes a beneficial use of surface water is generally specified and limited by individual states, and these uses are primarily limited to mining, agricultural (including the maintenance of livestock), and domestic or municipal uses. The category of "beneficial use" generally does not include habitat conservation or recreational use, although some states have enacted laws to allow for those uses. Some states now statutorily define recreation and aesthetics as beneficial uses of surface waters, but these terms are narrowly interpreted and the real applications of the laws are limited by the states' constitutional parameters. Additionally, some states allow for nonconsumptive "use" of surface water (leaving water in the stream) by noting that this use serves the public interest, either economically through recreation or by protecting scenic values, a development known as the instream flows doctrine (*Idaho Department of Parks v. Idaho Department of Water Administration* 1974). Arizona's definition of beneficial uses was statutorily amended to include wildlife purposes, recreation, and artificial groundwater recharge (Ariz. Rev. Stat. Ann. 45-151A). The public-interest instream flows doctrine also has been extended to allow nonconsumptive water rights to be established to serve ecological as well as aesthetic and recreational purposes (Colo. Rev. Stat. § 37-92-102(3)). The current parameters of the instream flow doctrine in each western state, however, are beyond the scope of this work (but see Leshy 1991).

Some western states have specific statutes or separate court systems for appropriating (or decreeing) water rights from surface water sources among users. These same statutes or courts determine the appropriator's ability to change his use of his allocation of the waters from one beneficial use to another or to change the geographic location of his diversion mechanism. After proceeding through the established state process to perfect an appropriation, the appropriator obtains a decreed water right. In Arizona, after going through the required process to establish a right to a surface stream, one receives a permit to appropriate surface waters of the state.

In all western states, if an appropriator does not use the decreed appropriation for a specified period of time, the right to that water is forever forfeited, or "abandoned," by the holder of the appropriation. The origin of this law is the belief that water left in a stream is wasted and that such waste should be punished. This same belief, that in a region of scarce water no one should be able to claim a right to water and not use it, is the concept that generally bars leaving water in a stream for ecological purposes.

An important distinction between the law governing surface waters and that applied to groundwater is that the right to appropriate water from a surface stream is completely detached from the ownership of land adjoining that stream. One may live in a different state and own land hundreds of miles from that adjacent to the water source yet still have a valid water right to appropriate from that source. Many decreed water rights are to surface waters located on federal and state lands, held by those who graze livestock on those same lands or by those who live adjacent to the public lands.

Groundwater

Historically, and in contrast to the doctrines governing surface water, an individual's rights to use groundwater are attached to the ownership of, or other rights to, the land overlying the water source. Another distinction between groundwater and surface water rights is that the measure of an individual's right to use groundwater is measured by the reasonable use doctrine. Additionally, the law limits individual rights to use their groundwater to the land overlying the groundwater.

Some western states have statutory schemes defining the types and restrictions of water uses that pertain to the groundwaters located within those states. In other states, common law—or judge-made law, rather than statutes—applies to govern the use of groundwater resources. Desert springs may be classified as groundwater or surface water, depending on the particular legal doctrines developed by statute or common law by the state in which the springs are located. The distinction is important because some states do not allow for the appropriation of groundwater by an individual, and the laws governing use of the water are different for each category.

The definition of *groundwater* varies from state to state and often is illogical. A state's definition of what constitutes groundwater may be based on how the water is captured. In Nebraska, if one must dig to reach the water, the water is groundwater. This allows that if one digs a well in an island in the middle of a river, the water drawn from the river through the well in the island will be classified as groundwater, although the water is obviously drawn from a surface water source.

Arizona courts still use an arcane concept to define what constitutes groundwater—in Arizona, percolating waters are those waters that are not contained in a defined underground channel (see Ariz. Rev. Stat. § 45-131). All percolating waters are defined by the courts as groundwater and are not appropriable (see *Collier v. Arizona Department of Water Resources* 1986). Generally, in Arizona, evidence that water is contained in a defined underground channel is established by evidence indicating a line of vegetation growing above the surface of, and obviously watered by, the defined underground channel. However, this evidence alone has not always been enough to establish that one's water was removed from a defined underground channel. The Arizona courts have been increasingly willing to acknowledge the very real hydrologic connection between percolating water and surface water. More recently, if one can prove to the court that the water in question is subflow (although the legal test for this category of water remains uncertain), the water is classified as surface water. If there is insufficient evidence that the water in question is subflow, the water is classified as groundwater. The Arizona statute that provides for the adjudication of claims and priorities concerning the waters of the state river system defines surface water as including "springs on the surface." The statute also defines groundwater as "water under the surface of the earth regardless of the geologic structure in which it is standing or moving."

The amount of water that one is entitled to use from a groundwater source also varies from state to state. Some states, including Texas, have legal doctrines granting "absolute rights" to groundwater, giving the owner of land over a groundwater source the right to pump as much water from that source as the landowner wishes—short of the landowner's "malicious damage" of another's property or "willful waste" of any water (*Houston & T.C. Ry. Co. v. East* 1904). Some states limit groundwater use by the doctrine of correlative rights, dictating that the amount of groundwater to which one is entitled is determined by the amount of the overlying land one owns. California laws are a mixture of concepts governing groundwater law—the doctrine of correlative rights applies if one uses the groundwater on land abutting the water source, but prior appropriation doctrines govern the use if the holder of water rights wants to transport the water off the appurtenant land for remote use. Some states have, pursuant to statute, adopted a system of groundwater law that uses the doctrine of prior appropriation, as evidenced by a permit system governing use of the water. Some states (e.g., Colorado and New Mexico) also adjudicate groundwater rights in the same judicial system as their surface waters. In Utah, there is virtually no law governing the use of groundwater, except for programs designed to recharge groundwater sources. Some states limit the use of groundwater by the reasonable use doctrine. "Reasonable use" as the measure of a groundwater right is usually an unfettered right to any amount of water, and this category of uses is broader than beneficial use.

Rights to use groundwater are also subject to forfeiture for nonuse. If one does not exercise the right to use the water, one loses all rights to that water. Rights to groundwater are usufructuary—one owns the right to use the water, not the water itself. In Arizona, water users cannot own a right to groundwater, as they can own a decreed right to surface water; one can, however, capture groundwater and put it to reasonable use. Arizonans therefore have a right only to the use of the groundwater, not to the water itself. (Arizona has amended its statutes so that now groundwater must be put to "reasonable and beneficial" use before a right to the use of the water attaches, and the use must ostensibly be on the land from which it was pumped.)

Thus, if one owns land above or surrounding desert springs that are classified by the state as groundwater, one has a right to use an amount of water from the springs equal to that which one could reasonably use on land abutting the springs. Nonetheless, it is questionable whether a state court would find that cause of action existed if another user dewatered this spring, if that use of the water was reasonable. It is also questionable whether the category of reasonable use would encompass leaving water in the spring to provide for fish and wildlife habitat.

Additionally, in a hydrologic-related system, the dewatering of one's springs might occur as a result of a surface water diversion. Although Arizona courts have increasingly become willing to see a connection between groundwater and surface water, the Arizona legislature has enacted statutes that allow for the adjudication of rights to all waters, including surface water and groundwater, within a particular watershed or groundwater active management area, as that term is defined by Arizona statutes.

The Arizona Groundwater Management Act

Arizona's Groundwater Management Act (GMA) of 1980 was amended in 1994. The GMA established Active Management Areas (AMAs) and Irrigation Non-expansion Areas (INEAs) within the state. These areas cover the southern two-thirds of the state, encompass the portions of Arizona containing 80 percent of Arizona's population, and are coterminous with the areas of the state responsible for the vast majority of the groundwater overdraft occurring in the state. It is important to note that the restrictions and rules of the GMA only apply to AMAs and INEAs. Groundwater in all other parts of the state is still governed by the reasonable use doctrine, requiring that it be put to "reasonable and beneficial uses" (Ariz. Rev. Stat. § 45-453; *Bristor v. Cheatham* 1953).

One of the primary purposes of Arizona's GMA was to eliminate the overdraft of the state's most important aquifers by the year 2025. The GMA also was aimed at preventing any increase in the use of state waters for irrigation, which was historically a large percentage of the state's use. Another purpose of the GMA was to ensure that the state would have an assured future water supply for municipal and industrial uses (the "assured water supply" provisions). Some experts believe that the GMA acts or will act, in effect, as a growth management act by requiring residential subdivision developers to provide for an assured water supply for their development. Eventually the GMA will limit development because water supplies will become less available and more expensive. Additionally, the GMA limits the uses to which groundwater in AMAs and INEAs may be put; these limitations include conservation measures, including provisions by which water rights may be purchased and retired. The GMA also imposed pump taxes on private wells and required the registration and permitting of private wells. The GMA provides for grandfathered water rights and recognizes rights in the existing pumping of groundwater while essentially eliminating any new pumping in AMAs or INEAs.

The GMA is a set of statutes moving Arizona law toward the integration of groundwater and surface water as sources of water supply, while simultaneously encouraging the integration of these sources as a legal matter. Various provisions of the GMA acknowledge that groundwater and surface water supplies are related and need to be allocated within the same legal system. The GMA includes a statement that it does not impact "decreed and appropriative water rights," implying that surface water rights are to be managed separately from the groundwater rights covered by the GMA. However, John Leshy and James Belanger argue that "numerous parts of the Act specifically and deliberately restrict surface water uses" (Leshy and Belanger, 20 Ariz. St. L.J. 657, 714). Arizona also has statutes mandating that the director of water resources analyze the effects of groundwater pumping on surface water supplies. Additionally, the GMA provides that, outside of AMAs, groundwater may be transported for use on land not overlying the water source without a presumption of legal liability attaching as a matter of law (Ariz. Rev. Stat. Ann. § 45-544). Recent Arizona court opinions adjudicating Gila River flows also determined that subflow water rights (i.e., rights to water in the underground zone where groundwater pumping impacts the flow of surface streams) may be appropriated and adjudicated in the same proceeding as the surface water rights they impact.

Water Quality

Arizona has given the state director of water resources the mandatory duty to set water quality standards for all waters of the state. Arizona's "unique waters" statutes provide some protection for water quality and ensuring the ecological integrity and environmental values of the state's waters; these statues require the director of the Department of Water Resources to maintain the water quality of each source for the designated purpose of that water source, including unique habitat and biological values. The "unique waters" statutes apply to desert springs that are headwaters of a surface water source, but arguably the statutes do not apply to springs unattached to riparian areas.

Federal Environmental and Water Resources Protection Laws

Federal law, at least, acknowledges that surface water and groundwater systems and resources are connected, and the ESA provides secondary protection for the ecological values of some desert springs defined as critical habitat pursuant to the ESA. Although federal law protects the water quality of some surface waters through the Clean Water Act (CWA), observers generally note that it offers little protection regarding the downdraft of groundwater resources and no protection regarding the pollution of groundwater if the groundwater is not related to a source of drinking water. Some desert springs may be subject to protection via federal wetlands regulations; however, various federal statutes that attempt to protect wetlands utilize definitions of the term that do not include most desert springs environments.

Wetlands

Scientific definitions of wetlands are not necessarily equivalent to jurisdictional definitions (i.e., whether a particular land mass falls within the provisions of a given federal statute). Most federal statutes define the wetlands over which they grant jurisdiction by analyzing one, two, or all of three of the following factors: (1) the area's hydrology (the degree and duration of its saturation and/or inundation); (2) the prevalence of hydrophilic plants in the area; and (3) the presence of hydrophilic soils in the area. To define an area as a wetland, the U.S. Fish and Wildlife Service requires one of these three factors to be present. In contrast, the Army Corps of Engineers, which is responsible for CWA Section 404 permitting, requires that all three qualities be present before an area may be considered a wetland. Most desert springs found in the American West have sandy soils and infrequent saturation and hydrophytic species seldom dominate; as a result, few desert springs are defined as wetlands under the corps' definition.

Federal Groundwater Acts

The CWA prohibits the public or private destruction of wetlands by requiring a permit from the Army Corps of Engineers for any dredge and fill (discharge) activity, and the CWA includes sanctions for those violating its provisions. The CWA covers "waters of the U.S.," a definition that includes surface waters, tributaries, and interstate waters. The CWA requires that the effects of the proposed dredge and fill action be analyzed, allowing for a wetland to be depleted only if "no practicable alternative" exists and mitigating measures are taken.

However, the CWA exempts most agricultural and ranching activity, the types of activity most likely to affect southwestern desert springs. These activities are exempt from the CWA because the "Swampbuster" statutes provide that farmers and ranchers will lose their agricultural subsidy if they convert wetland to agricultural use; however, Swampbuster's definition of wetlands requires a predominance of hydrophilic soils and plants, a more stringent standard than the CWA. Additionally, springs included in the CWA's provisions are defined as isolated if they cover more than one acre or if they lie in an abandoned surface water channel where the ground is usually saturated. Even if a particular desert spring falls within the CWA's jurisdiction requiring a permit, the Army Corps of Engineers does not deny many Section 404 permit applications and the Environmental Protection Agency seldom exercises its veto power over Section 404 permits issued by the Corps.

Another means by which springs may be protected is pursuant to a state designation of "unique waters" under the requirements of the Clean Water Act. In the case of Sycamore Spring, the National Wildlife Federation successfully challenged a Bureau of Land Management decision to allow grazing at Sycamore Spring, which forms the headwaters of Peeples Canyon Creek. All of the length of Peeples Canyon Creek, including Sycamore Spring, was designated as "unique waters," and the bureau was required to comply with state water laws protecting unique waters when determining whether to approve a grazing permit at the site (National Wildlife Federation et al. 1999).

The Endangered Species Act

Section 9 of the Endangered Species Act (ESA) subjects both private property owners and public agencies to penalties if they "take" a single member of a species listed as endangered by the U.S. Fish and Wildlife Service, pursuant to the ESA. The definition of the taking of a listed species includes destruction of critical habitat, which is that established by the Fish and Wildlife Service as habitat necessary to the perpetuation of the species and listed as such by notice published in the Federal Register.

The case of the endangered fountain darter and the Edwards Aquifer in Texas illustrates the potential impact of the ESA and its interface with state water laws. The Edwards Aquifer provides the sole source of municipal and industrial water for San Antonio (see Votteler 2002). The aquifer underlies the area east from the Texas-Mexico border under San Antonio and feeds into the Guadalupe River through Comal and San Marcos springs. Both Comal and San Marcos springs are home to the endangered fountain darter (*Etheostoma fonticola*); however, critical habitat for the fountain darter is designated only within San Marcos Spring. During periods of low rainfall, pumping from the aquifer increases and flow from the springs diminishes, sometimes to levels that alter the aquatic habitat of the springs and cause "takes" of the fountain darter. In 1991, a catfish farm began pumping up to forty million gallons per day (45,000 gallons per minute) of water from the system, one-fourth of the amount of water used by San Antonio in the same amount of time, causing a dewatering of the springs below the level necessary to support the fountain darter. Because of the adoption in Texas of a legal rule that provides no remedy for the unlimited pumping of groundwater put to use, state law could not be applied to halt the pumping.

In 1991, the Sierra Club sued the U.S. Department of the Interior and the Fish and Wildlife Service, alleging that the Interior Department had violated the ESA by allowing take of the fountain darter through the dewatering of the springs. As a result, in 1993 a federal court mandated that the Texas Legislature protect the springs' water flow by establishing a water-authority body to regulate groundwater withdrawals and prevent dewatering of the springs. The Edwards Aquifer Authority was formed; however, Texas residents brought suit in state court to prevent any attempts by the Edwards Aquifer Authority to limit their withdrawal of groundwater. The plaintiffs in the state suit maintained that any regulations enforced by the aquifer authority would violate the Texas state constitutional guarantee of an absolute right to withdraw groundwater. The Texas Supreme Court held that the Edwards Aquifer Authority was constitutional in 1996. However, a federal court-ordered plan to reduce pumping from the aquifer was voided, and litigation regarding pumping from the aquifer has continued (Texas 1996).

In Arizona, the loach minnow (*Rhinichthys cobitis*) has been listed as federally endangered, and critical habitat was established within the state along portions of the Gila River, the San Francisco and Blue rivers, and Aravaipa Creek (see Federal Register 1994). Theoretically, if any groundwater pumping adversely impacts habitat designated as critical to the loach minnow, citizens could sue to stop this pumping, or to limit it, pursuant to the ESA. The designation of the minnow's critical habitat within Arizona by the U.S. Fish and Wildlife Service notes that inclusion of the entire watershed was not necessary to protect the species.

The Administrative Procedures Act

The Administrative Procedures Act (APA) provides for judicial review of many federal executive agency decisions regarding the issuance of grazing or mining leases or permits on federal lands, as well as other decisions potentially affecting desert springs. In an APA suit, the plaintiff generally asserts that the adoption of a particular management plan, the granting of a particular permit, or some other agency decision was "arbitrary and capricious or contrary to law." If a plaintiff succeeds on such a claim, the relief granted is generally that the agency must reexamine its decision or conduct further inquiry and/or collect further data prior to issuing or reissuing a decision on the contested matter. An APA suit may be successful if the plaintiff claims that the challenged decision violates the science or other purpose of the land's initial designation as federal land, such as wilderness. For example, in the Sycamore Spring/Peeples Canyon Creek example cited above, the National Wildlife Federation challenged the decision of the Bureau of Land Management granting the grazing permit at the waterway designated by the state as a unique water (National Wildlife Federation et al. 1999). The National Wildlife Federation asserted that the bureau's decision to allow grazing was arbitrary, capricious, and contrary to the law because bureau personnel knew that such action would degrade unique waters.

Federal Reserved Water Rights

The doctrine of federal reserved water rights states that when the federal government formally withdraws land from the public's ability to use the land for settlement or extractive purposes, it also implicitly or explicitly reserves that amount of unappropriated waters on or under the land that are necessary to fulfill the federal government's purpose in creating the national forest, Indian reservation, or other form of federal land enclave. For example, when the federal government created Indian reservations as federal enclaves of land, it is legally presumed to have also reserved as much water as necessary from the available local water supply to support the Indians on the land. Similarly, if the federal government withdrew land from mineral entry and dedicated it as a national monument for its scientific value as unique wildlife habitat, then the courts would likely agree that water sufficient to sustain that habitat had also been claimed by the federal government at the time the land was withdrawn. Additionally, in practice the doctrine of federal reserved rights has presumed that surface water and groundwater resources are connected, if only when the courts find that the surface waters are inadequate to fulfill the governments' purposes.

The federal courts have held that the federal government might reserve not only surface water rights but also rights to underground water for the use of a federal enclave in *United States v. Cappaert* (1974). In the *Cappaert* case, the land surrounding Devils Hole was withdrawn by presidential proclamation and added to Death Valley National Monument in 1952. The land was withdrawn not only for the purpose of preserving Devils Hole but also to protect the Devils Hole pupfish (*Cyprinodon diabolis*), an endemic endangered species living in the pool. In 1968, groundwater pumping by the Cappaerts, the owners of land adjacent to the monument, caused the water level in the pool to fall below the level necessary to sustain the pupfish population. The United States sued the adjacent landowners to prohibit their pumping of water when this adversely affected the pupfish. In 1974 the Ninth Circuit Court of Appeals held that the United States had implicitly reserved enough surface and/or groundwater from the local appropriable sources to sustain the desert pupfish and that the Cappaerts could not pump water when this threatened the water level in the pool and the pupfish population. The court determined that, although the pool itself was surface water, it was adversely affected by groundwater pumping.

Following this doctrine, the Arizona state courts have held in the recent Gila River stream adjudication proceedings that the federal reserved water rights to surface waters supersede any state-mandated water rights. This means that the federal reserved water rights would limit any private rights to groundwater beyond the "reasonable use" limitations currently dictating the use of groundwater outside AMAs in the state. However, most withdrawals of land are made specifically subject to prior existing rights, such as grazing or mineral permits; the purpose of the withdrawal justifying the water right must be explicit in the withdrawal document and there must be appropriable water available at the time the withdrawal occurred before a legitimate reserved right may be found.

Federal and State Incentives

In addition to the common law and statutes cited above, some federal and state measures provide incentives for the conservation of ecological qualities of desert springs. For example, the federal and state governments have enacted laws allowing private landowners to receive tax benefits for dedicating land, potentially including desert springs habitat, to conservation purposes in perpetuity. Additionally, Arizona has established the Arizona Water Protection Fund (AWPF), dedicated to the protection and restoration of rivers and streams and their associated riparian habitat. Although the AWPF statutes do not explicitly mention desert springs projects as being eligible for funding from the so-called Riparian Trust Fund, some desert springs science projects have been funded from this source.

One jointly coordinated federal and state program is the Environmental Protection Agency's (2000) State Wetland Grant Program. The program was implemented to assist states with preservation of wetlands. Desert springs projects appear to be eligible for funding through this program.

Conclusions

The legal and cultural systems in most western states are gradually moving to evaluate and allocate all water resources in a comprehensive manner that synthesizes both groundwater and surface water availability and quality. However, in many respects, the existing statutory and common law system does not allow for comprehensive protection of western desert springs. Of the many factors confronting the conservation of these unique ecosystems and the species they sustain, legal limitations are among those in most serious and immediate need of reconsideration and rectification.

Chapter 17

Every Last Drop

Future Springs Ecosystem
Ecology and Management

LAWRENCE E. STEVENS

Odum's (1957) seminal work on ecosystem ecology was conducted at Silver Springs in Florida, but prior to this volume, little subsequent synthesis of springs ecology has been attempted (Sada et al. 2001). Florida is one of the few states that now has a relatively thorough inventory of its springs; however, in the absence of scientific and management attention, many of its nearly seven hundred known limnocrene (pool-forming) springs have been negatively affected by groundwater pollution or other human activities (Scott et al. 2004). In more arid regions, where demands on limited water resources are often severe, the health and persistence of springs ecosystems are even more jeopardized (e.g., Grand Canyon Wildlands Council 2002). With the presentation, analysis, and review of case studies and concepts, this volume provides insight into numerous basic and applied questions about springs and their conservation. However, this is only a small step toward more comprehensive protection of springs ecosystems. Such conservation efforts will do much to preserve ecological processes, cultural and biological diversity, and evolutionary trajectories.

At the conclusion of the 2000 Arizona-Sonora Desert Museum Symposium, we convened a panel of experts to explore the next steps in springs ecological research and conservation. Many participants voiced grave concerns regarding the fate of these ecosystems in arid regions, and many productive ideas were exchanged. The topics described by that panel and other concepts that were subsequently recognized during the preparation of this book include the development of (1) a springs ecosystem conceptual model, (2) testing of a comprehensive classification system (see chapter 4), (3) an overall approach to management and restoration, (4) inventory and monitoring protocols, and (5) ecosystem health assessment protocols.

Collectively, the panel recognized the need for conservation initiatives at regional, state, national, and international scales. In addition, the panel identified information gaps and research needs that, if filled, will help focus and advance springs ecosystem ecology and conservation.

A Springs Ecosystem Conceptual Model

A general conceptual model will help organize the diverse fields of inquiry associated with springs ecosystem research and management (Stevens and Springer 2005). Preliminary conceptual ecosystem models are not likely to accurately predict the outcomes of management actions but will be useful for identifying information gaps, relationships among physical and biological variables, and management options (e.g., Walters et al. 2000). This is particularly true for springs ecosystems, in which significant theoretical and data limitations presently exist. A springs ecosystem conceptual model should consist of two linked components: a controls model, and an anthropogenic impacts model. The controls component should seek to link static/descriptive, probabilistic, and dynamics submodels that govern natural ecosystem functions and characteristics. This component model should include eight submodels: (1) physical/aquifer dynamics, (2) geomorphology, (3) microclimate, (4) disturbance-productivity gradients (particularly whether terrestrial springs are dominated by spring flows or surface water flows), (5) microhabitat array, (6) biogeography, (7) trophic dynamics, and (8) ecosystem goods and services. The natural disturbance regime is a defining point in the springs conceptual model because the frequency and intensity of natural disturbances (e.g., rockfall, scouring floods, tidal exposure of littoral subaqueous springs, glacial advance/retreat cycles) exert powerful impacts on springs processes and biota. Naturally highly disturbed springs can only function as neorefugia (*sensu* Nekola 1999), supporting native and non-native weedy, generalist species or life stages, rather than microhabitat-adapted, rare, and endemic taxa. Expansion of the above submodels will be enhanced if they are described in terms of general theoretical issues and then related to physical driving variables and interactions. When refined, the controls model should produce measurable outputs, well-defined ecosystem consequences, and clear linkages among the constituent submodels as well as with the stressors model (below).

The conceptual controls model should then be linked to an anthropogenic stressors component model, which will help predict ecosystem changes in response to environmental alterations. This anthropogenic stressors model should focus on the major forms of human impacts to springs, particularly groundwater depletion, groundwater pollution, post-emergence flow alteration, geomorphic alteration of sources, livestock grazing, and nonnative species impacts. These stressors should be linked to the appropriate controls submodels.

Coupling the controls and stressor models will require many additional baseline data and analyses before the overall model can be used to predict springs ecosystem responses to natural and anthropogenic environmental changes. Some elements, such as groundwater modeling, already are theoretically robust, while other elements, particularly physical-biotic interactions (such as solar radiation budget impacts on plant assemblages) have scarcely been examined (but see chapter 9). Nonetheless, conceptual modeling will be quite useful for identifying the information needs, inventory, assessment, and monitoring priorities for springs managers and will help guide further springs ecosystem research.

Testing the Springs Classification Scheme

The symposium panel identified the need for a comprehensive springs classification system both to develop a common lexicon about these ecosystems and to allow managers to conduct strategic planning on the distribution and status of various types of springs under their jurisdiction. The classification system defined in chapter 4 is a significant contribution toward those goals; however, it requires rigorous testing to clarify its utility and to improve the description and assessment of springs (see below).

Improving Management

General Adaptive Management

As conceptual modeling and classification advance, a framework is needed for adaptive springs ecosystem management that focuses on the development and use of scientific information about ecosystem status, threats, protection, and restoration. Such a hierarchical system should help guide springs managers through a logical suite of activities including inventory, assessment, strategic planning, protection, and restoration of springs. Inventory is a primary information need, and assessment is an essential process to clarify and prioritize management strategies. Unfortunately, basic mapping and inventory of springs at regional landscape scales has been neglected, and few long-term monitoring data have been compiled on individual springs. Strategic planning is needed to frame discussion about what to manage for, how to accomplish those actions, and contingencies (i.e., management and administrative emergencies). Implementation of a springs management program requires initiation of monitoring at sites where actions are to be taken. Restoration requires credible pre- and post-treatment monitoring and analyses, information that should be archived and fed back into the adaptive management of the springs. Following these steps logically is likely to result in fulfillment of management goals and achievement of desired conditions at these unique ecosystems.

Management at Regional Scales

Grand Canyon Wildlands Council (2002) made several recommendations regarding springs management at regional scales. These recommendations emphasize the importance of planning and conducting spring inventories: springs ecosystems have yet to be systematically inventoried and assessed to any great extent by any state in arid North America. Second, GCWC (Grand Canyon Wildlands Council 2002) urged managers to consider the regional context of springs in their management jurisdiction by (1) describing the history of springs management in the region; (2) identifying how many and which kinds of springs exist; (3) conducting an ecologically based conservation inventory and assessment of springs ecosystem health; and (4) describing, prioritizing, and developing authorization for funding and implementation of springs ecosystem inventory, assessment, and, where possible, restoration. Also, GCWC (2002) recommended development of a regional springs vegetation model based on elevation, water quality, and slope-aspect. Such a model will be useful for springs restoration activities where prealteration data do not exist.

Management at Local Scales

At the local scale of individual springs, the Grand Canyon Wildlands Council (2002) made the following management recommendations:

1. Manage according to a plan that considers the uniqueness of each spring, with clear goals and priorities.
2. "Leave a little." Springs sources and moist backwalls are unique ecological and evolutionary settings that often support unusual and endemic plants and invertebrates. In addition, small pools of open water are much needed by wildlife, particularly birds, bats, predators, ungulates, and invertebrates. Piping all water from the source at such sites eliminates these functions and greatly reduces biodiversity and ecological integrity. Leaving a little flowing water at the source, particularly on backwall habitats, will disproportionately increase the spring's ecological integrity without significantly diminishing water supplies for human purposes.
3. If water is used for livestock watering, fence the source areas and maintain appropriate disturbance regimes (Kodrick-Brown and Brown 2007). Also, keep the pipe systems and tanks in good repair. Why dewater and destroy diverse and biologically productive springs only to let the abstracted water leak out of faulty pipes and tanks?
4. If water is piped from a spring, develop and archive a piping network diagram so that future generations will understand the extent and mechanisms of alteration.
5. If the spring is actively managed (for water supplies, livestock, etc.), create and maintain a discrete trail to the management area. This often involves a trail to the source. Construction of a trail will reduce erosion and the impacts of trampling on sensitive vegetation.
6. Install and maintain, or improve, wildlife rescue devices on all open water tanks to prevent needless loss of animals. Both small and large animals may fall in the tanks and drown. The Grand Canyon Wildlands Council (e.g., 2002) documented the deaths of birds of several species and mammals—including bats, rodents, and black bear—lost because no consideration of escape from open water tanks around springs was provided.

7. Seek to protect springs that harbor known populations of rare or endemic species (whether listed or not).

8. Develop and implement a plan for monitoring and restoring springs where applicable.

9. Conduct outreach to public stakeholders or form citizen action groups to discuss the importance of springs and what actions can be taken to make springs more desirable to native vegetation and wildlife.

10. Create partnerships with groups to provide on-the-ground assistance to overburdened agency staff and create funding opportunities for contract work on springs information management, inventory, assessment, and restoration.

Inventory Protocols

Inventories of spring flow, water quality, and biota are needed to classify each spring, determine notable characteristics, assess ecological health, and reveal threats; however, rigorous, efficient inventory protocols for springs have yet to be developed or tested. Two levels of inventory are appropriate.

A Level I inventory, generally performed by trained technicians, involves accurate georeferencing, description of site access, site photography, rough estimation of flow (for estimation of sampling equipment needs), and human impacts (Sada and Pohlmann 2003). Although unusual plants and animals may be recorded, not all Level I inventory technicians need to be thoroughly familiar with all elements of regional natural history. The Level I inventory should take less than half an hour per spring.

A Level II springs ecosystem inventory protocol should be more thorough and include (1) a detailed site description, including a sketch map, highly accurate GPS readings, and photographs of the site; (2) geologic/geomorphologic analysis; (3) hydrogeochemical analysis (discharge and water chemistry); (4) initial biological survey (vegetation and macroinvertebrate composition and cover or abundance, and vertebrates detected); and (5) a detailed description of cultural resources and human impacts. Level II inventories should be designed so as to be conducted by three to four individuals with expertise in geohydrology, aquatic and terrestrial biology (including vegetation mapping), and cultural resources. A Level II inventory should require no more than four hours to complete during a single site visit, depending on the size of the springs, but is likely to require at least an additional day of associated laboratory and analysis time. Level II protocols were tested by the Grand Canyon Wildlands Council (2002) on 103 Arizona Strip and Grand Canyon springs, by Springer et al. (2006) on 76 springs on the Colorado Plateau, and by our colleagues. The protocols were efficient at springs smaller than two hectares in area; they required two to three hours per site for a three-person team, but cultural expertise was not included on these teams. Level II inventory of relatively remote individual springs had a total cost of $3,500 per site in 2005. To maximize the biological information gathered, it is recommended that the Level II inventory should take place in the middle of the growing season. However, flow measurements in midsummer are likely to be reduced by transpiration and therefore may underestimate base flow volume and overestimate the concentration of base flow water chemistry.

Some of the proposed Level II variables eventually may be demonstrated to be redundant. For example, vegetation cover and composition at neorefugial springs are likely to be highly dynamic: Grand Canyon Wildlands Council (2004) demonstrated a range of interannual variation of 10–70 percent for plant ground cover at flood- and rockfall-prone Monument Creek Spring in Grand Canyon between 2000 and 2002, a level of variation comparable to that described by Meretsky (chapter 10, this vol.). Resolution of which variables are most pertinent and the consequent streamlining and refining of the Level II inventory await a sufficiently large database in which to test for the strength of various relationships. The Level II inventory data also should be designed to contribute toward an ecosystem health assessment.

To facilitate refinement of the conceptual ecosystem model, testing the classification system, and archiving inventory information requires development of a comprehensive springs inventory database. This database should embrace a large array of different springs types, across latitude, elevation, and different biomes. Such a database can be used to distinguish multivariate forcing mechanisms in springs ecosystems, such as the importance of aspect or microhabitat interactions in biological assemblage composition. The database should be developed with rigorous quality control and metadata standards. Analyses needed for testing the classification system and the assessment process (below) include cluster, ordination, and other multivariate community analyses to relate physical variables to springs types, as well as discriminant function analyses to determine which suites of variables most reliably identify springs types and ecological functioning.

Ecosystem Health and Risk Assessment Protocols

In contrast to stream-riparian ecosystems, springs that have been subjected to dewatering or other extreme anthropogenic alterations appear to have relatively low resiliency. Dewatering of springs inevitably means the decline or loss of associated wetland plant and invertebrate populations and assemblages, some of which may be genetically distinctive or endemic (Bunn and Arthington 2002; see also chapter 14, this vol.). Once perennial springs have been dewatered, restoration of the former habitats and biotic assemblages becomes a dubious undertaking, particularly if pre-impact monitoring data are not available. In cases of aquifer depletion, little possibility of recharge and restoration may exist; therefore, springs should be regarded as nonrenewable resources. Ecosystem health assessment protocols are needed to understand the status, priorities, and options for springs management and restoration. However, such criteria have not been developed or tested. Trends assessment also is needed, requiring reliably collected and archived monitoring information: unfortunately, few springs have been subjected to comprehensive, long-term monitoring. Assessment protocols should be scientific and efficient, using quantified information from the inventory, and should incorporate spatiotemporal considerations of site uniqueness. While some aspects of stream-riparian ecosystem health assessment (e.g., Brinson 1993; Richter et al. 1996; Rosgen 1996; Beard et al. 1999; Karr 1999; Dale et al. 2000; Busch and Trexler 2002; Rapport et al. 2003) may be appropriate to use in springs assessment, the vastly different role of geomorphology at springs limits the applicability of stream hydrogeomorphic approaches (Brinson 1993) to springs assessment (Stevens et al. 2005).

Four elements are needed for springs ecosystem health assessment: (1) compilation of diverse inventory information on the ecosystem characteristics and processes of individual springs across a region; (2) a rating system to evaluate individual ecosystem characteristics or processes at each spring that provides a rating system for whole individual springs; (3) prioritization of management actions by evaluating a spring's ecological importance against competing socioeconomic variables; and (4) ecosystem health variables that are soundly related to conditions in natural springs ecosystems at a wide array of relatively pristine reference sites (Brinson and Rheinhardt 1996; Stevens et al. 2005). This latter point indicates the need for the development of a network of reference sites or surrogate regional ecosystem models to guide management in regions where no natural springs remain. This comprehensive ecosystem health assessment system is fundamental to improving understanding of springs ecology, management, and conservation. Ultimately, any springs ecosystem health assessment should be related to a global biodiversity intactness index (e.g., Scholes and Biggs 2005).

Monitoring Protocols

Monitoring involves collection and analysis of repeated qualitative or quantitative measurements to assess changes in resource condition and the extent of progress in meeting one or more management objectives (Elzinga et al. 1998). The overall purpose of springs monitoring is to acquire information needed to periodically assess ecosystem integrity. Presentation of monitoring information alerts managers to changes in resource conditions, reduces uncertainty about management decisions, and reduces threats and operational inefficiently. However, the collection and archiving of long-term monitoring data are expensive and time-consuming and requires permanent commitment to staff and information management. Therefore, planning for springs monitoring should focus initially on current and relevant future management issues, scientific credibility, and cost-effectiveness.

Springs ecosystem monitoring objectives can usually be classified as one of two types: target/threshold objectives and change/trend objectives (Elzinga et al. 1998). Target/threshold objectives include such directives as maintaining flow and water quality at springs, increasing a target population size to a desired level, and keeping a site free of non-native invasive species. Change/trend objectives include directives such as increasing the mean density of a given species by a certain percentage within a given time period and decreasing the frequency of invasive weed species by a certain percentage at a given site over a given time period.

Fancy (2000) used a peer-review panel to develop the following recommendations on monitoring plan design: (1) use probabilistic sampling; (2) design study plot distribution from a statistical standpoint to include the landscapes of interest; (3) avoid judgment-based sampling; (4) distribute samples across the entire landscape, and then increase sampling intensity in areas of special interest; (5) refrain from generalizing natural conditions from a single reference site; (6) avoid stratifying habitats derived from vegetation maps; (7) delineate areas of special interest based on physical characteristics, not habitat or biological characteristics; (8) revisit permanent plots over time, as opposed to sampling from random transects; (9) determine *a priori* the sample size needed to significantly reduce the uncertainty about status or trend; (10) maintain the statistical spatial integrity of samples or plots; (11) improve the precision of repeated measurements by increasing the duration of sampling; (12) determine whether visiting all sites each year is necessary if a large suite of sites is included in the monitoring program; and (13) co-locate samples (e.g., within springs microhabitats) to facilitate comparisons among ecosystem components.

Springs Information Management

A key element of ecosystem management is the development of a comprehensive information management (IM) system. Springs information management systems should be developed at local, regional, national, and whenever possible, international administrative scales. For example, a comprehensive springs database and library should be established in each national park and national forest, in each national park or forest network or region, for the National Park Service and the National Forest Service overall, and for all parks and forests worldwide.

Springs information management systems at each administrative scale should include several elements. First, a comprehensive, searchable bibliography is needed. Second, a facile, efficient, and easily queried data archive should be established, with well-developed metadata standards. Third, a library should be established to archive hard-copy elements, such as maps, photographs, field notes, and so forth. Such elements should be scanned and archived electronically as soon as possible. Fourth, a well-curated museum of collected specimens should be established, and the information archived electronically. The last element involves the establishment of a geographic information system to relate georeferencing and inventory data to aquifer, surface flow, ecosystem, and climate models. Databases on regional, national, and international scales are needed to allow managers to assess conditions and conservation priorities at those scales and to answer some of the research questions posed below.

Information Gaps and Research Needs

The symposium panel of experts identified numerous basic and applied information and research issues that need to be addressed to advance springs ecosystem ecology and conservation. For ease of consideration, the synopsis of these topics is divided into research (physical, bio-ecosystem, sociocultural) and management/administration categories.

PHYSICAL SCIENCE RESEARCH. Many physical sciences, particularly aquifer studies, contribute to springs ecology. However, the depth of present studies is insufficient to predict basic patterns of flow, water quality, microclimate, and habitat template development. Research on the following topics will help clarify these uncertainties:

- Develop basic and applied predictive capability of springs flow and water chemistry responses to aquifer and global climate change
- Analyze potential evapotranspiration and other microclimate patterns at springs, across elevation, slope, and aspect
- Test and refine the springs classification system defined in chapter 4
- Conduct water quality and paleontological analyses of long-term persistence and ecosystem changes of springs

BIOLOGICAL AND ECOSYSTEM RESEARCH. This volume demonstrates the complexities and multidisciplinary challenges facing those who attempt to conduct comprehensive springs research and management, as well as some of the ecological and evolutionary insights that integrated springs research has to offer. Major ecosystem questions remain to be explored, including the following:

- Disturbance-productivity relationships and impacts on diversity in the context of slope, aspect, elevation, and other ecological gradients
- Biogeography of species abundance and density as a function of springs habitat size
- The role of springs as keystone ecosystems
- How, where, and at what rate endemism arises at springs
- Testing Nekola's (1999) neorefuge-paleorefuge concept at springs with different disturbance histories
- Determining regional relationships between water quality and aquatic biota diversity, particularly for plants and invertebrates across other ecological gradients (e.g., elevation, slope, aspect)
- Understanding the role of top-down trophic cascades in neorefugial versus paleorefugial springs
- Understanding the population dynamics of rare species at springs, particularly in relation to flows and climate change
- Developing, refining, conceptually advancing, and empirical testing of springs ecosystem models

SOCIOCULTURAL RESEARCH. Sociocultural research topics of interest at springs include the fields of archaeology, anthropology, ethnogeology and ethnobiology, history, and economics, to address the following questions:

- The extent and manner of use of springs by aboriginal peoples
- The extent to which indigenous cultural use of springs was or is sustainable
- The role of springs in cultural historical development, particularly within arid regions
- Ethnology of springs, including paleontology, biology, and sacredness
- Analyses of ecosystem goods and services in relation to resource sustainability

SPRINGS ADMINISTRATION AND MANAGEMENT. Research on springs management and administration and related issues remains largely uncoordinated across agencies and spatial scales. Analyses of springs ecosystem planning and management will improve coordination; guide development of clear, well-conceived, and scientifically credible management strategies; and improve efficiency and reduce economic-environmental conflicts. Numerous areas, including the following, require further consideration:

- Develop, test, and refine springs inventory and monitoring protocols
- Analyze and test springs ecosystem health assessment protocols
- Assess springs rehabilitation and restoration decision making and implementation
- Develop regional (aquifer-based) groundwater protection needs and strategies
- Establish a springs ecosystems research center, providing a well-administered archive for comprehensive springs data and analyses
- Develop a springs ecosystem ecology study area at which to experimentally test physical and biological hypotheses
- Develop regional sets of reference sites for springs, a process that will likely involve forging new relationships between public and private land-managing agencies
- Publish a springs ecosystem research journal
- Coordinate multiagency regional, national, and international springs inventory and conservation initiatives
- Resolve polarization between the public and government over the use of water derived from springs and the watersheds and aquifers from which they originate
- Develop adaptive ecosystem management models for springs

Conclusions

The anthropogenic threats facing springs are real, imminent, and ever accelerating. The above topics are some, but by no means all, of the challenges facing the nascent field of springs ecosystem ecology; addressing these issues is critical for improving our knowledge and the conservation of springs. Most of the issues described here require new levels of integrative, creative, and coordinated scientific and conservation thought, and all involve a renewed and vigorous intent on the part of the scientific community, the public, conservation groups and funders, governmental agencies, and policy makers to improve the quality of knowledge and the application of that knowledge to contribute to springs management and conservation. In North America, private and public management of springs is terribly and unnecessarily polarized, and this conflict needs to be resolved. If these scientific and social issues are left unattended, springs ecosystems are likely to continue to rapidly degrade over the coming decades, taking with them much of our most remarkable biological diversity. However, with more focused research and conservation attention on springs ecosystems, and with the various parties working through these social challenges, springs may be better understood, managed, and protected. Speaking for the entire panel, we fervently hope that the recommendations here outline a course for that effort and contribute to the preservation of these most remarkable ecosystems.

ACKNOWLEDGMENTS

Some of the research presented here was funded through National Park Service Cooperative Agreement Number CA 1200-99-009. I thank Bianca S. Perla, Angie Evenden, Heidi Kloeppel, Donald W. Sada, Abe Springer, Lisa Thomas, and the Arizona-Sonora Desert Museum Symposium panel of experts for their thoughts on topics described herein. I thank the University of Arizona Press editorial staff for their assistance in focusing the scope of this chapter.

References

Abbott, R. T. 1973. Spread of *Melanoides tuberculata*. Nautilus 87:29.

Abell, R. A., D. M. Olson, E. Dinerstein, W. Eichbaum, S. Walters, P. T. Hurley, W. Wet-tengell, C. J. Loucks, and P. Hedao. 1998. A conservation assessment of the freshwater ecoregions of North America. World Wildlife Fund, Washington, D.C.

Abell, R. A., D. M. Olson, E. Dinerstein, P. T. Hurley, J. Doggs, W. Eichbaum, S. Walters, W. Wettengell, T. Allnutt, C. J. Loucks, and P. Hedao. 2000. Freshwater ecoregions of North America: A conservation assessment. Island Press, Washington, D.C.

Acuna-Soto, R., D. W. Stahle, M. K. Cleaveland, and M. D. Therrel. 2002. Megadrought and megadeath in 16th Century Mexico. Emerging Infectious Diseases 8 (4): 360–362.

Adkins, W. S. 1920. Cuatro Ciénegas, Coahuila—Water Resources. Pages 1–19. Unpublished manuscript in Walter Scott Adkins Papers, 1892–1963, Box CDL2/M329 (04438863)-AR89-194. Archived at the Center for American History, University of Texas at Austin.

Agenbroad, L. D. 1984. Hot Springs, South Dakota: Entrapment and taphonomy of Colum-bian mammoth. Pages 113–127 *in* P. S. Martin and R. G. Klein, editors. Quaternary extinctions: A prehistoric revolution. University of Arizona Press, Tucson.

Alcock, J. 1975. Animal behavior: An evolutionary approach. Sinauer Associates, Sunderland, Mass.

Alcorn, J. B. 1994. Noble savage or noble state? Northern myths and southern realities in biodiversity conservation. Ethnoecológica 2:6–19.

Alexander, S. E., S. H. Schneider, and K. Lagerquist. 1997. The interaction of climate and life. Pages 71–92 *in* G. C. Daily, editor. Nature's services: Societal dependence on natu-ral ecosystems. Island Press, Washington, D.C.

Alfaro, C., and M. Wallace. 1994. Origin and classification of springs and historical review with current applications. Environmental Geology 24:112–124.

Al Kahem, H. R., and R. J. Behnke. 1983. Freshwater fishes of Saudi Arabia. Fauna of Saudi Arabia 5:545–567.

Allphin, L., and K. T. Harper. 1994. Habitat requirements for *Erigeron kachinensis*, a rare endemic of the Colorado Plateau. Great Basin Naturalist 54:193–203.

Allphin, L., M. D. Windham, and K. T. Harper. 1996. A genetic evaluation of three potential races of the rare Kachina daisy. Pages 68–76 *in* J. Maschinski, H. D. Hammond, and L. Holter, editors. Southwestern rare and endangered plants. Proceedings of the Second Conference. USDA, Forest Service Technical Report RM-GTR-283.

Amin, O. M., and W. L. Minckley. 1996. Parasites of some fish introduced into an Arizona reservoir, with notes on introductions. Journal of the Helminthological Society of Wash-ington 63:193–200.

Anderson, A. B., and D. A. Posey. 1989. Management of a tropical shrub savanna by the

Gorotire Kayapo of Brazil. Pages 159–173 *in* D. A. Posey and W. Balée, editors. Resource management in Amazonia: Indigenous and folk strategies, advances in economic botany. New York Botanical Garden Press, New York.

Anonymous. 1948. They found the bottom. Desert 11:31.

———. 1999. En riesgo de extinción pozas. Zocalo de Monclova, November 8.

———. 2003. Extremophiles: Some like it hotter. Science News 163:366.

Arabagali, D. 2001. They trampled on our taboos. Pages 213–220 *in* D. Rothenberg and M. Ulvaeus, editors. The world and the wild: Expanding wilderness conservation beyond its American roots. University of Arizona Press, Tucson.

Arnold, E. T. 1971. Behavioral ecology of two pupfishes (Cyprinodontidae, genus *Cyprinodon*) from northern Mexico. Ph.D. dissertation, Arizona State University, Tempe.

Ashley, G. M. 2001. Archaeological sediments in springs and wetlands. Pages 183–210 *in* J. K. Stein and W. R. Farrand, editors. Sediments in archaeological context. University of Utah Press, Salt Lake City.

Atran, S. 1999. Managing the Maya commons: The value of local knowledge. Pages 190–214 *in* V. D. Nazarea, editor. Ethnoecology: Situated knowledge/located lives. University of Arizona Press, Tucson.

Avery, C. C., and H. Helmke. 1993. Montezuma Well: A dynamic hydrologic system. Pages 200–207 *in* P. G. Rowlands, C. van Riper III, and M. K. Sogge, editors. Proceedings of the first biennial conference on research in Colorado Plateau national parks. U.S. Department of the Interior, National Park Service Transactions and Proceedings Series NPS/NRNAU/NRTP-93/10.

Axelrod, D. I. 1950. Evolution of desert vegetation in western North America. Studies in Late Tertiary Paleobotany. Carnegie Institution of Washington Publication 590. Washington, D.C.

———. 1979. Age and origin of Sonoran Desert vegetation. Occasional Papers of the California Academy of Sciences 132:1–74.

Axelrod, D. I., and P. H. Raven. 1985. Origins of the cordilleran flora. Journal of Biogeography 12:21–47.

Back, W. 1966. Hydrochemical facies and groundwater flow patterns in northern part of Atlantic coastal plain. U.S. Geological Survey Professional Paper 498-C. U.S. Government Printing Office, Washington, D.C.

Baker, N. W. 1971. Observations on the biology of the giant palm-boring beetle, *Dinapate wrightii* (Horn; Coleoptera: Bostrichidae). Journal of the New York Entomological Society 79:31–42.

Baldinger, A. J., W. D. Shepard, and D. L. Threloff. 2000. Two new species of *Hyalella* (Crustacea: Amphipoda: Hyalellidae) from Death Valley National Park, California, U.S.A. Proceedings of the Biological Society of Washington 113:443–457.

Balée, W., and A. Gély. 1989. Managed forest succession in Amazonia: The Ka'apor case. Pages 129–158 *in* D. A. Posey and W. Balée, editors. Resource management in Amazonia: Indigenous and folk strategies, advances in economic botany. New York Botanical Garden Press, New York.

Barnett, A., and A. Brazel. 1975. A topoclimatic survey of Montezuma Well, Arizona. Arizona Academy of Science 10:40–41.

Batchelder, G. L., and G. A. Cole. 1978. Evidence of post-pluvial water level fluctuation at

Montezuma Well, Coconino County, Arizona, U.S.A. Report to National Park Service, U.S. Department of the Interior.

Bauer, S. B., and T. A. Burton. 1990. Monitoring protocols to evaluate water quality effects of grazing management on western rangeland streams. EPA 910/R-93-017. U.S. Environmental Protection Agency, Seattle, Washington.

Beard, G. R., W. A. Scott, and J. K. Adamson. 1999. The value of consistent methodology in long-term environmental monitoring. Environmental Monitoring and Assessment 54:239.

Beare, M. H., R. W. Parmelee, P. F. Hendrix, W. Cheng, D. C. Coleman, and D. A. Crossley Jr. 1992. Microbial and faunal interactions and effects on litter nitrogen and decomposition in agroecosystems. Ecological Monographs 62 (4): 569–591.

Beckman, J. E. 1991. A history of Montezuma Well. Report to National Park Service. Archived at the headquarters of Montezuma Castle National Monument, McGuireville.

Belnap, J. 1995. Surface disturbances: Their role in accelerating desertification. Environmental Monitoring and Assessment 37:39–57.

Ben-Tuvia, A. 1981. Man-induced changes in the freshwater fish fauna of Israel. Fish Management 12:139–148.

Betancourt, J. L. 1984. Late Quaternary plant zonation and climate in southeastern Utah. Great Basin Naturalist 44:1–35.

Betancourt, J. L., T. R. Van Devender, and P. S. Martin. 1990. Packrat middens: The last 40,000 years of biotic change. University of Arizona Press, Tucson.

Blair, J. M., R. W. Parmelee, and M. H. Beare. 1990. Decay rates, nitrogen fluxes, and decomposer communities of single- and mixed-species foliar litter. Ecology 71:1976–1985.

Blair, J. M., T. R. Seastedt, C. W. Rice, and R. A. Ramundo. 1998. Terrestrial nutrient cycling in tallgrass prairie. Pages 222–243 in A. K. Knapp, J. M. Briggs, D. C. Hartnett, and S. L. Collins, editors. Grassland dynamics: Long-term ecological research in tallgrass prairie. Oxford University Press.

Blinn, D. W., and R. W. Davies. 1989a. The evolutionary importance of mechanoreception in three erpobdellid leech species. Oecologia 79:6–9.

———. 1989b. New distribution records for leeches (Hirudinoidea) in Arizona. Southwestern Naturalist 34:431–432.

———. 1990. Concomitant vertical migration of a predatory leech and its amphipod prey. Freshwater Biology 24:401–407.

Blinn, D. W., R. W. Davies, and B. Dehdashti. 1987. Specialized pelagic feeding by *Erpobdella montezuma* (Hirudinea). Holarctic Ecology 10:235–240.

Blinn, D. W., B. Dehdashti, C. Runck, and R. W. Davies. 1990. The importance of prey size and density in an endemic predator-prey couple (leech *Erpobdella montezuma*–amphipod *Hyalella montezuma*). Journal of Animal Ecology 59:187–192.

Blinn, D. W., N. E. Grossnickle, and B. Dehdashti. 1988b. Diel vertical migration of a pelagic amphipod in the absence of fish predation. Hydrobiologia 160:165–171.

Blinn, D. W., R. H. Hevly, and O. K. Davis. 1994. Continuous Holocene record of diatom stratigraphy, paleohydrology, and anthropogenic activity in a spring-mound in southwestern United States. Quaternary Research 42:197–205.

Blinn, D. W., and D. B. Johnson. 1982. Filter-feeding of *Hyalella montezuma*, an unusual behavior for a freshwater amphipod. Freshwater Invertebrate Biology 1:48–52.

Blinn, D. W., and G. E. Oberlin. 1996. Montezuma Castle National Monument integrated inventory and monitoring studies: Aquatic biota, invertebrates and algae. Colorado Plateau Research Station, Flagstaff, Ariz.

Blinn, D. W., C. Pinney, and M. W. Sanderson. 1982. Nocturnal planktonic behavior of *Ranatra montezuma* Polhemus (Nephidae: Hemiptera) in Montezuma Well, Arizona. Journal of the Kansas Entomological Society 53:481–484.

Blinn, D. W., C. Pinney, and V. T. Wagner. 1988a. Intraspecific discrimination of amphipod prey by a freshwater leech through mechanoreception. Canadian Journal of Zoology 66:427–430.

Blinn, D. W., and C. Runck. 1989. Substratum requirements for oviposition, seasonal egg densities, and conditions for egg eclosion in *Ranatra montezuma* (Heteroptera: Nepidae). Annals of the Entomological Society of America 82:707–711.

Blinn, D. W., C. Runck, and R. W. Davies. 1993. The impact of prey behavior and prey density on the foraging ecology of *Ranatra montezuma* (Heteroptera): A serological examination. Canadian Journal of Zoology 71:387–391.

Blinn, D. W., and M. Sanderson. 1989. Aquatic insects in Montezuma Well, Arizona, USA: A travertine spring mound with high alkalinity and dissolved carbon dioxide. Great Basin Naturalist 49:85–88.

Blinn, D. W., V. T. Wagner, and J. N. Grim. 1986. Surface sensilla on the predaceous freshwater leech *Erpobdella montezuma*: Possible importance in feeding. Transactions of the American Microscopical Society 105:21–30.

Blombery, A., and T. Rodd. 1982. Palms. Angus and Robertson, London, England.

Boldurian, A. T., and J. L. Cotter. 1999. Clovis revisited: New perspectives on Paleoindian adaptations from Blackwater Draw, New Mexico. University Museum Monograph 1003. University Museum, University of Pennsylvania, Philadelphia.

Bolton, H. E. 1936. The rim of Christendom: A biography of Eusebio Francisco Kino, Pacific Coast pioneer. Macmillan, New York.

———. 1984. The rim of Christendom: A biography of Eusebio Francisco Kino. Reprint, University of Arizona Press, Tucson.

Bonham, C. D. 1989. Measurements for terrestrial vegetation. John Wiley and Sons, New York.

Botosaneanu, L., editor. 1998. Studies in crenobiology: The biology of springs and springbrooks. Backhuys, Leiden, Netherlands.

Boucher, P. 1980. The phytoplankton ecology of Montezuma Well, Arizona. M.S. thesis, Northern Arizona University, Flagstaff.

Boucher, P., D. W. Blinn, and D. B. Johnson. 1984. Phytoplankton ecology in an unusually stable environment (Montezuma Well, Arizona, U.S.A.). Hydrobiologia 119:149–160.

Boulinier, T., J. D. Nichols, J. R. Sauer, J. E. Hines, and K. H. Pollock. 1998. Estimating species richness: The importance of heterogeneity in species detectability. Ecology 79:1018–1028.

Bowers, J. E. 1980. Flora of Organ Pipe Cactus National Monument. Journal of the Arizona-Nevada Academy of Science 15:1–11.

Bowers, W. E. 1990. Geologic map of Bryce Canyon National Park and vicinity, southwestern Utah. U.S. Geological Survey Miscellaneous Investigations Series Map I-2108. U.S. Government Printing Office, Washington, D.C.

Bowles, D. E., and T. L. Arsuffi. 1993. Karst aquatic ecosystems of the Edwards Plateau region of central Texas, USA: A consideration of their importance, threats to their existence, and efforts for their conservation. Aquatic Conservation: Marine and Freshwater Ecosystems 3:317–329.

Boyd, W. E. 1990a. Mound springs. Pages 107–118 in M. J. Tyler, C. R. Twidale, M. Davis, and C. B. Wells, editors. Natural history of the northeastern deserts. Royal Society of South Australia, Adelaide.

———. 1990b. Quaternary pollen analysis in the arid zones of Australia: Dalhousie Springs, central Australia. Review of Palaeobotany and Palynology 64:331–341.

———. 1992. A pollen flora of the native plants of South Australia and southern Northern Territory, Australia. South Australian Geographical Papers 3:1–212.

———. 1994. Quaternary pollen analysis in the arid zone of Australia: Further results from Dalhousie Springs, central Australia. Australian Geographical Studies Research Note 32:274–280.

Brain, C. K., and W. Koste. 1991. Rotifers of the genus *Proales* from saline springs in the Namib Desert, with the description of a new species. Hydrobiologia 255/256:449–454.

Brakenridge, G. R. 1981. Late Quaternary floodplain sedimentation along the Pomme de Terre River, southern Missouri. Quaternary Research 15:62–76.

Brame, A. H., Jr. 1970. A new species of *Batrachoseps* (slender salamander) from the desert of southern California. Los Angeles County Museum Contributions in Science 200:1–11.

Brazel, A. J. 1976. A preliminary water and energy budget analysis of Montezuma Well, Arizona. Journal of the Arizona Academy of Science 11:9–15.

Brinson, M. M. 1993. Hydrogeomorphic classification for wetlands: Final report. Report No. WESTRWRPDE4. East Carolina University, Greenville, N.C.

Brinson, M. M., and R. D. Rheinhardt. 1996. The role of reference wetlands in functional assessment and mitigation. Ecological Applications 16:69–76.

Bristor v. Cheatham. 1953. 75 Ariz. 227, 238, 255 P. 2d 173, 178–179.

Brooks, R. H. 1967. A comparative analysis of bone from Locality 2, Tule Springs, Nevada. Pages 403–411 in H. M. Wormington and D. Ellis, editors. Pleistocene studies in southern Nevada. Nevada State Museum Anthropological Papers 13. Carson City.

Brower, J. E., J. H. Zar, and C. N. von Ende. 1990. Field and laboratory methods for general ecology. 3rd ed. Wm. C. Brown Publishers, Dubuque, Iowa.

———. 1998. Field and laboratory methods for general ecology. 4th ed. McGraw-Hill, Boston.

Brown, D. E., N. B. Carmony, C. H. Lowe, and R. M. Turner. 1976. A second locality for native California fan palms in Arizona. Journal of the Arizona Academy of Science 11:37–41.

Brown, D. S., B. L. Petri, and G. M. Nalley. 1992. Compilation of hydrologic data for the Edwards Aquifer, San Antonio area, 1991, with 1934–91 summary. Edwards Underground Water District Bulletin 51. San Antonio, Texas.

Brown, J. H. 1971. The desert pupfish. Scientific American 225:104–110.

———. 1984. On the relationship between abundance and distribution of species. American Naturalist 124:255–279.

———. 1995. Macroecology. University of Chicago Press, Chicago.

Brown, J. H., and M. Lomolino. 1998. Biogeography. 2nd ed. Sinauer Associates, Sunderland, Mass.

Brown, W. S. 1971. Morphometrics of *Terrapene coahuila* (Chelonia, Emydidae), with comments on its evolutionary status. Southwestern Naturalist 16 (2): 171–184.

———. 1974. Ecology of the aquatic box turtle, *Terrapene coahuila* (Chelonia, Emydidae) in northern Mexico. Bulletin of the Florida State Museum, Biological Sciences 19 (1): 1–67.

Brues, C. T. 1928. Studies on the fauna of hot springs in the western United States and the biology of thermophilous animals. Proceedings of the American Academy of Arts and Sciences 63:139–228.

———. 1932. Further studies on the fauna of North American hot springs. Proceedings of the American Academy of Arts and Sciences 67:183–303.

Brune, G. 1975. Major and historical springs of Texas. Report of the Texas Water Development Board 189. Austin.

———. 1981. Springs of Texas, vol. 1. Privately published by Gunnar Brune, printed by Branch-Smith, Fort Worth, Texas.

Brussard, P. F., D. A. Charlet, and D. Dobkin. 1999. The Great Basin–Mojave Desert region, *in* M. J. Mac, P. A. Opler, C. E. Puckett-Haecker, and P. D. Doran, editors. The status and trends of the nation's biological resources. U.S. Department of the Interior, U.S. Geological Survey, Reston, Va.

Bryan, K. 1919. Classification of springs. Journal of Geology 27:522–561.

Bull, W. B. 1991. Geomorphic responses to climatic change. Oxford University Press, New York.

Bunn, S. E., and A. H. Arthington. 2002. Basic principles and ecological consequences of altered flow regimes for aquatic biodiversity. Environmental Management 30:492–507.

Burk, J. H. 1977. Sonoran desert vegetation. Pages 869–889 *in* M. G. Barbour and J. Major, editors. Terrestrial vegetation of California. John Wiley and Sons, New York.

Burkle, L. H. 1995. Current issues in Pliocene paleoclimatology. Pages 3–7 *in* E. S. Vrba, G. H. Denton, T. C. Partridge, and L. H. Burckle, editors. Paleoclimate and evolution, with emphasis on human origins. Yale University Press, New Haven, Conn.

Busch, D., and J. Trexler, editors. 2002. Interdisciplinary approaches for evaluating ecoregional initiatives. Island Press, New York.

Cafaro, P., and M. Verma. 2001. For Indian wilderness. Pages 58–63 *in* D. Rothenberg and M. Ulvaeus, editors. The world and the wild: Expanding wilderness conservation beyond its American roots. University of Arizona Press, Tucson.

Campbell, C. J., and W. Green. 1968. Perpetual succession of stream-channel vegetation in a semiarid region. Journal of the Arizona Academy of Science 5:86–98.

Carney, R. S. 1994. Consideration of the oasis analogy for chemosynthetic communities at Gulf of Mexico hydrocarbon vents. Geo-Marine Letters 14:149–159.

Carson, E. W., and T. E. Dowling. 2006. Influence of hydrogeographic history and hybridization on the distribution of genetic variation in the pupfishes *Cyprinodon atrorus* and *C. bifasciatus*. Molecular Ecology 15:667–679.

Castetter, E. F., and W. H. Bell. 1942. Pima and Papago Indian agriculture. University of New Mexico Press, Albuquerque.

Castetter, E. F., and R. M. Underhill. 1935. The ethnobiology of the Papago Indians. Ethnobiological Studies in the American Southwest. University of New Mexico Bulletin 275:3–84.

Chaney, E. W., W. Elmore, and W. S. Platts. 1990. Livestock grazing on western riparian areas. U.S. Environmental Protection Agency, Denver, Colo.

Clarholm, M. 1981. Protozoan grazing of bacteria in soil: Impact and importance. Microbial Ecology 7:343–350.

Clark, A., and R. L. Burgess. 1966. The persistent perennial vegetation on the rim of Montezuma Well, Arizona. Journal of the Arizona Academy of Science 4:35–42.

Clarke, F. W. 1924. Mineral wells and springs. Pages 181–217 in U.S. Geological Survey, editor. The data of geochemistry. U.S. Government Printing Office, Washington, D.C.

Clarkson, R. W., A. T. Robinson, and T. L. Hoffnagle. 1997. Asian tapeworm (*Bothriocephalus acheilognathi*) in native fishes from the Little Colorado River, Grand Canyon, Arizona. Great Basin Naturalist 57:66–69.

Clover, E. U., and L. Jotter. 1944. Floristic studies in the canyons of the Colorado and tributaries. American Midland Naturalist 32:591–642.

Coad, B. W. 1980. Environmental change and its impact on the freshwater fishes of Iran. Biological Conservation 19:51–80.

Coad, B. W., and A. Abdoli. 1993. Exotic fish species in the fresh waters of Iran. Zoology of the Middle East 9:65–80.

Cohen, A. E., D. A. Hendrickson, and J. C. Marks. 2001. Cuatro Ciénegas yesterday and today: A look at historic and modern photographs. Proceedings of the Desert Fishes Council 33. Bishop, Calif.

Cole, G. A. 1965. Final report to Montezuma Castle National Monument of investigations of Montezuma Well. Archived at the headquarters of Montezuma Castle National Monument, McGuireville.

———. 1981. Habitats of North American desert fishes. Pages 477–492 in R. J. Naiman and D. L. Soltz, editors. Fishes in North American deserts. Wiley Interscience, New York.

———. 1983. The uniqueness and value of the Montezuma Well–ditch system as a research resource. Unpublished report. Archived at the headquarters of Montezuma Castle National Monument.

———. 1984. Crustacea from the bolson of Cuatro Ciénegas, Coahuila, Mexico. Journal of the Arizona-Nevada Academy of Science 19:3–12.

Cole, G. A., and W. T. Barry. 1973. Montezuma Well, Arizona, as a habitat. Journal of the Arizona Academy of Science 8:7–13.

Cole, G. A., and G. L. Batchelder. 1969. Dynamics of an Arizona travertine forming stream. Journal of the Arizona Academy of Science 5:271–283.

Cole, G. A., and R. W. Watkins. 1977. *Hyalella montezuma*, a new species (Crustacea: Amphipoda) from Montezuma Well, Arizona. Hydrobiologia 52:175–184.

Cole, K. L. 1990. Reconstruction of past desert vegetation along the Colorado River using packrat middens. Palaeogeography, Palaeoclimatology, Palaeoecology 76:349–366.

Cole, K. L., and L. K. Murray. 1997. Middle and late Holocene packrat middens from Capitol Reef National Park. Pages 3–24 in C. van Riper III and E. T. Deshler, editors. Proceedings of the third biennial conference on research in Colorado Plateau national parks. U.S. Department of the Interior, National Park Service, Transactions and Proceedings Series NPS/NRNAU/NRTP-95/11.

Coleman, D. C., E. R. Ingham, and J. C. Moore. 1990. An across ecosystem analysis of seasonal effects and faunal reduction on decomposition in a semiarid prairie, meadow, and lodgepole pine forest. Pedobiologia 34:207–219.

Coleman, R. L., and C. N. Dahm. 1990. Stream geomorphology effects on periphyton standing crop and primary production. Journal of the North American Benthological Society 9:293–302.

Collier v. Arizona Department of Water Resources. 1986. 150 Ariz. 195, 197, 722 P. 2d 363, 364.

Colorado State University Cooperative Extension Tri River Area. 2002. Tamarisk Symposium, 26–27 September 2001, Holiday Inn, Grand Junction, Colo. Available from http://www.coopext.colostate.edu/TRA/tamarisksymposium.html.

Compton-O'Brien, A., R. D. Foust Jr., M. E. Ketterer, and D. W. Blinn. 2003. Total arsenic in a fishless desert spring: Montezuma Well, Arizona. Pages 200–209 *in* Y. Cai and O. C. Braids, editors. Biogeochemistry of environmentally important trace elements. American Chemical Society Symposium Series 835. Washington, D.C.

CONABIO. 1998. Actualización de la Norma Oficial Mexicana Nom-Ecol-059-1994, que determina las especies y subespecies de flora y fauna silvestres terrestres y acuaticas en peligro de extincion, amenazadas, raras y las sujetas a proteccion especial y que establece especificaciones para su proteccion. Available from http://www.conabio.gob.mx/informacion/nom-059.htm.

Contreras-Arquieta, A. 1998. New records of the snail *Melanoides tuberculata* (Müller, 1774) (Gastropoda: Thiaridae) in the Cuatro Cienegas basin, and its distribution in the state of Coahuila, Mexico. Southwestern Naturalist 43 (2): 283–286.

Contreras-Arquieta, A., G. Guajardo-Martinez, and S. Contreras-Balderas. 1995. *Thiara (Melanoides) tuberculata* (Müller, 1774) (Gastropoda: Thiaridae), su probable impacto ecológico en México. Publicaciones Faculdad Ciencas Biologicos, Universidad Autonoma de Nuevo León, México 8:17–24.

Contreras-Balderas, S. 1977. Biota endemica de Cuatro Ciénegas, Coahuila, Mexico. Memoria del Primer Simposio Nacional de Ornitiologia, pp. 106–113.

———. 1984. Environmental impacts in Cuatro Cienegas, Coahuila, Mexico: A commentary. Journal of Arizona-Nevada Academy of Science 19:85–88.

———. 1991. Conservation of Mexican freshwater fishes: Some protected sites and species, and recent federal legislation. Pages 191–197 *in* W. L. Minckley and J. E. Deacon, editors. Battle against extinction: Native fish management in the American West. University of Arizona Press, Tucson.

———. 2000. The valley of Cuatro Ciénegas, Coahuila: Its biota and its future. Pages 94–96 *in* R. A. Abell et al., editors. Freshwater ecoregions of North America: A conservation assessment. Island Press, Washington, D.C.

———. 2002. Letter to presidents Clinton and Fox (2001) regarding the status of México's desert aquatic habitats and associated aquifers. In possession of senior author.

Contreras-Balderas, S., and M. de L. Lozano-Vilano. 1994. Water, endangered fishes, and development perspectives in arid lands of México. Conservation Biology 8:379–387.

———. 1996. Extinction of most Sandia and Potosí valleys (Nuevo León, México) endemic pupfishes, crayfishes, and snails. Ichthyological Explorations of Freshwater 7:33–40.

Cooper, D. J., J. S. Sanderson, D. I. Stannard, and D. P. Groeneveld. 2006. Effects of long-term water table drawdown on evapotranspiration and vegetation in an arid region phreatophyte community. Journal of Hydrology 325:21–34.

Cornett, J. W. 1981. *Batrachoseps major* (Amphibia: Caudata: Plethodontidae) from the Colorado Desert. Bulletin of the Southern California Academy of Sciences 80 (2): 94–95.

———. 1984. The desert palm oasis. Education Bulletin 84-1. Education Foundation, Desert Protective Council, Spring Valley, Calif.

———. 1985a. A new locality for desert fan palms in California. Desert Plants 7 (3): 164.

———. 1985b. Notes on the use of spadices of *Washingtonia filifera* (Wendl) by *Xylocopa californica* (Cresson) (Hymenoptera: Apoidae). Pan-Pacific Entomologist 61 (3): 251–252.

———. 1985c. Germination of *Washingtonia filifera* seeds eaten by coyotes. Principes 29 (1): 19.

———. 1986a. The largest desert fan palm oases. Principes 30 (2): 82–84.

———. 1986b. Palm burning and increased spadix production. Southwestern Naturalist 31 (4): 552–553.

———. 1986c. Arthropod visitors at *Washingtonia filifera* (Wendl) flowers. Pan-Pacific Entomologist 62 (3): 224–225.

———. 1987a. Indians and the desert fan palm. Masterkey 60 (4): 12–17.

———. 1987b. Cold tolerance in the desert fan palm, *Washingtonia filifera* (Arecaceae). Madroño 34 (1): 57–62.

———. 1988a. Naturalized populations of the desert fan palm, *Washingtonia filifera*, in Death Valley National Monument. Pages 167–174 in C. A. Hall Jr. and V. Doyle-Jones, editors. Plant biology of eastern California. University of California Press, Los Angeles.

———. 1988b. The occurrence of the desert fan palm, *Washingtonia filifera*, in southern Nevada. Desert Plants 8 (4): 169–171.

———. 1989a. Another new locality for the desert fan palm in California. Crossosoma 15 (2): 1–4.

———. 1989b. The desert fan palm, not a relict. San Bernardino County Museum Association Quarterly 36 (2): 56–58.

———. 1989c. Desert palm oasis. Palm Springs Desert Museum, Palm Springs, Calif.

———. 1989d. The impact of rodents on desert fan palm (*Washingtonia filifera*) populations. Page 5 *in* Symposium on the scientific value of the desert. Anza-Borrego Foundation, Borrego Springs, Calif.

———. 1991. Population dynamics of the palm, *Washingtonia filifera*, and global warming. San Bernardino County Museum Association Quarterly 38 (2): 46–47.

———. 1993. Factors determining the occurrence of the desert fan palm, *Washingtonia filifera*. San Bernardino County Museum Association Special Publication 93 (1): 37–38.

Cornett, J. W., and J. Zabriskie. 1981. Fire in a desert oasis. Fremontia 8 (4): 18–21.

Cotera, M. C. 2001. Tenencia de la tierra y legislación en los derechos de uso de agua en el área de protección de flora y fauna de Cuatro Ciénegas. Pronatura A. C., Monterrey, Nuevo León, Mexico.

Cotter, J. L. 1937. The occurrence of flints and extinct animals in pluvial deposits near Clovis, New Mexico, part 4: Report on the excavations at the gravel pit in 1936. Proceedings Philadelphia Academy Natural Sciences 89:2–16.

————. 1938. The occurrence of flints and extinct animals in pluvial deposits near Clovis, New Mexico, part 6: Report on field season of 1937. Proceedings Philadelphia Academy Natural Sciences 90:113–117.

Courtenay, W. R., Jr., and G. K. Meffe. 1989. Small fishes in strange places: A review of introduced poeciliids. Pages 319–332 in G. K. Meffe and F. F. Snelson Jr., editors. Ecology and evolution of livebearing fishes (Poeciliidae). Prentice Hall, Englewood Cliffs, N.J.

Courtenay, W. R., Jr., and P. B. Moyle 1992. Crimes against biodiversity: The lasting legacy of fish introductions. Pages 365–372 in J. E. Williams and R. J. Neves, editors. Biological diversity in aquatic management. Transactions of the 57th North American Wildlife and Natural Resources Conference, Charlotte, North Carolina. Wildlife Management Institute, Washington, D.C.

Courtenay, W. R., Jr., and J. R. Stauffer Jr. 1984. Distribution, biology, and management of exotic fishes. Johns Hopkins University Press, Baltimore.

Cox, C. F., I. N. Healey, and P. D. Moore. 1976. Biogeography: An ecological and evolutionary approach. Blackwell Scientific Publications, Oxford, England.

Crowe, J. C., and J. M. Sharp. 1997. Hydrogeologic delineation of habitats for endangered species: The Comal Springs/River system. Environmental Geology 30 (1–2): 17–28.

Crutzen, P. J., and E. F. Stoermer. 2000. The "Anthropocene." Global Change Newsletter 41:12–13.

Curtin, L. S. M. 1949. By the prophet of the earth. San Vicente Foundation, Sante Fe, N.Mex.

Czarnecki, D. B. 1979. Epipelic and epilithic diatom assemblages in Montezuma Well National Monument, Arizona. Journal of Phycology 15:346–352.

Czarnecki, D. B., and D. W. Blinn. 1979. Observations on southwestern diatoms, part 2: *Caloneis latiuscula* var. *Reimeri* n. var., *Cyclotella pseudostelligera* f. *parva* n. f. and *Gomphonema montezumense* n. sp., new taxa from Montezuma Well National Monument. Transactions of the American Microscopical Society 98:110–114.

Dahl, T. E. 1990. Wetland losses in the United States, 1780s to 1980s. U.S. Department of Interior, Fish and Wildlife Service, Washington, D.C.

Dale, V. H., S. Brown, R. A. Haeuber, N. T. Hobbs, N. Huntly, R. J. Naiman, W. E. Riebsame, M. G. Turner, and T. J. Valone. 2000. Ecological principles and guidelines for managing the use of land. Ecological Applications 10:639–670.

Damon, P. E., C. V. Haynes, and A. Long. 1964. Arizona radiocarbon dates, part 5. Radiocarbon 6:91–107.

Daubenmire, R. 1959. A canopy-coverage method of vegetational analysis. Northwest Science 33:43–64.

Davies, R. W., D. W. Blinn, and B. Dehdashti. 1987. The comparative ecology of three species of Erpobdellidae (Annelida: Hirudinoidea). Archiv für Hydrobiologie 111:601–614.

Davies, R. W., and F. R. Govedich. 2001. Annelida: Euhirudinea and Acanthobdellidae. Pages 465–504 in J. H. Thorpe and A. P. Covich, editors. Ecology and classification of North American freshwater invertebrates. Academic Press, San Diego, Calif.

Davies, R. W., R. N. Singhal, and D. W. Blinn. 1985. *Erpobdella montezuma* (Hirudinoidea: Erpobdellidae), a new species of freshwater leech from North America (Arizona, USA). Canadian Journal of Zoology 63:965–969.

Davies, R. W., F. J. Wrona, and V. Kalarani. 1992. Assessment of activity-specific metabolism of aquatic organisms: An improved system. Canadian Journal of Fisheries and Aquatic Science 49:1142–1148.

Davies-Colly, R. J., and D. G. Smith. 2001. Turbidity, suspended sediment, and water clarity: A review. Journal of the American Water Resources Association 37:1–17.

Davis, E. H. 1920. The Papago ceremony of Vikita. Indian Notes and Monographs [Museum of the American Indian, Heye Foundation, New York] 4:157–177.

———. 1921. The Papago ceremony of the Vikita. Indian Notes and Monographs [Museum of the American Indian, Heye Foundation, New York] 3 (4): 1–24.

Davis, O. K., and D. S. Shafer. 1992. A Holocene climatic record for the Sonoran Desert from pollen analysis of Montezuma Well, Arizona, USA. Palaeogeography, Palaeoclimatology, Palaeoecology 92:107–119.

Deacon, J. E., and W. L. Minckley. 1974. Desert fishes. Pages 385–488 in G. W. Brown Jr., editor. Desert biology, vol. 2. Academic Press, New York.

Deacon, J. E., and C. D. Williams. 1991. Ash Meadows and the legacy of the Devils Hole pupfish. Pages 69–87 in W. L. Minckley and J. E. Deacon, editors. Battle against extinction: Native fish management in the American West. University of Arizona Press, Tucson.

DeAngelis, D. L. 1992. Dynamics of nutrient cycling and food webs. Chapman and Hall, London.

Dehdashti, B. 1987. Population dynamics of Hyalella montezuma (Amphipoda) and interactions with Erpobdella montezuma (Hirudinoidea) in Montezuma Well, Arizona. Ph.D. dissertation, Northern Arizona University, Flagstaff.

Dehdashti, B., and D. W. Blinn. 1986. A bryozoan from an unexplored cave at Montezuma Well, Arizona. Southwestern Naturalist 31:557–558.

———. 1991. Population dynamics and production of the pelagic amphipod Hyalella montezuma in a thermally constant system. Freshwater Biology 25:131–141.

DeLoach, C. J., R. I. Carruthers, J. E. Lovich, T. L. Dudley, and S. D. Smith. 2000. Ecological interactions in the biological control of saltcedar (Tamarix spp.) in the United States: Toward a new understanding. Pages 819–873 in N. R. Spencer, editor. Proceedings of the X International Symposium on Biological Control of Weeds. 4–14 July 1999, Montana State University, Bozeman.

Dettinger, M. D., J. R. Harrill, D. L. Schmidt, and J. W. Hess. 1995. Distribution of carbonate-rock aquifers in southern Nevada and adjacent parts of California, Arizona, and Utah. U.S. Geological Survey Water-Resources Investigations Report 91-4146. U.S. Government Printing Office, Washington, D.C.

Devereux, P. 2000. The sacred place: The ancient origin of holy and mystical sites. Cassell, London.

Diegues, A. C. 2001. Recycled rainforest myths. Pages 155–170 in D. Rothenberg and M. Ulvaeus, editors. The world and the wild: Expanding wilderness conservation beyond its American roots. University of Arizona Press, Tucson.

Dinerstein, E., D. Olson, J. Atchley, C. Loucks, S. Contreras-Balderas, R. A. Abell, E. Iñigo, E. Enkerlin, C. Williams, and G. Castilleja. 2000. Ecoregion-based conservation in the Chihuahuan Desert: A biological assessment. Island Press, Washington, D.C.

Dinger, E. 2001. Aquatic invertebrates of Cuatro Ciénegas, Coahuila, México, and effects of fish on stromatolite invertebrate assemblages. M.S. thesis, Northern Arizona University, Flagstaff.

Dinius, S. H. 1987. Design of an index of water quality. Water Resources Bulletin 23:833–843.

Douce, G. K., and D. P. Webb. 1978. Indirect effects of soil invertebrates on litter decomposition: Elaboration via analysis of a tundra model. Ecological Modelling 4:339–359.

Douglas, M. E., P. C. Marsh, and W. L. Minckley. 1994. Indigenous fishes of western North America and the hypothesis of competitive displacement: *Meda fulgida* (Cyprinidae) as a case study. Copeia 1994 (1), 9–19.

Dove, A. D. M., T. H. Cribb, S. P. Mockler, and M. Lintermans. 1997. The Asian fish tapeworm, *Bothriocephalus acheilognathi*, in Australian freshwater fishes. Marine and Freshwater Research 48:181–183.

Dryer, J. D. 1994. Late Pleistocene vegetation change at Stanton's Cave, Colorado River, Grand Canyon National Park, Arizona. M.S. thesis, Quaternary Studies, Northern Arizona University, Flagstaff.

Drysdale, R. N., and S. J. Gale. 1997. The Indarri Falls travertine dam, Lawn Hill Creek, northwest Queensland, Australia. Earth Surface Processes and Landforms 22:413–418.

Duan, Y., S. I. Guttman, J. T. Oris, and A. J. Bailer. 2000. Genetic structure and relationships among populations of *Hyalella azteca* and *Hyalella montezuma* (Crustacean: Amphipods). Journal of the North American Benthological Society 19:308–320.

DuBois, S. M., and A. W. Smith. 1980. The 1887 earthquake in San Bernardino Valley, Sonora: Historic accounts and intensity patterns in Arizona. Arizona Bureau of Geology and Mining Technology Special Paper 3. Tucson.

Dudley, W. W., and J. D. Larson. 1976. Effect of irrigation pumping on desert pupfish habitats in Ash Meadows, Nye County, Nevada. U.S. Geological Survey Professional Paper 927. U.S. Government Printing Office, Washington, D.C.

Duffy, J. E. 1993. Genetic population structure in two tropical sponge-dwelling shrimps that differ in dispersal potential. Marine Biology 116:691–695.

Dumont, H. J. 1982. Relict distribution patterns of aquatic animals: Another tool in evaluating Late Pleistocene climate changes in the Sahara and Sahel. Paleoecology of Africa 14:1–32.

Duvernell, D. D., and B. J. Turner. 1999. Variation and divergence of Death Valley pupfish populations at retrotransposon-defined loci. Molecular Biology and Evolution 16:363–371.

Eastwood, A. 1896. Report of a collection of plants from San Juan County, in southeastern Utah. Proceedings of the California Academy of Science II, 6:271–329.

Echelle, A. A., and P. J. Conner. 1989. Rapid, geographically extensive genetic introgression after secondary contact between two pupfish species (Cyprinodon, Cyprinodontidae). Evolution 43:717–727.

Echelle, A. A., and A. F. Echelle. 1994. Assessment of genetic introgression between two pupfish species, *Cyprinodon elegans* and *C. variegatus* (Cyprinodontidae), after more than 20 years of secondary contact. Copeia 1994 (1), 590–597.

———. 1997. Genetic introgression of endemic taxa by non-natives: A case study with Leon Springs pupfish and sheepshead minnow. Conservation Biology 11:153–161.

Edwards, C. A. 2000. Soil invertebrate controls and microbial interactions in nutrient and organic matter dynamics in natural and agroecosystems. Pages 141–160 *in* D. C. Coleman and P. F. Hendrix, editors. Invertebrates as webmasters in ecosystems. CABI Publishing, New York.

Edwards, J. R., G. Longley, R. Moss, R. Ward, R. Matthews, and B. Stewart. 1989. A classification of Texas aquatic communities with special consideration toward the conservation of endangered and threatened taxa. Texas Journal of Science 41:231–240.

Ellsworth, P. M., and D. W. Blinn. 2003. Distribution and biomass of *Tropocyclops prasinus mexicanus* (Cyclopoida) in a near-thermally constant environment, Montezuma Well, Arizona. Southwestern Naturalist 48:341–346.

Elzinga, C. L., D. W. Salzer, and J. W. Willoughby. 1998. Measuring and monitoring plant populations. BLM Technical Reference 1730-1. BLM/RS/ST-98/005+1730. Available from Bureau of Land Management, National Business Center BC-650B, PO Box 25047, Denver, CO 80225-0047.

English, D., T. M. Alam, and D. W. Blinn. 1986. Electrophoretic characterization of *Hyalella montezuma*: An endemic filter feeding amphipod. Journal of Heredity 77:284–285.

Environmental Protection Agency. 2000. Office of Water, Wetland Program Development Grants, FY2000 Grant Guidance.

Erman, N. A. 1992. Factors determining biodiversity in Sierra Nevada cold spring systems. Pages 119–127 *in* C. A. Hall Jr., V. Doyle-Jones, and B. Widawski, editors. The history of water: Eastern Sierra Nevada, Owens Valley, White-Inyo Mountains. University of California White Mountains Research Station, Los Angeles.

Ezcurra, E., R. S. Felger, A. Russell, and M. Equihua. 1988. Freshwater islands in a desert sand sea: The hydrology, flora and phytogeography of the Gran Desierto oases of northwestern Mexico. Desert Plants 9:35–44, 55–63.

Ezcurra, E., and V. Rodrigues. 1986. Rainfall patterns in the Gran Desierto, Sonora, Mexico. Journal of Arid Environments 10:13–28.

Fancy, S. G. 2000. Guidance for the design of sampling schemes for inventory and monitoring of biological resources in national parks. U.S. National Park Service. Available at http://science.nature.nps.gov/im/monitor/docs/nps_sg.doc.

Federal Register. 1994. 59 F. R. 10898, Mar. 8.

Felger, R. S. 2000. Flora of the Gran Desierto and Rio Colorado of northwestern Mexico. University of Arizona Press, Tucson.

Felger, R. S., P. L. Warren, S. A. Anderson, and G. P. Nabhan. 1992. Vascular plants of a desert oasis: Flora and ethnobotany of Quitobaquito, Organ Pipe Cactus National Monument, Arizona. Proceedings of the San Diego Society of Natural History 8:1–39.

Fenelon, J. M., and M. T. Moreo. 2002. Trend analysis of ground-water levels and spring discharge in the Yucca Mountain region, Nevada and California, 1960–2000. U.S. Geological Survey Water-Resources Investigations Report 02-4178. U.S. Government Printing Office, Washington, D.C.

Ferrington, L. C., Jr. 1995. Biodiversity of aquatic insects and other invertebrates in springs: Introduction. Journal of the Kansas Entomological Society (supplement) 68:1–3.

Fetter, C. W. 1994. Applied hydrogeology. Macmillan College Publishing Company, New York.

Fewkes, J. W. 1900. A theatrical performance at Walpi. Proceedings of the Washington Academy of Sciences 2:605–629.

Fitzgerald, J. 1996. Subsurface residence time and geochemical evolution of spring waters issuing from the South Rim aquifer in the eastern Grand Canyon, Arizona. M.S. thesis, University of Nevada, Las Vegas.

Flanagan, L. B., C. S. Cook, and J. R. Ehleringer. 1997. Unusually low carbon isotope ratios in plants from hanging gardens in southern Utah. Oecologia 111:481–489.

Fleischner, T. L. 1994. Ecological costs of livestock grazing in western North America. Conservation Biology 8:629–644.

Fondo Mundial para la Naturaleza and Protección de la Fauna Mexicana. 1999. Planeación del uso del agua en el Valle de Cuatrociénegas. Proteccion de la Fauna Mexicana (PROFAUNA), Saltillo, Coahuila, Mexico.

Foust, R. D., Jr., P. Mohapatra, A. Compton, and J. Reifel. 2000. Surface water and groundwater arsenic in the Verde Valley in central Arizona. Proceedings of the 31st International Geological Congress, Pre-Congress Workshop on Arsenic in Groundwaters of Sedimentary Aquifers, Rio de Janeiro, Brazil.

Fowler, J. F. 1995. Biogeography of hanging gardens on the Colorado Plateau. Ph.D. dissertation, Department of Zoology and Physiology, University of Wyoming, Laramie.

Fowler, J. F., N. L. Stanton, R. L. Hartman, and C. L. May. 1995. Level of endemism in hanging gardens of the Colorado Plateau. Pages 215–223 in C. van Riper III, editor. Proceedings of the second biennial conference on research in Colorado Plateau national parks. U.S. Department of the Interior, National Park Service, Transactions and Proceedings Series NPS/NRNAU/NRTP-97/12.

Freeze, R. A., and J. A. Cherry. 1979. Groundwater. Prentice-Hall, Englewood Cliffs, N.J.

Fritze, H., T. Pennanen, and J. Pietikäinen. 1993. Recovery of soil microbial biomass and activity from prescribed burning. Canadian Journal of Forest Research 23:1286–1290.

Fritze, H., A. Smolander, T. Levula, V. Kitunen, and E. Mälkönen. 1994. Wood-ash fertilization and fire treatments in a Scots pine forest stand: Effects on the organic layer, microbial biomass, and microbial activity. Biology and Fertility of Soils 17:57–63.

Frumkin, A., A. Shimron, and J. Rosenbaum. 2003. Radiometric dating of the Siloam Tunnel, Jerusalem. Nature 425:169–171.

Fry, B. 1988. Food web structure on Georges Bank from stable C, N, and S isotopic compositions. Limnology and Oceanography 33:1182–1190.

———. 1991. Stable isotope diagrams of freshwater food webs. Ecology 72:2293–2297.

Fuller, M. L. 1904. Underground waters of eastern United States. U.S. Geological Survey Water Supply Paper 114. Washington, D.C.

Fuller, P. L., L. G. Nico, and J. D. Williams. 1999. Nonindigenous fishes introduced into inland waters of the United States. American Fisheries Society Special Publication 27. Bethesda, Md.

Gage, J. D., and T. A. Tyler. 1991. Deep sea biology: A natural history of organisms of the deep sea floor. Cambridge University Press, Cambridge, England.

Gamradt, S. C., and L. B. Kats. 1996. Effect of introduced crayfish and mosquitofish on California newts. Conservation Biology 10 (4): 1155–1162.

Garate, D.T. 1999. Who named Arizona? The Basque connection. Journal of Arizona History 40:35–57.

Garcia, F. O. 1992. Carbon and nitrogen dynamics and microbial ecology in tallgrass prairie. Kansas State University, Manhattan.

Garcia, F. O., and C. W. Rice. 1994. Microbial biomass dynamics in tallgrass prairie. Soil Science Society of America Journal 58:816–823.

Garrett, G. P. 1999. The renovation of Lake Balmorhea. Proceedings of the Desert Fishes Council 30:9–12.

Gastil, G., J. Minch, and R. P. Phillips. 1983. The geology and ages of islands. Pages 13–25 in T. J. Case and M. L. Cody, editors. Island biogeography in the Sea of Cortez. University of California Press, Berkeley.

Gibson, D. J., T. R. Seastedt, and J. M. Briggs. 1993. Management practices in tallgrass prairie: Large- and small-scale experimental effects on species composition. Journal of Applied Ecology 30:247–255.

Glennon, R. 2002. Water follies: Groundwater pumping and the fate of America's fresh waters. Island Press, Washington, D.C.

Glinski, R. 1977. Regeneration and distribution of sycamore and cottonwood trees along Sonoita Creek, Santa Cruz County, Arizona. Pages 116–124 in R. R. Johnson and D. A. Jones, editors. Importance, preservation and management of riparian habitat: A symposium. U.S.D.A., National Forest Service General Technical Report RM-43.

Goren, M., and R. Ortal. 1999. Biogeography, diversity, and conservation of the inland water fish communities in Israel. Biological Conservation 89:1–9.

Govedich, F. R. 1996. Evolutionary ecology of erpobdellid leeches (Hirudinoidea: Erpobdellidae) in northern Arizona. M.S. thesis, Northern Arizona University, Flagstaff.

Govedich, F. R., D. W. Blinn, R. H. Hevly, and P. S. Keim. 1999. Cryptic radiation in erpobdellid leeches in xeric landscapes: A molecular analysis of population differentiation. Canadian Journal of Zoology 77:52–57.

Govedich, F. R., D. W. Blinn, P. Keim, and R. W. Davies. 1998. Phylogenetic relationships of three genera of Erpobdellidae (Hirudinoidea), with a description of a new genus, *Motobdella*, and species, *Motobdella sedonensis*. Canadian Journal of Zoology 76:2164–2171.

Graf, W. L. 1978. Fluvial adjustments to the spread of tamarisk in the Colorado Plateau region. Geological Society of America Bulletin 89:1491–1501.

Graham, T. B. 1997. The Knowles Cañon hanging garden, Glen Canyon National Recreation Area, five years after burning: Vegetation and soil biota patterns. Pages 173–190 in C. van Riper III and E. T. Deshler, editors. Proceedings of the third biennial conference on research in Colorado Plateau national parks. U.S. Department of the Interior, National Park Service, Transactions and Proceedings Series NPS/NRNAU/NRTP-95/11.

Grall, G. 1995. Mexico's desert aquarium. National Geographic 188 (4): 85–97.

Gramly, R. M. 1988. Palaeo-Indian sites south of Lake Ontario, western and central New York state. Pages 265–280 in R. S. Laub, N. G. Miller, and D. W. Steadman, editors. Late Pleistocene and early Holocene paleoecology and archaeology of the eastern Great Lakes region. Bulletin of the Buffalo Society of Natural Sciences 33.

Grand Canyon Wildlands Council. 2002. Inventory of 100 Arizona Strip springs, seeps and natural ponds: Final project report. Grand Canyon Wildlands Council, Inc., Flagstaff.

———. 2004. Biological inventory of 10 South Rim Springs, Grand Canyon National Park. Final report. Grand Canyon Wildlands Council, Inc., Flagstaff, Ariz.

Green, F. E. 1992. Comments on the report of a worked mammoth tusk from the Clovis site. American Antiquity 57 (2): 331–337.

Gregory, H. E. 1951. The geology and geography of the Paunsaugunt region, Utah. U.S. Geological Survey Water Supply Paper 226. USGPO, Washington, D.C.

Griffith, J. 1992. Beliefs and holy places: A spiritual geography of the Pimería Alta. University of Arizona Press, Tucson.

Griffiths, R. P., J. A. Entry, and W. H. Emmingham. 1997. Chemistry and microbial activity of forest and pasture riparian-zone soils along three Pacific Northwest streams. Plant and Soil 190:169–178.

Grim, J. N., and C. A. Manganaro. 1985. Form of the extrusomes and secreted material of the ciliated protozoan *Pseudourostyla cristata*, with some phylogenetic interpretations: A light, scanning electron, and transmission electron microscopic study. Transactions of the American Microscopical Society 104:350–359.

Grime, J. P. 1979. Plant strategies and vegetation processes. John Wiley and Sons, Chichester, U.K.

Grimm, N. B., A. Chacon-Torres, C. N. Dahm, O. T. Lind, P. L. Starkweather, and W. W. Wurtsbaugh. 1996. Sensitivity of aquatic ecosystems to climatic and anthropogenic changes: The Basin and Range, American Southwest, and México. Hydrological Processes 11:1023–1041.

Grossnickle, N. E., T. N. Beversdorf, and T. A. Sobat. 1985. Predation by *Erpobdella punctata* (Hirudinoidea) during nocturnal migrations in Montezuma Well, Arizona. Page 39, abstract, *in* 48th Annual Meeting of the American Society of Limnology and Oceanography. University of Minnesota, Minneapolis.

Guerra, L. V. 1952. Ichthyological survey of the Rio Salado, Mexico. M.S. thesis, University of Texas, Austin.

Habermehl, M. A. 1980. The Great Artesian Basin, Australia. Journal of Australian Geology and Geophysics 5:9–38.

———. 1982. Springs in the Great Artesian Basin, Australia: Their origin and nature. Australian Bureau of Mineral Resources, Geology, and Geophysics Report 235:1–50.

Hanski, I. 1982. Dynamics of regional distribution: The core and satellite hypothesis. Oikos 38:210–221.

Harbaugh, A. W., and M. G. McDonald. 1996. User's documentation for MODFLOW-96, an update to the U.S. Geological Survey modular finite-difference ground-water flow model. U.S. Geological Survey Open-File Report 96-485. USGPO, Washington.

Harper, K. T., L. L. St. Clair, K. H. Thorne, and W. M. Hess, editors. 1994. Natural history of the Colorado Plateau and Great Basin. University Press of Colorado, Niwot.

Harris, C. 1981. Oases in the desert: The mound springs of northern South Australia. Proceedings of the Royal Geographic Society of Australasia. South Australian Branch 81:26–39.

———. 1992. Mound springs: South Australian conservation initiatives. Rangeland Journal 14:157–173.

Hastings, J. R., and R. R. Humphrey. 1969. Climatological data and statistics for Sonora and northern Sinaloa. Technical Reports on the Meteorology and Climatology of Arid Regions 19. University of Arizona Institute of Atmospheric Physics, Tucson.

Haury, L. R. 1982. Crustacean collections from Montezuma Well: Unpublished report to the National Park Service. Montezuma Castle National Monument, McGuireville.

Haynes, C. V., Jr. 1967. Quaternary geology of the Tule Springs area, Clark County, Nevada. Pages 15–104 *in* H. M. Wormington and D. Ellis, editors. Pleistocene studies in southern Nevada. Nevada State Museum Anthropological Papers 13. Carson City.

———. 1985. Mastodon-bearing springs and Late Quaternary geochronology of the lower Pomme de Terre valley, Missouri. Geological Society of America, Special Paper 204. Boulder, Colo.

———. 1987. Curry Draw, Cochise County, Arizona: A Late Quaternary stratigraphic record of Pleistocene extinction and Paleoindian activities. Geological Society of America Centennial Field Guide, Cordilleran Section 1:23–28.

———. 1991. Geoarchaeological and paleohydrological evidence for a Clovis-age drought in North America and its bearing on extinction. Quaternary Research 35 (3): 438–450.

———. 1995. Geochronology of paleoenvironmental change, Clovis type site, Blackwater Draw, New Mexico. Geoarchaeology 10 (5): 317–388.

———. 1998. Geochronology of the stratigraphic manifestations of paleoclimatic events at Paleoindian sites. Page 138 *in* Abstracts of the 63rd Annual Meeting of the Society for American Archaeology. Seattle, Washington.

Haynes, C. V., Jr., and G. A. Agogino. 1966. Prehistoric springs and geochronology of the Clovis site, New Mexico. American Antiquity 31 (6): 812–821.

Haynes, C. V., Jr., P. E. Damon, and D. C. Grey. 1966. Arizona radiocarbon dates, part 6. Radiocarbon 8:1–21.

Haynes, C. V., Jr., and E. Hemmings. 1968. Mammoth-bone shaft wrench from Murray Springs, Arizona. Science 159:186–187.

Haynes, C. V., Jr., and B. B. Huckell (editors). 2007. Murray Springs: A Clovis site with multiple activity areas in the San Pedro Valley, Arizona. Anthropological Paper 71, University of Arizona Press, Tucson.

Haynes, C. V., Jr., J. J. Saunders, D. Stanford, and G. A. Agogino. 1992. Reply to Green's comments on the Clovis site. American Antiquity 57:338–344.

Haynes, C. V., Jr., D. J. Stanford, M. Jodry, J. Dickenson, J. L. Montgomery, P. H. Shelley, I. Rovner, and G. A. Agogino. 1999. A Clovis well at the type site 11,500 B.C.: The oldest well in America. Geoarchaeology 14 (4): 455–470.

Haynes, C. V., Jr., M. Stuiver, H. Haas, J. E. King, F. B. King, and J. J. Saunders. 1983. Mid-Wisconsinan radiocarbon dates from mastodon- and mammoth-bearing springs, Ozark Highland, Missouri. Radiocarbon 25 (2): 381–391.

Haynes, R. C., and U. T. Hammer. 1978. The saline lakes of Saskatchewan, part 4. Primary production by phytoplankton in selected saline ecosystems. Internationale Revue der Gesamten Hydrobiologie 63:337–351.

Heath, R. C. 1984. Ground-water regions of the United States. U.S. Geological Survey Water-Supply Paper 2220. U.S. Government Printing Office, Washington, D.C.

———. 1989. Basic ground-water hydrology. U.S. Geological Survey Water-Supply Paper 2242. U.S. Government Printing Office, Washington, D.C.

Hecht, S., and D. A. Posey. 1989. Preliminary results on soil management techniques of the Kayapo Indians. Pages 174–188 in D. A. Posey and W. Balée, editors. Resource management in Amazonia: Indigenous and folk strategies, advances in economic botany. New York Botanical Garden Press, New York.

Heckman, R. A., P. D. Greger, and R. C. Furtek. 1993. The Asian tapeworm, *Bothriocephalus acheilognathi*, in fishes from Nevada. Journal of the Helminthological Society of Washington 60:127–128.

Heino, J., T. Muotka, H. Mykra, R. Paavola, H. Hamalainen, and E. Koskenniemi. 2003. Defining macroinvertebrate assemblage types of headwater springs: Implications for bioassessment and conservation. Ecological Applications 13:843–852.

Hemmings, E. T. 1970. Early man in the San Pedro Valley, Arizona. Ph.D. dissertation, University of Arizona, Tucson.

Henderson, J. 1933. Caverns, ice caves, sinkholes, and natural bridges, part 2. University of Colorado Studies 20:130.

Henderson, R. 1961. On desert trails today and yesterday. Westernlore Press, Los Angeles.

Hendrickson, D. A. 1993. Evaluation of the razorback sucker (*Xyrauchen texanus*) and Colorado squawfish (*Ptychocheilus lucius*) reintroduction programs in central Arizona based on surveys of fish populations in the Salt and Verde rivers from 1987 to 1990. Report to the Arizona Game and Fish Department (Project E5-2, Job 7, Section VI of the Endangered Species Act) to U.S. Fish and Wildlife Service, Albuquerque, N.Mex.

Hendrickson, D. A., and W. L. Minckley. 1985. Cienegas—Vanishing climax communities of the American Southwest. Desert Plants 6:103–175.

Hershler, R. 1984. The Hydrobiid snails (Gastropoda: Rissoacea) of the Cuatro Cienegas Basin: Systematic relationships and ecology of a unique fauna. Journal of the Arizona-Nevada Academy of Science 19 (1): 61–76.

———. 1985. Systematic revision of the Hydrobiidae (Gastropoda: Rissoacea) of the Cuatro Cienegas basin, Coahuila, Mexico. Malacologia 26 (1–2): 31–123.

———. 1994. A review of the North American freshwater snail genus *Pyrgulopsis* (Hydrobiidea). Smithsonian Contributions to Zoology 554. Smithsonian Institution Press, Washington, D.C.

———. 1998. A systematic review of the hydrobiid snails (Gastropoda: Rissooidea) of the Great Basin, western United States: part 1, Genus *Pyrgulopsis*. Veliger 41:1–132.

Hershler, R., and L. A. C. Hayek. 1988. Shell variation of springsnail populations in the Cuatro Cienegas Basin, Mexico: Preliminary analysis of limnocrene fauna. Nautilus 102 (2): 56–64.

Hershler, R., and J. J. Landye. 1988. Arizona Hydrobiidae (Prosobranchia: Rissoacea). Smithsonian Contributions in Zoology 459 (I, I–IV): 1–63. Washington, D.C.

Hershler, R., H.-P. Liu, and M. Mulvey. 1999. Phylogenetic relationships within the aquatic snail genus *Tryonia*: Implications for biogeography of the North American Southwest. Molecular Phylogenetics and Evolution 13:377–391.

Hershler, R., and W. L. Minckley. 1986. Microgeographic variation in the banded spring snail (Hydrobiidae: *Mexipyrgus*) from the Cuatro Cienegas basin, Coahuila, Mexico. Malacologia 27 (2): 357–374.

Hershler, R., and D. W. Sada. 2002. Biogeography of Great Basin aquatic snails of the genus *Pyrgulopsis*. Smithsonian Contributions to the Earth Sciences 33:255–276.

Hester, J. J. 1972. Blackwater Locality No. 1: A stratified early man site in eastern New Mexico. Fort Burgwin Research Center, Publication 8. Ranchos de Taos, N.Mex.

Hevly, R. H. 1974. Recent paleoenvironments and geological history at Montezuma Well. Journal of the Arizona Academy of Science 9:66–75.

Hevly, R. H., J. D. Nations, and B. J. Szabo. 1992. Age and deposition environment of tufa at Montezuma Well and Red Tank Draw, Arizona. Proceedings of the SWARM Division of AAAS. Tucson.

Hibbard, C. W., and E. S. Riggs. 1949. Upper Pliocene vertebrates from Keefe Canyon, Meade County, Kansas. Geological Society of America Bulletin 60:829–860.

Hickman, J. C., editor. 1993. The Jepson manual. University of California Press, Berkeley.

Hill, A. M., and D. M. Lodge. 1995. Multi-trophic-level impact of sublethal interactions between bass and omnivorous crayfish. Journal of the North American Benthological Society 14 (2): 306–314.

Hillis, D. M., and R. L. Mayden. 1985. Spread of the Asiatic clam *Corbicula* (Bivalvia: Cobiculacea) in the New World tropics. Southwestern Naturalist 30 (3): 454–456.

Hitchcock, F. H. 1927. The phantom of the desert. Grafton Press, New York.

Hoffman, G. L. 1970. Intercontinental and transcontinental dissemination and transfaunation of fish parasites with emphasis on whirling disease (*Myxosoma cerebralis*). Pages 69–81 *in* S. F. Sniesko, editor. A symposium on diseases of fishes and shellfishes. American Fisheries Society Special Publication 5. Washington, D.C.

Hoffman, G. L., and G. Schubert. 1984. Some parasites of exotic fishes. Pages 233–261 *in* W. R. Courtenay Jr. and J. R. Stauffer Jr., editors. Distribution, biology, and management of exotic fishes. Johns Hopkins University Press, Baltimore, Md.

Holliday, V. T. 2000. Folsom drought and episodic drying on the southern High Plains from 10,900–10,200 ^{14}C yr B.P. Quaternary Research 53:1–12.

Holmes, J. W., A. F. Williams, J. W. Hall, and C. J. Henschke. 1981. Measurement of discharges from some of the mound springs in the desert of northern South Australia. Journal of Hydrology 49:329–339.

Holmgren, A. H., L. M. Schultz, and T. K. Lowrey. 1976. *Sphaeromeria*, a genus closer to *Artemisia* than to *Tanacetum* (Asteraceae: Anthemideae). Brittonia 28:255–262.

Houston & T. C. Ry. Co. v. East. 1904. 98 Tex. 146, 82 S.W. 279.

HRS Water Consultants. 1998. Groundwater reconnaissance of the Cuatrocienegas area. Unpublished report prepared for the Nature Conservancy. HRS Water Consultants, Lakewood, Colo.

Hubbard, H. G. 1899. The home of *Dinapate wrightii* Horn. Entomological News 10 (4): 83–89.

Hubbs, C. 1957. *Gambusia heterochir*, a new poeciliid fish from Texas, with an account of its hybridization with *G. affinis*. Tulane Studies in Zoology 5:1–16.

———. 1959. Population analysis of a hybrid swarm between *Gambusia affinis* and *G. heterochir*. Evolution 13:236–246.

———. 1971. Competition and isolation mechanisms in the *Gambusia affinis* × *G. heterochir* hybrid swarm. Bulletin of the Texas Memorial Museum 19:1–47.

———. 1980. The solution to the *Cyprinodon bovinus* problem: Eradication of a pupfish genome. Proceedings of the Desert Fishes Council 10 (1978): 9–18.

———. 1995. Springs and spring runs as unique aquatic systems. Copeia 1999 (4): 989–991.

Hubbs, C., and H. J. Broderick. 1963. Current abundance of *Gambusia gaigei*, an endangered fish. Southwestern Naturalist 8:46–48.

Hubbs, C., G. Hoddenbach, and C. M. Fleming. 1986. An enigmatic population of *Gambusia gaigei*, an endangered fish species. Southwestern Naturalist 32:121–123.

Hubbs, C. L., and R. R. Miller. 1948. The zoological evidence: Correlation between fish distributions and hydrographic history in the desert basins of western United States. Pages 17–166 *in* The Great Basin with emphasis on glacial and postglacial times. Bulletin of the University of Utah 38, Biological Series 10. Salt Lake City.

Hubbs, C. L., R. R. Miller, and L. C. Hubbs. 1974. Hydrographic history and relict fishes of the north-central Great Basin. Memoirs of the California Academy of Sciences 7:1–259.

Hudson, L. E., R. A. Coleman, and S. Charles. 2000. A preliminary population study of *Platanthera zothecina* (Higgins & Welsh) *Kartesz* & *Gandhi* (Orchidaceae) at Navajo National Monument, Arizona. North American Native Orchid Journal 6:103–118.

Hunt, C. B., and D. R. Mabey. 1966. Stratigraphy and structure, Death Valley, California. U.S. Geological Survey Professional Paper 494A. U.S. Government Printing Office, Washington, D.C.

Hunt, H. W., E. R. Ingham, D. C. Coleman, E. T. Elliott, and C. P. P. Reid. 1988. Nitrogen limitation of decomposition and primary production in shortgrass, mountain meadow and lodgepole pine forest. Ecology 69:1009–1016.

Hurd, E. G., N. L. Shaw, J. Mastrogiuseppe, L. C. Smithman, and S. Goodrich. 1998. Field guide to intermountain sedges. General Technical Report RMRS-GTR-10. Rocky Mountain Research Station, U.S. Forest Service.

Hurlbert, S. H. 1987. Pseudoreplication and the design of ecological field experiments. Ecological Monographs 54 (2): 187–211.

Huston, M. 1979. A general hypothesis of species diversity. American Naturalist 113:81–101.

———. 1994. Biological diversity: The coexistence of species on changing landscapes. Cambridge University Press, Cambridge, England.

Hynes, H. B. N. 1970. The ecology of running waters. University of Toronto Press, Toronto.

Idaho Department of Parks v. Idaho Department of Water Administration. 1974. 530 P. 2d 924.

INEGI. 1974a. Carta Topográfica 1:50,000, Cuatro Ciénegas G13B59, Coahuila. Instituto Nacional de Estadística, Geografía e Informática, Mexico City.

———. 1974b. Carta Topográfica 1:50,000, El Venado G14A51, Coahuila. Instituto Nacional de Estadística, Geografía e Informática, Mexico City.

Ingham, E. R., C. Cambardella, and D. C. Coleman. 1986. Manipulation of bacteria, fungi and protozoa by biocides in lodgepole pine forest soil microcosms: Effects on organism interactions and nitrogen mineralization. Canadian Journal of Soil Science 66:261–272.

Ingham, E. R., D. C. Coleman, and J. C. Moore. 1989. Analysis of food-web structure and function in a shortgrass prairie, a mountain meadow and lodgepole pine forest. Biology of the Fertility of Soils 8:29–37.

Ingham, E. R., and K. A. Horton. 1987. Bacterial, fungal and protozoan responses to chloroform fumigation in stored prairie soil. Soil Biology and Biochemistry 19:545–550.

Ingham, R. E., J. A. Trofymow, E. R. Ingham, and D. C. Coleman. 1985. Interactions of bacteria, fungi and their nematode grazers: Effects on nutrient cycling and plant growth. Ecological Monographs 55:119–140.

Insam, H., and K. Haselwandter. 1989. Metabolic quotient of the soil microflora in relation to plant succession. Oecologia 79:174–178.

Jameson, E., Jr., and H. J. Peeters, editors. 1988. California mammals. University of California Press, Berkeley.

Jenkins, M. 2005. Squeezing water from a stone. High Country News 37:8–13.

Johannesson, K. H., K. J. Stetzenbach, V. F. Hodge, D. K. Kreamer, and X. Zhou. 1997. Delineation of ground-water flow systems in the southern Great Basin using aqueous rare earth element distributions. Ground Water 35 (5): 807–819.

Johnson, D. B. 1982. The biology of a freshwater planktonic amphipod, *Hyalella montezuma*. M.S. thesis, Northern Arizona University. Flagstaff.

Johnson, R. R. 1991. Historic changes in vegetation along the Colorado River in the Grand Canyon. Pages 178–206 *in* National Research Council, editor. Colorado River ecology and dam management. National Academy Press, Washington, D.C.

Johnson, R. R., C. D. Ziebell, D. R. Patton, P. F. Folliott, and R. H. Hamre. 1985. Riparian ecosystems and their management: Reconciling conflicting issues. Proceedings of the First North American Riparian Conference. U.S. Department of Agriculture National Forest Service General Technical Report RM-120. Washington, D.C.

Johnson, S. R., and A. K. Knapp. 1993. The effect of fire on gas exchange and aboveground biomass production in annually vs biennially burned *Spartina pectinata* wetlands. Wetlands 13:299–303.

Kalarani, V., D. C. Reddy, D. W. Blinn, and R. W. Davies. 1993. Interspecific differences in respiration and energy storage reserves in two freshwater predatory leeches from ecosystems of contrasting stability. Comparative Biochemistry and Physiology 104A:239–242.

Karr, J. R. 1991. Biological integrity: A long-neglected aspect of water resource management. Ecological Applications 1:66–84.

———. 1999. Defining and measuring river health. Freshwater Biology 41:221–234.

Kauffman, J. B., and W. C. Krueger. 1984. Livestock impacts on riparian ecosystems and streamside management implications. Journal of Range Management 37:430–438.

Kaufman, L. S. 1992. Catastrophic changes in species-rich freshwater ecosystems: The lessons of Lake Victoria. BioScience 42:846.

Keate, N. S. 1996. Quantitative analysis of hanging garden endemic plant species richness in the Moab region, Utah. Ph.D. dissertation, University of Utah, Salt Lake City.

Keilhack, K. 1912. Lehrbuch der grundwasser and quellenkunde. 3rd ed. Geb. Borntraeger, Berlin.

Kelso, S. 1991. Taxonomy of Primula sects: Aleuritia and Armerina in North America. Rhodora 93:67–99.

Kidd, D. E., and W. E. Wade. 1963. Algae of Montezuma Well, Arizona and vicinity. Plateau 36:63–71.

Killham, K. 1995. Soil ecology. Cambridge University Press, Cambridge, England.

Kimball, B. A., and P. K. Christensen. 1996. Residence time of water discharging from the hanging gardens of Zion Park. Water Resources Bulletin 32:531–539.

Kimberling, D. N., A. R. Ferreira, S. M. Shuster, and P. Keim. 1996. RAPD marker estimation of genetic structure among isolated northern leopard frog populations in the southwestern USA. Molecular Ecology 5:521–529.

King, J. E. 1973. Late Pleistocene palynology and biogeography of the western Missouri Ozarks. Ecological Monographs 43:539–565.

Knopf, F. L., and R. W. Cannon. 1982. Structural resilience of a willow riparian community to changes in grazing practices. Pages 198–207 *in* J. M. Peek and P. D. Dalke, editors. Wildlife-livestock relationships symposium: Proceedings 10. University of Idaho, Forest, Wildlife, and Range Experimental Station, Moscow.

Knopf, F. L., R. R. Johnson, T. Rich, F. B. Samson, and R. C. Szaro. 1988. Conservation of riparian ecosystems in the United States. Wilson Bulletin 100:272–284.

Kodrick-Brown, A., and J. H. Brown. 2007. Native fishes, exotic mammals, and the conservation of desert springs. Frontiers in Ecology and the Environment. 5:549–553.

Konieczki, A. D., and S. A. Leake. 1997. Hydrogeology and water chemistry of Montezuma Well in Montezuma Castle National Monument and surrounding area, Arizona. U.S. Geological Survey Water Resources Investigations Report 97-4156. U.S. Government Printing Office, Washington, D.C.

Kornfield, I. L., and R. K. Koehn. 1975. Genetic variation and speciation in New World cichlids. Evolution 29:427–437.

Kornfield, I. L., D. C. Smith, P. S. Gagnon, and J. N. Taylor. 1982. The cichlid fish of Cuatro Cienegas, Mexico: Direct evidence of conspecificity among distinct trophic morphs. Evolution 36 (4): 658–664.

Kornfield, I. L., and J. N. Taylor. 1983. A new species of polymorphic fish, *Cichlasoma minckleyi*, from Cuatro Cienegas, Mexico (Teleostei: Cichlidae). Proceedings of the Biological Society of Washington 96 (2): 253–269.

Kreamer, D. K., V. F. Hodge, I. Rabinowitz, K. H. Johannesson, and K. J. Stetzenbach. 1996. Trace element geochemistry in water from selected springs in Death Valley National Park, California. Ground Water 34:95–103.

Krupp, F., and W. Schneider. 1989. The fishes of the Jordan River drainage basin and Azraq Oasis. Fauna of Saudi Arabia 10:347–416.

Krupp, F., W. Schneider, L. A. Nader, and O. Kushaim. 1990. Zoological Survey in Saudi Arabia, Spring 1990. Fauna of Saudi Arabia 11:3–9.

Laity, J. E., and M. C. Malin. 1985. Sapping processes and the development of theater-headed valley networks on the Colorado Plateau. Geological Society of America Bulletin 96:203–217.

Langdon, J. S. 1988. Prevention and control of fish diseases in the Murray-Darling basin. Pages 163–172 *in* Anonymous, editor. Proceedings of the workshop on native fish management. Murray-Darling Basin Commission, Canberra, Australia.

Lange, A. 1957. Studies on the origin of Montezuma Well and Cave, Arizona. Cave Studies 1:31–45.

Larmer, P. 1998. Tackling tamarisk. High Country News, April 25, 1.

Laub, R. S. 1994. The Pleistocene/Holocene transition in western New York state: Fruits of interdisciplinary studies of the Hiscock site. Pages 155–167 *in* R. I. McDonald, editor. Great Lakes archaeology and paleoecology: Exploring interdisciplinary initiatives in the nineties. Quaternary Sciences Institute, University of Waterloo, Waterloo, Ontario.

———. 2000. A second dated mastodon bone artifact from Pleistocene deposits at the Hiscock site (western New York state): Archaeology of Eastern North America 28:141–154.

Laub, R. S., M. F. DeRemer, C. A. Dufort, and W. L. Parsons. 1988. The Hiscock site: A rich

Late Quaternary locality in western New York state. Pages 67–81 *in* R. S. Laub, N. G. Miller, and D. W. Steadman, editors. Late Pleistocene and early Holocene paleoecology and archeology of the eastern Great Lakes region. Buffalo Society of Natural Sciences 33.

Laub, R. S., and G. Haynes. 1998. Fluted points, mastodons, and evidence of late-Pleistocene drought of the Hiscock site, western New York state. Current Research in the Pleistocene 15:32–34.

Laury, R. L. 1980. Paleoenvironments of a Late Quaternary mammoth-bearing sinkhole deposit, Hot Springs, South Dakota. Geological Society of America Bulletin (Part 1) 91:465–475.

———. 1990. Geological history of the Mammoth Site and surrounding region, Hot Springs area, Fall River and Custer counties, South Dakota: An Overview. Pages 15–21 *in* L. D. Agenbroad, J. I. Mead, and L. W. Nelson, editors. Megafauna and man: Discovery of America's heartland. Mammoth Site of Hot Springs, South Dakota, Inc., Scientific Papers 1.

Lepesme, P. 1947. Les insectes des palmiers. Lechevalier, Paris.

Leshy, J. 1991. Instream flow rights: The private and public roles. C616 ALI-ABA 163. Western Water Law in the Age of Reallocation.

Leshy, J., and J. Belanger. 1988. Arizona water law: Where ground and surface water meet. Arizona State Law Journal 20:657.

Lévêque, C. 1990. Relict tropical fish fauna in the central Sahara. Ichthyological Exploration of Freshwaters 1:39–48.

Liem, K. F., and L. S. Kaufman. 1984. Intraspecific macroevolution: Functional biology of the polymorphic cichlid species *Cichlasoma minckleyi*. Pages 203–215 *in* A. A. Echelle and I. L. Kornfield, editors. Evolution of fish species flocks. University of Orono Press, Orono, Maine.

Linke, P., E. Suess, M. Torres, V. Martens, W. D. Rugh, W. Ziebis, and L. D. Kulm. 1994. In situ measurement of fluid flow from cold seeps at active continental margins. Deep-Sea Research I 41:721–739.

Loiselle, P. V. 1979. A revision of the genus *Hemichromis* (Teleostei: Cichlidae). Part 3, The *Hemichromis guttatus* species group. Koninklijk Museum voor Midden-Afrika Tervuren Belgie Annalen Reeks in Octavo Zoologische Wetenschappen o (228): 69–98.

Lomolino, M. V., B. R. Riddle, and J. H. Brown. 2006. Biogeography. 3rd ed. Sinauer Associates, Sunderland, Mass.

Long, B. A., and R. A. Smith. 1996. Water quality data analysis and interpretation for spring monitoring sites: Southeast Utah group. Technical Report NPS/NRWRD/NRTR-96/77. National Park Service.

Longley, G. 1981. The Edwards Aquifer: Earth's most diverse groundwater ecosystem? International Journal of Speleology 11:123–128.

———. 1992. The subterranean aquatic ecosystem of the Balcones Fault Zone Edwards Aquifer in Texas—threats from over-pumping. Pages 291–300 *in* J. A. Stanford and J. J. Simons, editors. Proceedings of the first international conference on ground water ecology. American Water Resources Association, Bethesda, Md.

Loope, W. L. 1977. Relationships of vegetation to environment in Canyonlands National Park. Ph.D. dissertation, Utah State University, Logan.

Lozano-Vilano, M. de L., and S. Contreras-Balderas. 1993. Four new species of *Cyprinodon* from southern Nuevo León, México, with a key to the *C. eximius* complex (Teleostei: Cyprinodontidae). Ichthyological Exploration of Freshwaters 4:295–308.

Lumholtz, C. S. 1912. New trails in Mexico. Charles Scribner's Sons, New York.

Lyle, D. A. 1878. The springs of southern Nevada. American Naturalist 12:18–27.

MacArthur, R. H., and E. O. Wilson. 1963. An equilibrium theory of insular zoogeography. Ecology 17:373–387.

MacAvoy, S. E., R. S. Carney, C. R. Fisher, and S. A. Macko. 2002. Use of chemosynthetic biomass by large mobile benthic predators in the Gulf of Mexico. Marine Ecology Progress Series 225:65–78.

Madsen, H., and F. Frandsen. 1989. The spread of freshwater snails, including those of medical and veterinary importance. Acta Tropica 46:139–146.

Magurran, A. E. 2004. Measuring biological diversity. Blackwell Publishing, Maldan, Mass.

Malanson, G. P. 1980. Habitat and plant distributions in hanging gardens of the Narrows, Zion National Park, Utah. Great Basin Naturalist 40:178–182.

———. 1982. The assembly of hanging gardens: Effects of age, area, and location. American Naturalist 119:145–150.

———. 1993. Riparian landscapes. Cambridge University Press, New York.

Malanson, G. P., and J. Kay. 1980. Flood frequency and the assemblage of dispersal types in hanging gardens of the Narrows, Zion National Park, Utah. Great Basin Naturalist 40:365–371.

Mangerud, J., S. T. Anderson, B. E. Berglund, and J. J. Donner. 1974. Quaternary stratigraphy of Norden, a proposal for terminology and classification. Boreas 4:109–128.

Marcuson, P. E. 1977. Overgrazed streambanks depress fishery production in Rock Creek, Montana. Fish and Game Federation Aid Project, Special Project Report F-20-R-21-11a. Montana Department of Fish and Game, Helena.

Marine, W. 1963. Ground-water resources of the Bryce Canyon National Park area, Utah. U.S. Geological Survey Water Supply Paper 1475-M. U.S. Government Printing Office, Washington, D.C.

Marsh, P. C., and W. L. Minckley. 1990. Management of endangered Sonoran topminnow at Bylas Springs, Arizona: Description, critique, and recommendations. Great Basin Naturalist 50:265–272.

Martin, P. S. 1967. Prehistoric overkill. Pages 75–120 *in* P. S. Martin and H. E. Wright Jr., editors. Pleistocene extinctions: The search for a cause. Yale University Press, New Haven, Conn.

Mason, I. L. 1939. Studies on the fauna of an Algerian hot spring. Journal of Experimental Biology 16:487–498.

Mather, M. E., and R. A. Stein. 1993. Direct and indirect effects of fish predation on the replacement of a native crayfish by an invading congener. Canadian Journal of Fisheries and Aquatic Sciences 50 (6): 1278–1288.

Mathiot, M. 1973. A dictionary of Papago usage. Vol. 2, Ku'u. Language Science Monographs 8 (2). Indiana University Publications, Bloomington.

Mawby, J. E. 1967. Fossil vertebrates of the Tule Springs site, Nevada. Pages 105–129 *in* H. M. Wormington and D. Ellis, editors. Pleistocene studies in southern Nevada. Nevada State Museum Anthropology Papers 13. Carson City.

Maxey, G. B. 1968. Hydrogeology of desert basins. Ground Water 6:10–22.

May, C. L., J. F. Fowler, and N. L. Stanton. 1995. Geomorphology of the hanging gardens of the Colorado Plateau. Pages 3–24 *in* C. van Riper III, editor. Proceedings of the second biennial conference on research in Colorado Plateau national parks. U.S. Department of the Interior, National Park Service, Transactions and Proceedings Series NPS/NRNAU/NRTP-95/11.

McArthur, E. D., R. Van Buren, S. C. Sanderson, and K. T. Harper. 1998. Taxonomy of *Sphaeromeria*, *Artemisia*, and *Tanacetum* (Compositae, Anthemideae) based on randomly amplified polymorphic DNA (RAPD). Great Basin Naturalist 58:1–11.

McClenaghan, L. R., and A. C. Beauchamp. 1986. Low genic differentiation among isolated populations of the California fan palm. Evolution 40 (2): 315–322.

McCoy, C. J., Jr. 1984. Ecological and zoogeographic relationships of amphibians and reptiles of the Cuatro Cienegas basin. Journal of the Arizona-Nevada Academy of Science 19 (1): 45–59.

McCune, B., and M. J. Mefford. 1995. PC-ORD: Multivariate analysis of community data. Version 2.0. MjM Software Design, Gleneden Beach, Oreg.

McDougall, W. B., and H. S. Haskell. 1960. Seed plants of Montezuma Castle National Monument. Museum of Northern Arizona Bulletin 35. Flagstaff

McLaughlin, S. P. 1989. Natural floristic areas of the western United States. Journal of Biogeography 16:239–248.

McLaughlin, S. P., and J. E. Bowers. 1999. Diversity and affinities of the flora of the Sonoran floristic province. Pages 12–35 *in* R. Robichaux, editor. Ecology of Sonoran Desert plants and plant communities. University of Arizona Press, Tucson.

McLoughlin, N. J., D. W. Blinn, and R. W. Davies. 1999. An energetic evaluation of a predator (leech)–prey (amphipod) couple in Montezuma Well, Arizona. Functional Ecology 13:45–50.

McMillan, R. B. 1976. Man and mastodon: A review of Koch's 1840 Pomme de Terre expeditions. Pages 81–96 *in* W. R. Wood and R. B. McMillan, editors. Prehistoric man and his environments: A case study in the Ozark Highland. Academic Press, New York.

Mead, J., R. Hevly, and L. Agenbroad. 1990. Late Pleistocene invertebrates and plant remains, Mammoth Site, Black Hills, South Dakota. Pages 9–10 *in* L. D. Agenbroad, J. I. Mead, and L. W. Nelson, editors. Megafauna and man: Discovery of America's heartland. Mammoth Site of Hot Springs, South Dakota, Inc., Scientific Papers 1.

Medellín, M. S., and M. C. Cotera. 2001. Plan comunitario de manejo de recursos naturales del ejido antiguos mineros del norte Cuatro Ciénegas, Coahuila, México. Pronatura A. C., Monterrey, Nuevo León, Mexico.

Medellín, M. S., M. C. Cotera, and C. Treviño. 2001. Diagnóstico social y diseño de estratégias operativas para el área de protección de flora y fauna de Cuatro Ciénegas. Pronatura A. C., Monterrey, Nuevo León, Mexico.

Meffe, G. K. 1985. Predation and species replacement in American southwestern fish: A case study. Southwestern Naturalist 30:173–187.

———. 1989. Fish utilization of springs and cienegas in the arid southwest. Pages 475–485 *in* R. R. Sharitz and J. W. Gibbons, editors. Freshwater wetlands and wildlife: Perspectives on natural, managed, and degraded ecosystems. U.S. Department of Energy, Office of Science, Technology, and Information, Oak Ridge, Tenn.

Meffe, G. K., and F. F. Snelson Jr. 1989. Ecology and evolution of livebearing fishes (Poecilii-dae). Pages 13–32 in G. K. Meffe and F. F. Snelson Jr., editors. Ecology and evolution of livebearing fishes (Poeciliidae). Prentice Hall, Englewood Cliffs, N.J.

Mehringer, P. J., Jr. 1967. Pollen analysis of the Tule Springs site, Nevada. Nevada State Museum Anthropology Papers 13:130–200. Carson City.

Meinzer, O. E. 1911. Ground waters of Juab, Millard, and Iron counties, Utah. U.S. Geological Survey Water Supply Paper 277:44–45. U.S. Government Printing Office, Washington, D.C.

———. 1923. Outline of ground-water hydrology, with definitions. U.S. Geological Survey Water Supply Paper 494. U.S. Government Printing Office, Washington, D.C.

Meinzer, O. E., and R. F. Hare. 1915. Geology and water resources of Tularosa Basin, New Mexico. U.S. Geological Survey Water Supply Paper 343:52–53. U.S. Government Printing Office, Washington, D.C.

Mellink, E. 1985. Agricultural disturbance and rodents: Three farming systems in the Sonoran Desert. Journal of Arid Environments 8:207–222.

Meyer, E. R. 1973. Late-Quaternary paleoecology of the Cuatro Ciénegas basin, Coahuila, México. Ecology 54:982–995.

Miller, M. P., D. N. Kimberling, and P. Keim. 1999. Re-analysis of genetic structure among populations of Rana pipiens in Arizona and Utah. Southwestern Naturalist 44:527–530.

Miller, R. R. 1948. The cyprinodont fishes of the Death Valley system of eastern California and southwestern Nevada. Miscellaneous Publications of the Museum of Zoology, University of Michigan 68:1–155.

———. 1950. Speciation in fishes of the genera Cyprinodon and Empetrichthys inhabiting the Death Valley Region. Evolution 4:155–163.

———. 1961. Man and the changing fish fauna of the American Southwest. Papers of the Michigan Academy of Science, Arts, and Letters 46:365–404.

———. 1968. Two new fishes of the genus Cyprinodon from the Cuatro Cienegas basin, Coahuila, Mexico. Occasional Papers of the Museum of Zoology, University of Michigan 659:1–15.

———. 1981. Coevolution of deserts and pupfishes (genus Cyprinodon) in the American Southwest. Pages 39–94 in R. J. Naiman and D. L. Soltz, editors. Fishes in North American deserts. John Wiley and Sons, New York.

Miller, R. R., and C. L. Hubbs. 1960. The spiny-rayed cyprinid fishes (Plagopterini) of the Colorado River system. Miscellaneous Publications of the Museum of Zoology, University of Michigan 115:1–39.

Miller, R. R., and E. P. Pister. 1971. Management of the Owens pupfish, Cyprinodon radiosus, in Mono County, California. Transactions of the American Fisheries Society 100:502–509.

Minckley, W. L. 1969. Environments of the Bolsón of Cuatro Ciénegas, Coahuila, México. Science Series, University of Texas, El Paso 2:1–65.

———. 1973. Fishes of Arizona. Arizona Game and Fish Department, Phoenix.

———. 1974. Endemic fishes of the Cuatro Ciénegas Basin, northern Coahuila, México. Pages 383–404 in R. H. Wauer and D. H. Riskind, editors. Transactions of the symposium of the biological resources of the Chihuahuan Desert region, United States and

Mexico. U.S. Department of the Interior, National Park Service, Transactions and Proceedings Series 3.

————. 1984. Cuatro Ciénegas fishes: Research review and a local test of diversity versus habitat size. Journal of the Arizona-Nevada Academy of Science 19:13–21.

————. 1992. Three decades near Cuatro Ciénegas, México: Photographic documentation and a plea for area conservation. Journal of the Arizona-Nevada Academy of Science 26 (2): 89–118.

————. 1994. A bibliography for natural history of the Cuatro Ciénegas basin and environs, Coahuila, México. Proceedings of the Desert Fishes Council 25. Bishop, Calif.

————. 1999a. Fredric Morton Chamberlain's 1904 survey of Arizona fishes, with annotations. Journal of the Southwest 41:177–237.

————. 1999b. Ecological review and management recommendations for recovery of the endangered Gila topminnow. Great Basin Naturalist 59:230–244.

Minckley, W. L., and D. E. Brown. 1994. Part 6, wetlands. Pages 222–287, 333–351, and literature cited in D. E. Brown, editor. Biotic communities: Southwestern United States and northwestern Mexico. University of Utah Press, Salt Lake City.

Minckley, W. L., and J. E. Deacon. 1968. Southwestern fishes and the enigma of endangered species. Science 159:1424–1433.

————. 1991. Battle against extinction: Native fish management in the American West. University of Arizona Press, Tucson.

Minckley, W. L., G. K. Meffe, and D. L. Soltz. 1991. Conservation and management of short-lived fishes: The cyprinodontoids. Pages 247–282 in W. L. Minckley and J. E. Deacon, editors. Battle against extinction: Native fish management in the American West. University of Arizona Press, Tucson.

Minckley, W. L., and C. O. Minckley. 1986. *Cyprinodon pachycephalus*, a new species of pupfish (Cyprinodontidae) from the Chihuahuan Desert of northern México. Copeia 1986 (1), 184–192.

Minckley, W. L., and P. Unmack. 2000. Western springs: Their faunas and threats to their existence. Pages 52–53 in R. A. Abell et al., editors. Freshwater ecoregions of North America: A conservation assessment. Island Press, Washington, D.C.

Mollemans, F. H. 1989. Terrestrial and semi-aquatic plants. Pages 57–77 in W. Zeidler and W. F. Ponder, editors. Natural history of Dalhousie Springs. South Australian Museum, Adelaide, Australia.

Monroe, S. A., R. C. Antweiler, R. J. Hart, H. E. Taylor, M. Truini, J. R. Rihs, and T. J. Felger. 2005. Chemical characteristics of ground-water discharge along the South Rim of Grand Canyon in Grand Canyon National Park, Arizona, 2000–2001. U.S. Geological Survey Scientific Investigations Report 2004-5146. USGPO, Washington.

Montgomery, C. W., and E. C. Perry Jr., 1982. Isotopic methods in hydrologic studies — An introduction. In E. C. Perry and C. W. Montgomery, editors. Isotope studies of hydrologic processes. Northern Illinois University Press, DeKalb.

Moran, R. 1977. Palms in Baja California. Environment Southwest 478:10–14.

Moyle, P. B. 1976. Inland fishes of California. University of California Press, Berkeley.

————. 1986. Fish introductions into North America: Patterns and ecological impact. Pages 27–43 in H. A. Mooney and J. A. Drake, editors. Ecology of biological invasions of North America and Hawaii. Springer-Verlag, New York.

———. 1995. Conservation of native freshwater fishes in the Mediterranean-type climate of California, USA: A review. Biological Conservation 72 (2): 271–279.

Moyle, P. B., H. W. Li, and B. A. Barton. 1986. The Frankenstein effect: Impact of introduced fishes on native fishes in North America. Pages 415–426 *in* R. H. Stroud, editor. Fish culture in fisheries management. American Fisheries Society, Bethesda, Md.

Muli, J. R. 1996. Environmental problems of Lake Victoria (East Africa): What the international community can do. Lakes and Reservoirs: Research and Management 2 (1–2): 47–53.

Muller, E. H., and P. E. Calkin. 1988. Late Pleistocene and Holocene geology of the eastern Great Lakes region: Geologic setting of the Hiscock paleontological site, western New York. Pages 53–63 *in* R. S. Laub, N. G. Miller, and D. W. Steadman, editors. Late Pleistocene and early Holocene paleoecology and archaeology of the eastern Great Lakes region. Buffalo Society of Natural Sciences 33.

Mullineaux, L. S., and S. C. France. 1995. Dispersal of deep-sea hydrothermal vent fauna. Pages 408–424 *in* S. E. Humphris, R. A. Zierenberg, L. S. Mullineaux, and R. E. Thomson, editors. Seafloor hydrothermal systems: Physical, chemical, biological, and geochemical interactions. Geophysical Monograph Series 91.

Mundorff, J. C. 1971. Non-thermal springs of Utah. Utah Geological and Mineralogical Society Water Resources Bulletin 16.

Myers, A. A., and P. S. Giller, editors. 1988. Analytical biogeography: An integrated approach to the study of animal and plant distributions. Chapman and Hall, New York.

Myers, G. S. 1965. *Gambusia*, the fish destroyer. Tropical Fish Hobbyist 15 (3): 31–32, 53–54, 61–65.

Nabhan, G. P. 1982. The desert smells like rain: A naturalist in Papago Indian country. Northpoint Press, San Francisco.

Nabhan, G. P., W. Hodgson, and F. Fellows. 1989. A meager living on lava and sand? Hia ced O'odham food resources and habitat diversity in oral and documentary histories. Journal of the Southwest 31 (4): 508–533.

Nabhan, G. P., A. M. Rea, K. L. Reichhardt, E. Mellink, and C. F. Hutchinson. 1982. Papago influences on habitat and biotic diversity: Quitovac oasis ethnoecology. Journal of Ethnobiology 2:124–143.

Nash, J. M. 2005. Western water wars: Las Vegas eyes rural Nevada's aquifers, triggering a debate about the future of this arid region. Time 166:56–57.

National Park Service Advisory Board. 2001. Rethinking the national parks for the 21st century. National Park Service Advisory Board Report, Washington, D.C.

National Wildlife Federation, Wilderness Society, Yuma Audubon Society, Palo Verde Group of the Sierra Club. 1999. 151 IBLA 66 (IBLA 96-535).

Nations, J. D., R. H. Hevly, J. J. Landy, and D. W. Blinn. 1981. Paleontology, paleoecology, and depositional history of the Miocene-Pliocene Verde Formation, Yavapai County, Arizona. Arizona Geological Society Digest 13:133–149.

Nature Conservancy Wildland Invasive Species Team. 2002. Invasives on the Web: *Tamarix* species (saltcedar, tamarisk). Available from http://tncweeds.ucdavis.edu/esadocs/tamaramo.html.

Naumburg, E., R. Mata-Gonzales, R. G. Hunter, T. Mclendon, and D. W. Martin. 2005. Phreatophytic vegetation and groundwater fluctuations: A review of current research and application of ecosystem response modeling with an emphasis on Great Basin vegetation. Environmental Management 35:726–740.

Nazarea, V. D. 1998. Cultural memory and biodiversity. University of Arizona Press, Tucson.

Nebeker, G. T., K. T. Harper, J. D. Brotherson, and S. L. Welsh. 1977. Characteristics of plants of common occurrence in hanging gardens of the Colorado Plateau. Unpublished report. Brigham Young University, Provo.

Nekola, J. C. 1999. Paleorefugia and neorefugia: The influence of colonization history on community pattern and process. Ecology 80:2459–2473.

Nelson, J. S. 1984. The tropical fish fauna in cave and basin hot-springs drainage, Banff National Park, Alberta. Canadian Field-Naturalist 97:255–261.

Netopil, R. 1971. The classification of water springs on the basis of the variability of yields. Studia Geographica 22:145–150.

Norris, S. M., and W. L. Minckley. 1997. Two new species of *Etheostoma* (Osteichthyes: Percidae) from central Coahuila, northern México. Ichthyological Exploration of Freshwaters 8 (2): 159–176.

Norusis, M. J. 1993. SPSS for windows: Base system user's guide. Release 6.0. SPSS, Inc., Chicago.

Noss, R. F., E. T. LaRoe, and M. J. Scott. 1995. Endangered ecosystems of the United States: A preliminary assessment of loss and degradation. Biological Report 28. National Biological Survey, Washington, D.C.

Nxomani, C. D., A. J. Bibbink, and R. Kirby. 1994. Differentiation of isolated, threatened fish populations in dolomitic waters of the Transvaal, South Africa, by polyacrylamide gel electrophoresis of total cellular proteins. Biological Conservation 69:185–189.

Oberlin, G. 1995. Aquatic invertebrates in Montezuma Castle National Monument, Arizona: Response to physicochemical parameters. M.S. thesis, Northern Arizona University, Flagstaff.

Oberlin, G. E., and D. W. Blinn. 1997. The effect of temperature on the metabolism and behavior of an endemic amphipod, *Hyalella montezuma*, from Montezuma Well, Arizona, USA. Freshwater Biology 37:55–59.

O'Brien, C., and D. W. Blinn. 1999. The endemic spring snail *Pyrgulopsis* habitats as refugia. Freshwater Biology 42:225–234.

Odum, H. T. 1957. Trophic structure and productivity of Silver Springs, Florida. Ecological Monographs 27:55–112.

Olin, G. 1977. House in the sun. Southwest Parks and Monuments Association, Tucson.

Olu, K., A. Duperret, M. Sibuet, J. P. Foucher, and A. Fiala-Medioni. 1996. Structure and distribution of cold seep communities along the Peruvian active margin: Relation to geological and fluid patterns. Marine Ecology Progress Series 132:109–125.

Ono, R. D., J. D. Williams, and A. Wagner. 1983. Vanishing fishes of North America. Stone Wall Press, Washington, D.C.

Orr, R. T. 1971. Vertebrate biology. W. B. Saunders, Philadelphia.

Parajuli, P. 2001. How can four trees make a jungle? Pages 3–20 *in* D. Rothenberg and M. Ulvaeus, editors. The world and the wild: Expanding wilderness conservation beyond its American roots. University of Arizona Press, Tucson.

Parsons, E. C. 1939. Pueblo Indian religion. University of Chicago Press, Chicago. Reprinted, Bison Books, University of Nebraska Press, Lincoln.

Paul, E. A., and F. E. Clark. 1996. Soil microbiology and biochemistry. Academic Press, San Diego.

Peden, A. E. 1973. Virtual extinction of *Gambusia amistadensis* n. sp., a new poeciliid fish from Texas. Copeia 1973 (2), 210–221.

Pellant, M., D. A. Pyke, P. Shaver, and J. E. Herrick. 2000. Interpreting indicators of rangeland health. Version 3. U.S. Bureau of Land Management National Science and Technology Center Technical Reference 1734-6. Denver.

Perkins, D. J., B. N. Carlson, and M. Fredstone. 1984. The effects of groundwater pumping on natural spring communities in Owens Valley. Pages 515–527 *in* R. E. Warner and K. M. Hendrix, editors. California riparian systems: Ecology, conservation, and productive management. University of California Press, Berkeley.

Peterson, B. J., and B. Fry. 1987. Stable isotopes in ecosystem studies. Annual Review of Ecology and Systematics 18:293–320.

Peterson, B. J., R. W. Howarth, and R. H. Garritt. 1986. Multiple stable isotopes used to trace the flow of organic matter in estuarine food webs. Science (Wash.) 227:1361–1363.

Petersen, K. L. 1994. Modern and pleistocene climatic patterns in the West. Pages 27–54 *in* K. T. Harper, L. L. St. Clair, K. H. Thorne, and W. M. Hess, editors. Natural history of the Colorado Plateau and Great Basin. University Press of Colorado, Niwot.

Phillips, B. G., R. R. Johnson, A. M. Phillips III, and J. E. Bowers. 1980. Resource values of the aquatic and riparian vegetation of Roaring Springs, Grand Canyon. U.S. Dept. of the Interior, National Park Service. Proceedings of the Second Conference on Research in the National Parks. National Park Service, Washington, D.C.

Phillips, B. G., A. M. Phillips III, and M. A. S. Bernzott. 1987. Annotated checklist of vascular plants of Grand Canyon National Park. Grand Canyon Natural History Association Monograph 7.

Pietikäinen, J., and H. Fritze. 1993. Microbial biomass and activity in humus layer following burning: Short term effects of two different fires. Canadian Journal of Forest Research 23:1275–1285.

Pinkava, D. J. 1984. Vegetation and flora of the bolson of Cuatro Cienegas region, Coahuila, Mexico: part 4, Summary, endemism and corrected catalogue. Journal of the Arizona-Nevada Academy of Science 19 (1): 23–47.

Pister, E. P. 1991. The Desert Fishes Council: Catalyst for change. Pages 55–68 *in* W. L. Minckley and J. E. Deacon, editors. Battle against extinction: Native fish management in the American West. University of Arizona Press, Tucson.

Plenet, S., J. Gibert, and P. Vervier. 1992. A floodplain spring: An ecotone between surface water and groundwater. Regulated Rivers: Research and Management 7:93–102.

Plume, R. W. 1996. Hydrogeologic framework of the Great Basin region of Nevada, Utah, and adjacent states. U.S. Geological Survey Professional Paper 1409. U.S. Government Printing Office, Washington, D.C.

Pohlmann, K. F., D. Campagna, J. B. Chapman, and S. Earman. 1998. Investigation of the origin of springs in the Lake Mead National Recreation Area. Publication 41161. Water Resources Center, Desert Research Institute, University of Nevada.

Pointier, J. P., R. N. Incani, C. Balzan, P. Chrosciechowski, and S. Prypchan. 1994. Invasion of the rivers of the littoral central region of Venezuela by *Thiara granifera* and *Melanoides tuberculata* (Mollusca: Prosobranchia: Thiaridae) and the absence of *Biomphalaria glabrata*, snail host of *Schistosoma mansoni*. Nautilus 107 (4): 124–128.

Pointier, J. P., W. L. Paraense, and V. Mazille. 1993. Introduction and spreading of *Biomphalaria straminea* (Dunker, 1848) (Mollusca: Pulmonata: Planorbidae) in Guadeloupe, French West Indies. Memorias do Instituto Oswaldo Cruz 88 (3): 449–455.

Polhemus, D. A. 1993. Conservation of aquatic insects: Worldwide crisis or localized threats? American Zoologist 33:588–598.

Polhemus, D. A., and J. T. Polhemus. 2002. The biogeography of aquatic true bugs (Insect: Heteroptera) in the Great Basin. Pages 235–254 *in* R. Hershler, D. B. Madsen, and D. R. Currey, editors. Great Basin aquatic systems history. Smithsonian Contributions to the Earth Sciences 33. Washington, D.C.

Polhemus, J. T. 1976. Notes on North American Nepidae (Hemiptera: Heteroptera). Pan-Pacific Entomologist 52:204–208.

Polhemus, J. T., and M. W. Sanderson. 1987. *Microvelia rasilis* Drake in Arizona: A species new to the United States (Heteroptera: Veliidae). Great Basin Naturalist 47:660.

Pollack, D. W. 1994. User's guide for MODPATH/MODPATH-PLOT, Version 3: A particle tracking post-processing package for MODFLOW. U.S. Geological Survey Open-File Report 94-464. U.S. Government Printing Office, Washington, D.C.

Ponder, W. F. 1986. Mound springs of the Great Artesian Basin. Pages 403–420 *in* P. DeDeckker and W. D. Williams, editors. Limnology of Australia. CSIRO, Australia, and W. Junk, The Hague, Netherlands.

Ponder, W. F., and G. A. Clark. 1990. A radiation of hydrobiid snails in threatened artesian springs in western Queensland. Records of the Australian Museum 42:301–363.

Por, F. D. 1984. An outline of the distribution patterns of the freshwater copepoda of Israel and surroundings. Hydrobiologia 113:151–154.

Por, F. D., H. J. Bromley, C. Dimentman, G. H. Herbst, and R. Ortal. 1986. River Dan, headwater of the Jordan, an aquatic oasis in the Middle East. Hydrobiologia 134:121–140.

Posey, D. A. 1983. Indigenous ecological knowledge and development of the Amazon. Pages 225–257 *in* E. Moran, editor. The dilemma of Amazonian development. Westview Press, Boulder, Colo.

———. 1985. Indigenous management of tropical forest ecosystems: The case of the Kayapo Indians of the Brazilian Amazon. Agroforestry Systems 3:139–158.

Postel, S. 1992. The last oasis: Facing water scarcity. W. W. Norton, New York.

Postel, S., and S. Carpenter. 1997. Freshwater ecosystem services. Pages 195–214 *in* G. C. Daily, editor. Nature's services: Societal dependence on natural ecosystems. Island Press, Washington, D.C.

Powell, J. W. 1895. The exploration of the Colorado River and its canyons. Reprinted 1961, Dover Reprint, New York.

Pringle, C. M. 2000. River conservation in tropical versus temperate latitudes. Pages 371–384 *in* P. J. Boon, B. R. Davies, and G. E. Petts, editors. Global perspectives on river conservation: Science, policy, and practice. John Wiley and Sons, London, England.

Pringle, C. M., and F. J. Triska. 2000. Emergent biological patterns and surface-subsurface interactions at landscape scales. Pages 167–193 *in* J. B. Jones and P. J. Molholland, editors. Stream and groundwaters. Academic Press, New York.

Prudic, D. E., J. R. Harrill, and T. J. Burbey. 1995. Conceptual evaluation of regional ground-water flow in the carbonate-rock province of the Great Basin, Nevada, Utah, and adjacent states. U.S. Geological Survey Professional Paper 1409-D. U.S. Government Printing Office, Washington, D.C.

Quade, J., R. M. Forester, W. L. Pratt, and C. Carter. 1998. Black mats, spring-fed streams, and late-glacial-age recharge in the southern Great Basin. Quaternary Research 49 (2): 129–148.

Raison, R. J. 1979. Modification of the soil environment by vegetation fires, with particular reference to nitrogen transformations: A review. Plant and Soil 51:73–108.

Raison, R. J., and J. W. McGarity. 1980. Effects of ash, heat, and the ash-heat interaction on biological activities in two contrasting soils: part 1, Respiration rate. Plant and Soil 55:363–376.

Rapport, D. J., N. O. Nielsen, B. L. Lasley, D. E. Rolston, C. O. Qualset, and A. B. Damania, editors. 2003. Managing for healthy ecosystems. Lewis Publishers, Boca Raton, Fla.

Raven, P. H., and D. I. Axelrod. 1978. Origin and relationships of the California flora. University of California Publications in Botany 72.

Rea, A. M. 1983. Once a river: Bird life and habitat changes on the Middle Gila. University of Arizona Press, Tucson.

———. 1997. At the desert's green edge: An ethnobotany of the Gila River Pima. University of Arizona Press, Tucson.

Rea, A. M., G. P. Nabhan, and K. Reichhardt. 1983. Sonoran desert oases: Plants, birds and native peoples. Environment Southwest 503:5–9.

Recio, C. G. 1999. Bajo caudal de agua en pozas. Zocalo de Monclova, December 18.

Redman, C. L. 1999. Human impact on ancient environments. University of Arizona Press, Tucson.

Reichhardt, K. L., E. Mellink, G. P. Nabhan, and A. Rea. 1994. Habitat heterogeneity and biodiversity associated with indigenous agriculture in the Sonoran Desert. Ethno-ecológica 2 (3): 21–34.

Rice, C. W., T. C. Todd, and others. 1998. Belowground biology and processes. Pages 244–264 in A. K. Knapp, J. M. Briggs, D. C. Hartnett, and S. L. Collins, editors. Grassland dynamics: Long-term ecological research in tallgrass prairie. Oxford University Press, New York.

Richter, B. D., J. V. Baumgartner, J. Powell, and D. P. Braun. 1996. A method for assessing hydrologic alteration within ecosystems. Conservation Biology 10:1163–1174.

Richter, B. D., D. P. Braun, M. A. Mendelson, and L. L. Master. 1997. Threats to imperiled freshwater fauna. Conservation Biology 11 (5): 1081–1093.

Ricklefs, R. E. 1997. The economy of nature. 4th ed. W. H. Freeman, New York.

Riggs, A. C., and J. E. Deacon. 2002. Connectivity in desert aquatic ecosystems: The Devils Hole story. Pages 1–38 in D. W. Sada and S. E. Sharpe, editors. Spring-fed wetlands: Important scientific and cultural resources of the intermountain region. Conference Proceedings, Spring-Fed Wetlands: Important Scientific and Cultural Resources of the Intermountain Region, 7–9 May 2002, Las Vegas, Nev. DHS Publication No. 41210. Also available at http://www.wetlands.dri.edu.

Roberts, C. R., and C. W. Mitchell. 1987. Spring mounds in southern Tunisia. Pages 321–334 *in* L. Frostick and I. Reid, editors. Desert sediments: Ancient and modern. Geological Society Special Publication 35. Blackwell Scientific Publications, Boston.

Rodriguez-Almaraz, G. A., and E. Campos. 1994. Distribution and status of the crayfishes (Cambaridae) of Nuevo Leon, Mexico. Journal of Crustacean Biology 14 (4): 729–735.

Rodriguez-Almaraz, G. A., J. A. Gonzalez-Aguilar, and R. Mendoza-Alfaro. 1997. Biological and ecological notes on *Paleomonetes suttkusi* (Crustacea: Palaemonidae) from the Cuatro Cienegas basin, Coahuila, Mexico. Southwestern Naturalist 42 (4): 501–503.

Rodriguez-Loubet, F., M. Antochiw, and E. Araux. 1993. Quitovac: Ethnoarqueologia del desierto de Sonora, Mexico. Editions Recherches sur les Civilisations, Paris.

Rolston, H. I. 1991. Fishes in the desert: Paradox and responsibility. Pages 93–108 *in* W. L. Minckley and J. E. Deacon, editors. Battle against extinction: Native fish management in the American West. University of Arizona Press, Tucson.

Romme, K. D., J. M. Porter, and R. Fleming. 1993. Plant communities of Capitol Reef National Park, Utah. U.S. Dept. of the Interior, National Park Service, Technical Report NPS/NAUCARE/NRTR-93/02. Cooperative Park Studies Unit, Northern Arizona University, Flagstaff.

Rosen, P. C., C. R. Schwalbe, D. A. Parizek Jr., P. A. Holm, and C. H. Lowe. 1995. Introduced aquatic vertebrates in the Chiricahua Region: Effects on declining native ranid frogs. Pages 251–261 *in* L. E. DeBano, G. J. Gottfried, R. H. Hamre, C. B. Edminster, P. F. Ffolliott, and A. Ortega-Rubio, technical coordinators. Biodiversity and management of the Madrean Archipelago: The sky islands of southwestern United States and northwestern Mexico. U.S. Forest Service General Technical Report RM-GTR-264. Fort Collins, Colo.

Rosenberg, D. M., P. McCully, and C. M. Pringle. 2000. Global-scale environmental effects of hydrological alterations: Introduction. BioScience 50:746–751.

Rosenzweig, M. L. 1995. Species diversity in space and time. Cambridge University Press, Cambridge, England.

Rosgen, D. L. 1996. Applied river morphology. Wildlands Hydrology, Pagosa Springs, Colo.

Ross, W. 1985. Oasis fishes of eastern Saudi Arabia. Fauna of Saudi Arabia 7:303–317.

Roth, F. 1987. Data on the distribution and faunal history of the genus *Theodoxus* in the Middle East (Gastropoda: Neritidae). Pages 73–79 *in* F. Krupp, W. Schneider, and R. Kinzelbach, editors. Proceedings of the symposium on the fauna and zoogeography of the Middle East, Mainz, 1985. Reichert, Wiesbaden, Germany.

Rowell, K., and D. W. Blinn. 2003. Herbivory on a chemically defended plant as a predation deterrent in *Hyalella azteca*. Freshwater Biology 48:247–254.

Runck, C. 1989. Secondary production, caloric ingestion, and laboratory rearing of *Ranatra montezuma* Polhemus (Heteroptera: Nepidae). M.S. thesis, Northern Arizona University, Flagstaff.

———. 1993. Influence of invertebrate predation on behavior and energy flow in a freshwater animal community. Ph.D. dissertation, Northern Arizona University, Flagstaff.

Runck, C., and D. W. Blinn. 1990. Population dynamics and secondary production by *Ranatra montezuma* (Heteroptera: Nepidae). Journal of the North American Benthological Society 9:262–270.

———. 1991. Effect of vegetational refuges of invertebrate predation rate. North American Benthological Society meeting, Santa Fe, N.Mex. Abstr.

————. 1992. The foraging ecology of *Ranatra montezuma* (Heteroptera): An optimal forager in Montezuma Well, Arizona. Journal of the Arizona-Nevada Academy of Science 26:119–129.

————. 1993. Secondary productivity by *Telebasis salva* (Odonata) in a thermally constant aquatic ecosystem. Journal of the North American Benthological Society 12:136–147.

————. 1994. Role of *Belostoma bakeri* (Heteroptera) in the trophic ecology of a fishless desert spring. Limnology and Oceanography 39:1800–1812.

————. 1995. Daily energy ingestion by the predatory *Ranatra montezuma* (Heteroptera) on two prey types: Importance of prey behavior. Ecoscience 2:280–285.

Rushforth, S. R., and G. S. Merkley. 1988. Comprehensive list by habitat of the algae of Utah. Great Basin Naturalist 48:154–179.

Sabine, B. J., editor. 1994. National list of plant species that occur in wetlands: Regions 4, 5, and 8. Resource Management Group, Inc., Grand Haven, Mich.

Sada, D. W., E. Fleishman, and D. D. Murphy. 2005. Associations among spring dependent aquatic assemblages and environmental and land use gradients in a Mojave Desert mountain range. Diversity and Distributions 11:91–99.

Sada, D. W., and K. F. Pohlmann. 2003. Draft U.S. National Park Service, Mojave Inventory and Monitoring Network spring survey protocols: Level I, 19 November 2003. Desert Research Institute, Inc., Reno, Nev.

Sada, D. W., J. E. Williams, J. C. Silvey, A. Halford, J. Ramakka, P. Summers, and L. Lewis. 2001. A guide to managing, restoring, and conserving springs in the western U.S. Technical Reference 1737-17. U.S. Bureau of Land Management, BLM/ST/ST-01/001+1737. Denver, Colo.

Sage, R. D., and R. K. Selander. 1975. Trophic radiation through polymorphism in cichlid fishes. Proceedings of the National Academy of Sciences, U.S.A. 72 (11): 4669–4673.

Sakamoto, K., and Y. Oba. 1994. Effect of fungal to bacterial biomass ratio on the relationship between CO_2 evolution and the total soil microbial biomass. Global Change Biology 11 (2): 266.

Sala, A., S. D. Smith, and D. A. Devitt. 1996. Water Use by *Tamarix ramosissima* and associated phreatophytes in a Mojave Desert floodplain. Ecological Applications 6 (3): 888–898.

Salick, J. 1989. Ecological basis of Amuesha agriculture, Peruvian Upper Amazon. Pages 189–212 *in* D. A. Posey and W. Balée, editors. Resource management in Amazonia: Indigenous and folk strategies, advances in economic botany. New York Botanical Garden Press, New York.

Sarkar, S. 2001. Restoring wilderness or reclaiming forests? Pages 37–55 *in* D. Rothenberg and M. Ulvaeus, editors. The world and the wild: Expanding wilderness conservation beyond its American roots. University of Arizona Press, Tucson.

Saunders, J. J. 1977. Late Pleistocene vertebrates of the western Ozark Highlands, Missouri. Illinois State Museum, Reports of Investigations 33:118.

————. 1988. Fossiliferous spring sites in southwestern Missouri. Pages 127–149 *in* R. S. Laub, N. G. Miller, and D. W. Steadman, editors. Late Pleistocene and early Holocene paleoecology and archaeology of the eastern Great Lakes region. Buffalo Society of Natural Sciences 33.

Saxton, D., and L. Saxton. 1973. O'othham Hoho'ok A'agitha: Legends and lore of the Papago and Pima Indians. University of Arizona Press, Tucson.

Scates, M. D. 1968. Notes on the hydrobiology of Azraq Oasis, Jordan. Hydrobiologia 31:73–89.

Schaefer, D. H., and J. R. Harrill. 1995. Simulated effects of proposed ground-water pumping in 17 basins of east-central and southern Nevada. U.S. Geological Survey Water-Resources Investigations Report 95-4173. U.S. Government Printing Office, Washington, D.C.

Schoenherr, A. A. 1981. The role of competition in the displacement of native fishes by introduced species. Pages 173–203 in R. J. Naiman and D. L. Soltz, editors. Fishes in North American deserts. John Wiley and Sons, New York.

Scholes, R. J., and R. Biggs. 2005. A biodiversity intactness index. Nature 434:45–49.

Schroeder, A. H. 1948. Montezuma Well. Plateau 20:37–40.

Schroeder, A. H., and H. F. Hastings. 1958. Montezuma Castle National Monument. National Park Service, Historical Handbook Series 27. Washington, D.C.

Schultz, B. W. 2001. Extent of vegetated wetlands at Owens Dry Lake, California, USA, between 1977 and 1992. Journal of Arid Environments 48:69–87.

Schultz, C. B., L. G. Tanner, F. C. Whitmore, L. L. Ray, and E. C. Crawford. 1963. Paleontological investigations at Big Bone Lick State Park, Kentucky: A preliminary report. Science 142 (29): 1167–1169.

Schütt, H. 1987. The molluscs of the Oasis Palmyra. Pages 62–71 in F. Krupp, W. Schneider, and R. Kinzelbach, editors. Proceedings of the 1985 symposium on the fauna and zoogeography of the Middle East, Mainz, 1985. Reichert, Wiesbaden, Germany.

Schwenkmeyer, D. 1986. The palm oasis: Our tropical vestige. Environment Southwest 514:18–23.

Scott, M. L., P. B. Shafroth, and G. T. Auble. 1999. Responses of riparian cottonwoods to alluvial water table declines. Environmental Management 23:347–358.

Scott, T. M., G. H. Means, R. P. Meegan, R. C. Means, S. B. Upchurch, R. E. Copeland, J. Jones, T. Roberts, and A. Willet. 2004. Springs of Florida. Florida Geologic Society Bulletin 66. Tallahassee.

Seaber, P. R. 1988. Hydrostratigraphic units. Pages 9–14 in W. Back, J. S. Rosenhein, and P. R. Seaber, editors. Hydrogeology O-2. Geological Society of America, Boulder, Colo.

Secretaría de Desarrollo Social. 1994a. Decreto por el que se declara como área protegida, con el carácter de protección de flora y fauna, la región conocida como Cuatrociénegas, municipio de Cuatro Ciénegas, Coahuila. Diario Oficial de la Federación, México 494 (5): 5–11.

———. 1994b. Norma Oficial Mexicana NOM-059-ECOL-1994, que determina las especies y subespecies de flora y fauna silvestres y acuáticas en peligro de extinción, amenazadas, raras y sujetas a protección especial, y que establece especificaciones para su protección. Diario Oficial de la Federación, México 488 (10): 2–60.

Secretaría del Medio Ambiente Recursos Naturales y Pesca. 1999. Programa de manejo del area de proteccion de flora y fauna Cuatrociénegas Mexico. Instituto Nacional de Ecología, Tlacopac, Mexico.

———. 2000. Ordenamiento ecológico para la región de Cuatro Ciénegas, Coahuila. Secretaría del Medio Ambiente, Recursos Naturales y Pesca, Instituto Nacional de Ecología, Mexico City.

Seehausen, O., J. J. M. van Alphen, and F. Witte. 1997a. Cichlid fish diversity threatened by eutrophication that curbs sexual selection. Science (Wash.) 277 (5333): 1808–1811.

Seehausen, O., F. Witte, E. F. Katunzi, J. Smits, and N. Bouton. 1997b. Patterns of the remnant cichlid fauna in southern Lake Victoria. Conservation Biology 11 (4): 890–904.

Sellards, E. H. 1952. Early man in America: A study in prehistory. University of Texas Press, Austin.

Setälä, H., M. Tyynismaa, E. Martikainen, and V. Huhta. 1991. Mineralization of C, N and P in relation to decomposer community structure in coniferous forest soil. Pedobiologia 35:285–296.

Shannon, C. E. 1948. A mathematical theory of communication. Bell System Technical Journal 27:379–423.

Sharpe, S. E. 1991. Late-Pleistocene and Holocene vegetation change in Arches National Park, Grand County, Utah, and Dinosaur National Monument, Moffat County, Colorado. M.S. thesis, Quaternary Studies, Northern Arizona University, Flagstaff.

———. 1993. Late-Wisconsin and Holocene vegetation in Arches National Park, Utah. Pages 109–122 in P. G. Rowlands, C. van Riper III, and M. K. Sogge, editors. Proceedings of the first biennial conference on research in Colorado Plateau National Parks. U.S. Department of the Interior, National Park Service, Transactions and Proceedings Series NPS/NRNAU/NRTP-93/10.

Shepard, W. D. 1992. *Microcylloepus formicoideus* (Coleoptera: Elmidae), a new riffle beetle from Death Valley National Monument, California. Entomological News 101:147–153.

———. 1993. Desert springs—both rare and endangered. Aquatic Conservation: Marine and Freshwater Ecosystems 3:352–359.

Shepard, W. D., D. W. Blinn, R. J. Hoffman, and P. T. Kantz. 2000. Algae of Devils Hole, Nevada, Death Valley National Park. Western North American Naturalist 60:410–419.

Shepard, W. D., and D. Threloff. 1997. Additional records of riffle beetles (Coleoptera: Elmidae) in Death Valley National Park, California. Southwestern Naturalist 42:196–197.

Shine, C., and C. de Klemm. 1999. Wetlands, water and the law: Using law to advance wetland conservation and wise use. IUCN Environmental Policy and Law Paper 38. World Conservation Union, Gland, Switzerland.

Shipton, Z. K., J. P. Evans, D. Kirchner, P. T. Kolesar, A. P. Williams, and J. Heath. 2004. Analysis of CO_2 leakage through "low-permeability" faults from natural reservoirs in the Colorado Plateau, east-central Utah. Pages 43–58 in S. J. Baines and R. H. Worden, editors. Geological storage of carbon dioxide. Geology Society of London Special Publication 233.

Shreve, F., and I. Wiggins. 1964. Vegetation and flora of the Sonoran Desert. Stanford University Press, Stanford, Calif.

Shutler, R., Jr. 1967. Archaeology of Tule Springs. Pages 298–303 in H. M. Wormington and D. Ellis, editors. Pleistocene studies in southern Nevada. Nevada State Museum Anthropological Papers 13. Carson City.

Sigler, J. W., and W. F. Sigler. 1992. Aquatic resources of the arid West: Perspectives on fishes and wilderness management. Pages 51–78 in S. I. Zeveloff, and C. M. McKell, editors.

Wilderness issues in the arid lands of the western United States. University of New Mexico Press, Albuquerque.

Silverberg, R. 1970. Mammoth, mastodons and man. McGraw-Hill, New York.

Skelton, P. H. 1990. The status of fishes from sinkholes and caves in Namibia. Journal of the Namibia Scientific Society/Namibia Wissenschaftliche Gesellschaft, Windhoek, Namibia 42:75–83.

Smalley, A. E. 1964. A new *Palaemonetes* from Mexico. Crustaceana 6:229–232.

Smith, D. C. 1982. Trophic ecology of the cichlid morphs of Cuatro Cienegas, Mexico. M.S. thesis, University of Maine, Orono.

Smith, M. J., W. R. Kay, D. H. D. Edward, P. J. Papas, K. St. J. Richardson, J. C. Simpson, A. M. Pinder, D. J. Cale, P. H. J. Horowitz, J. A. Davis, F. H. Jung, R. H. Norris, and S. A. Halse. 1999. AusRivAS: Using macroinvertebrates to assess ecological condition of rivers in Western Australia. Freshwater Biology 41:269–282.

Smith, M. L. 1981. Late Cenozoic fishes in the warm deserts of North America: A reinterpretation of desert adaptations. Pages 11–38 *in* R. J. Naiman and D. L. Soltz, editors. Fishes in North American deserts. John Wiley and Sons, New York.

Smith, M. L., and B. Chernoff. 1981. Breeding populations of cyprinodontoid fishes in a thermal stream. Copeia 1981 (3), 701–702.

Smith, M. L., and R. R. Miller. 1986. The evolution of the Rio Grande basin as inferred from its fish fauna. Pages 457–485 *in* C. Hocutt and E. O. Wiley, editors. The zoogeography of North American freshwater fishes. John Wiley and Sons, New York.

Smith, R. L. 1974. Ecology and field biology. Harper and Row, New York.

Solar Pathfinder. 1994. Instruction manual for the solar pathfinder. Solar Pathfinder, Inc., Hartford, S.Dak.

Soltz, D. L., and R. J. Naiman. 1978. The natural history of native fishes in the Death Valley system. Natural History Museum of Los Angeles County, California, Science Series 30:1–76.

Sorenson, S. K., P. D. Dileanis, and F. A. Branson. 1991. Soil water and vegetation responses to precipitation and changes in depth to groundwater in Owens Valley, California. U.S. Geological Survey Water-Supply Paper 2370-G. U.S. Government Printing Office, Washington, D.C.

Southwest Parks and Monuments Association. 1985. A checklist of birds from Tuzigoot, Montezuma Castle, and Montezuma Well, National Monument. Southwest Parks and Monuments Association, Tucson.

Spangel, P., and M. Sutton. 1949. The botany of Montezuma Well. Plateau 22: 11–19.

Spence, J. R. 1995. Characterization and possible origins of isolated Douglas fir stands on the Colorado Plateau. Pages 71–82 *in* W. J. Waugh, editor. Climate change in the Four Corners region. Proceedings of a symposium, Grand Junction, Colo., 12–14 September 1994. Department of Energy NTIS CONF-9409325.

———. 1996. A survey and classification of the riparian and spring vegetation in side canyons around Lake Powell, Glen Canyon National Recreation Area, Utah. Unpublished final report, Resource Management Division, Glen Canyon National Recreation Area.

———. 2001. Climate of the central Colorado Plateau, Utah and Arizona: Characterization and recent trends. Pages 187–203 *in* C. van Riper III, K. A. Thomas, and M. A. Stuart, editors. Proceedings of the fifth biennial conference of research on the Colorado Plateau. U.S. Department of the Interior, U.S. Geological Survey Report Series USGSFRESC/ COPL/2001/24. U.S. Government Printing Office, Washington, D.C.

———. 2002. Biotic, water and aquatic invertebrate components of springs along the Colorado River on the Colorado Plateau. Unpublished final report to the National Park Service, Water Resources Division, WASO. Resource Management Division, Glen Canyon National Recreation Area, Page, Arizona.

Spence, J. R., and N. R. Henderson. 1993. Tinaja and hanging garden vegetation of Capitol Reef National Park, southern Utah, U.S.A. Journal of Arid Environments 24:21–36.

Spencer, C. N., B. R. McClelland, and J. A. Stanford. 1991. Shrimp introduction, salmon collapse, and eagle displacement: Cascading interactions in a food web of a large aquatic ecosystem. BioScience 41:14–21.

Spicer, E. H. 1962. Cycles of conquest: The impact of Spain, Mexico, and the United States on the Indians of the Southwest, 1533–1960. University of Arizona Press, Tucson.

Springer, A. E., L. E. Stevens, and R. Harms. 2006. Inventory and classification of selected National Park Service springs on the Colorado Plateau. NPS Cooperative Agreement Number CA 1200-99-009. National Park Service, Flagstaff, Ariz.

Stafford, T. W., Jr. 1990. Late Pleistocene megafauna extinctions and the Clovis culture: Absolute ages based on accelerator ^{14}C dating of skeletal remains. Pages 118–122 in L. D. Agenbroad, J. K. Mead, and L. W. Nelson, editors. Megafauna and man: Discovery of America's heartland. Mammoth Site of Hot Springs, South Dakota, Inc., Scientific Papers 1.

Stanford, D., C. Haynes Jr., J. J. Saunders, G. A. Agogino, and A. T. Boldurian. 1990. Blackwater Draw Locality 1: History of current research and interpretations. Pages 105–128 in V. T. Holliday and E. Johnson, editors. Guidebook to the Quaternary history of the Llano Estacado. Lubbock Lake Landmark Quaternary Research Series 2. Lubbock, Texas.

Starkweather, P., and D. W. Blinn. 1986a. In situ feeding patterns of a planktonic freshwater amphipod: The importance of maximizing mass-specific filtering rates. 49th Annual Meeting of the American Society of Limnology and Oceanography. Kingston, R.I. Abstr.

———. 1986b. Feeding behavior of a unique planktonic freshwater amphipod, *Hyalella montezuma*: Exploitation of neustonic phytoplankton and copepod prey. Zooplankton Behavior Symposium, Skidway Institute Of Oceanography. University of Georgia, Savannah. Abstr.

Steadman, D. W. 1988. Vertebrates from the Late Quaternary Hiscock site, Genessee County, New York. Pages 95–113 in R. S. Laub, H. G. Miller, and D. W. Steadman, editors. Late Pleistocene and early Holocene paleoecology and archeology of the eastern Great Lakes region. Buffalo Society of Natural Sciences 33.

Stebbins, G. L. 1982. Floristic affinities of the high Sierra Nevada. Madroño 29:189–199.

Stebbins, R. C. 1985. A field guide to western reptiles and amphibians. Houghton Mifflin, Boston.

Stevens, L. E., and T. Ayers. 2002. The biodiversity and distribution of alien vascular plants and animals in the Grand Canyon region. Pages 241–265 in B. Tellman, editor. Invasive exotic species in the Sonoran region. University of Arizona Press, Tucson.

Stevens, L. E., and A. E. Springer. 2005. A conceptual model of springs ecosystem ecology. National Park Service, Flagstaff, Ariz. Available at http://www.nature.nps.gov/im/ units/scpn/phase2.htm.

Stevens, L. E., P. B. Stacey, A. Jones, D. Duff, C. Gourley, and J. C. Caitlin. 2005. Protocol for rapid assessment of southwestern stream-riparian ecosystems. Pages 397–420 *in* C. van Riper III and D. J. Mattson, editors. Fifth conference on research on the Colorado Plateau. University of Arizona Press, Tucson.

Stevenson, M. M., and T. M. Buchanan. 1973. An analysis of hybridization between the cyprinodont fishes *Cyprinodon variegatus* and *C. elegans*. Copeia 1973 (4): 683–692.

Stiny, J. 1933. Springs: The geological foundations of springs for engineers of all disciplines as well as students of natural science. Julius Springer, Vienna.

Stohlgren, T. J., D. T. Barnett, and J. T. Kartesz. 2003. The rich get richer: Patterns of plant invasions in the United States. Frontiers in Ecology and the Environment 1:11–14.

Stohlgren, T. J., D. Binkley, G. W. Chong, M. A. Kalkhan, L. D. Schell, K. A. Bull, Y. Otsuki, G. Newman, M. Bashkin, and Y. Son. 1999. Exotic plant species invade hot spots of native plant diversity. Ecological Monographs 69:25–46.

Stone, R. D. 1998. Inventory of sensitive species and ecosystems in Utah: Endemic and rare plants of Utah, an overview of their distribution and status. Utah Division of Wildlife Resources, Salt Lake City.

Stromberg, J. C., R. Tiller, and B. Richter. 1996. Effects of groundwater decline on riparian vegetation of semiarid Arizona. Ecological Applications 6 (1): 113–131.

Stromberg, J. C., J. A. Tress, S. D. Wilkens, and S. D. Clark. 1992. Response of velvet mesquite to ground water decline. Journal of Arid Environments 23:45–58.

Suess, E., G. Bohrman, R. von Huene, P. Linke, K. Wallmann, S. Lammers, and H. Sahling. 1998. Fluid venting in the eastern Aleutian subduction zone. Journal of Geophysical Research 103:2597–2614.

Sutton, M. 1954. Montezuma Well. Arizona Highways 30 (July): 30–35.

Swanson, G. J. 1989. Harvesting ground water: The West's newest crop. Water Well Journal 43:54–57.

Tate, R. L. 1995. Soil microbiology. John Wiley and Sons, New York.

Taylor, D. W. 1966. A remarkable snail fauna from Coahuila, Mexico. Veliger 9 (2): 152–228.

———. 1967. Late Pleistocene molluscan shells from the Tule Springs area. Pages 397–402 *in* H. M. Wormington and D. Ellis, editors. Pleistocene studies in southern Nevada. Nevada State Museum Anthropological Papers 13. Carson City.

Taylor, D. W., and W. L. Minckley. 1966. New world for biologists. Pacific Discovery 19 (5): 18–22.

Taylor, F. B., L. A. Gillman, and J. W. Pedretti. 1989. Impact of cattle on two isolated fish populations in Pahranagat Valley, Nevada. Great Basin Naturalist 49:491–495.

Teal, J. M. 1957. Community metabolism in a temperate cold spring. Ecological Monographs 27:283–302.

Team Arundo del Norte. 2002. Team Arundo del Norte. Available from http://ceres.ca.gov/ tadn/.

Texas. 1996. Barshop & Medina County Underground Water Conservation District. 925 SW 2d 618, 625–26.

Thomas, E. P. 1995. Evolutionary ecology of amphipods in xeric and mesic environments. Ph.D. dissertation, Northern Arizona University, Flagstaff.

Thomas, E. P., D. W. Blinn, and P. Keim. 1994. A test of an allopatric speciation model for congeneric amphipods in an isolated aquatic ecosystem. Journal of the North American Benthological Society 13:100–109.

———. 1997. Genetic and behavioral divergence among desert spring amphipod populations. Freshwater Biology 38:137–143.

———. 1998. Do xeric landscapes increase genetic divergence in aquatic ecosystems? Freshwater Biology 40:587–593.

Thomas, J. W., C. Maser, and J. E. Rodiek. 1979. Wildlife habitats in managed rangelands— The Great Basin of southeastern Oregon: Riparian zones. General Technical Report PNW-80. U.S. Forest Service, Portland, Oreg.

Tilly, L. J. 1968. The structure and dynamics of Cone Spring. Ecological Monographs 38:169–197.

Timmins, S. M. 1992. Wetland vegetation recovery after fire: Eweburn Bog, Te Anau, New Zealand. New Zealand Journal of Botany 30:383–399.

Tiner, R. W. 2003. Geographically isolated wetlands of the United States. Wetlands 23:494–516.

Tintoretto, E. 2004. Under the desert: The mysterious waters of Quatro Cienegas. Associacion La Venta, Instituto Coahuilense de Ecologia, Coahuila, Mexico.

Todd, T. C. 1996. Effects of management practices on nematode community structure in tallgrass prairie. Applied Soil Ecology 3:235–246.

Tryon, M. D., and K. M. Brown. Complex flow patterns through Hydrate Ridge and their impact on seep biota. Geophysical Research Letters 28:2863–2866.

Tryon, M. D., K. M. Brown, L. M. Dorman, and A. Sauter. 2001. A new benthic aqueous flux meter for very low to moderate discharge rates. Deep-Sea Research I 48:2121–2146.

Tsuk, T. 2000. Ancient water systems in settlements in Eretz-Israel. Tel Aviv University, Tel Aviv, Israel.

Tuhy, J. S., and J. A. MacMahon. 1988. Vegetation and relict communities of Glen Canyon National Recreation Area. Final report to the National Park Service. The Nature Conservancy and Department of Biology and Ecology Center, Utah State University, Logan.

Uhl, N. W., and J. Dransfield. 1987. Genera Palmarum: The L. H. Bailey Hortorium and International Palm Society. Allen Press, Lawrence, Kans.

United States v. Cappaert. 1974. 508 F. 2d 313, 9th Cir.

Unmack, P. J. 1995. Desert fishes down under. Proceedings of the Desert Fishes Council 26:71–95.

U.S. Census Bureau. 1999. Statistical abstract of the United States. 119th ed. U.S. Census Bureau, Washington, D.C.

U.S. Department of Agriculture. 1999. Final environmental impact statement for Tusayan growth: Hydrology appendix. Montgomery and Associates, Inc. U.S. Department of Agriculture, Forest Service, Kaibab National Forest, Ariz.

U.S. Department of Agriculture, NRCS. 2001. The PLANTS database. Version 3.5 (http://plants.usda.gov). Data compiled from various sources by Mark W. Skinner. National Plant Data Center, Baton Rouge, La.

U.S. Department of the Interior. 1993. Riparian area management: Process for assessing proper functioning condition. Bureau of Land Management Technical Reference 1737-9. Washington, D.C.

————. 1998. Riparian area management: A user guide to assessing proper functioning condition and the supporting science for lotic areas. Bureau of Land Management Technical Reference 1737-15. Denver, Colo.

U.S. Fish and Wildlife Service. 1984. Recovery plan for the Owens pupfish, *Cyprinodon radiosus*. U.S. Fish and Wildlife Service, Portland, Oreg.

————. 1989. Recovery plan for Owens tui chub, *Gila bicolor snyderi*. U.S. Fish and Wildlife Service, Portland, Oreg.

————. 1990. Recovery plan for the endangered and threatened species of Ash Meadows, Nevada. U.S. Fish and Wildlife Service, Portland, Oreg.

————. 1995. Rio Yaqui fishes recovery plan. U.S. Fish and Wildlife Service, Albuquerque, N.Mex.

U.S. Geological Survey. 1997. Ground water atlas of the United States—Segment 1, California-Nevada. *In* M. J. Mac, P. A. Opler, C. E. Puckett-Haecker, and P. D. Doran, editors. The status and trends of the nation's biological resources. U.S. Department of the Interior, U.S. Geological Survey, Reston, Va. Internet.

Van Der Kamp, G. 1995. The hydrogeology of springs in relation to the biodiversity of spring fauna: A review. Journal of the Kansas Entomological Society 68 (supplement): 4–17.

Vander Zanden, M. J., J. M. Cassleman, and J. B. Rasmussen. 1999. Stable isotope evidence for the food web consequences of species invasions in lakes. Nature 401:464–467.

Van Dover, C. L., C. R. German, K. G. Speer, L. M. Parson, and R. C. Vrijenhoek. 2002. Evolution and biogeography of deep-sea vent and seep invertebrates. Science 295:1253–1257.

Van Everdingen, R. O. 1991. Physical, chemical, and distributional aspects of Canadian springs. Memoirs of the Entomological Society of Canada 155:1–217.

Vannote, R. L., G. W. Minshall, K. W. Cummins, J. R. Sedell, and C. E. Cushing. 1980. The river continuum concept. Canadian Journal of Fisheries and Aquatic Sciences 37:130–137.

Van Pelt, N. S., C. D. Schelz, J. S. Tuhy, D. W. Johnson, and J. R. Spence. 1991. Survey and analysis of relict plant communities within national park system units of the Colorado Plateau, vols. 1–3. Final report to the National Park Service, Rocky Mountain Region. Nature Conservancy, Great Basin Field Office, Salt Lake City, Utah.

Vela-Coiffer, P., and D. F. Lozano Garcia. 2000. S.I.G. de Cuatro Ciénegas: Distribución vegetal en el area de proteccion de flora y fauna de Cuatrociénegas, Coahuila, México. Available at http://www-labsig.mty.itesm.mx/Proyectos/4cienegas/4cienegas.html.

Vogl, R. J., and L. T. McHargue. 1966. Vegetation of California fan palm oases on the San Andreas Fault. Ecology 47:532–540.

Votteler, T. H. 2002. The little fish that roared. Tulane Environmental Law Journal 15:257–270.

Wager, R., and P. J. Unmack. 2000. Fishes of the Lake Eyre catchment of central Australia. Queensland Department of Primary Industries and Queensland Fisheries Service, Brisbane, Australia.

Wagner, V. T., and D. W. Blinn. 1987a. A comparative study of the maxillary setae of two co-existing species of *Hyalella* (Amphipoda), a filter feeder and a detritus feeder. Archiv für Hydrobiologie 109:409–419.

———. 1987b. Montezuma Well: The living desert oasis. National Park Service brochure.

Wallace, A. R. 1863. On the physical geography of the Malay Archipelago. Journal of the Royal Geographic Society (London) 33:217–234.

Wallace, M. P., and C. Alfaro. 2001. Geologic/hydrogeologic setting and classification of springs. Pages 33–72 *in* P. E. LaMoreaux and J. T. Tanner, editors. Springs and bottled waters of the world: Ancient history, source, occurrence, quality and use. Springer, Berlin.

Walters, C., J. Korman, L. E. Stevens, and B. D. Gold. 2000. Ecosystem modeling for evaluation of adaptive management policies in the Grand Canyon. Conservation Ecology 4:1. Available at http://www.consecol.org/vol4/iss2/art1.

Waring, G. A. 1965. Thermal springs of the United States and other countries of the world: A summary. Geological Survey Professional Paper 492. U.S. Government Printing Office, Washington, D.C.

Warnica, J. M. 1966. New discoveries at the Clovis site. American Antiquity 31:345–357.

Warren, P. L., K. L. Reichhardt, D. A. Mouat, B. T. Brown, and R. R. Johnson. 1982. Vegetation of Grand Canyon National Park. Technical Report No. 9. National Park Service, Cooperative National Park Resources Studies Unit, University of Arizona, Tucson.

Weed, W. H. 1889. Formation of travertine and siliceous sinter by vegetation of hot springs. U.S. Geological Survey 9th Annual Report, 1887–1888: 613–676. U.S. Government Printing Office, Washington, D.C.

Weeks, G. F. 1918. Seen in a Mexican plaza: A summer's idyll of an idle summer, by El Gringo (Geo. F. Weeks). Fleming H. Revell, New York.

Welsh, S. L. 1984. Flora of Glen Canyon National Recreation Area. Unpublished report to Glen Canyon National Recreation Area, Page, Ariz.

———. 1989a. Hanging gardens of Zion National Park. Contract No. CX 1590-7-0001. Report prepared for Water Resource Division, National Park Service, Ft. Collins, Colo.

———. 1989b. On the distribution of Utah's hanging gardens. Great Basin Naturalist 49:1–30.

Welsh, S. L., N. D. Atwood, L. C. Goodrich, and L. C. Higgins. 1993. A Utah flora. 2nd ed. Brigham Young University Press, Provo, Utah.

Welsh, S. L., and C. A. Toft. 1981. Biotic communities of hanging gardens in southeastern Utah. National Geographic Society Reports 13:663–681.

Wendland, H. A. 1879. Botanische Zeitung 37: lxi, 68, 148.

Wendorf, F., and J. J. Hester, editors. 1975. Late Pleistocene environments of the southern High Plains. Publication of the Fort Burgwin Research Center, no. 9. Dallas, Texas.

West, N. E. 1988. Intermountain deserts, shrub steppes, and woodlands. Pages 209–230 *in* M. G. Barbour and W. D. Billings, editors. North American terrestrial vegetation. Cambridge University Press, New York.

Wetzel, M. J., G. E. Oberlin, and D. W. Blinn. 1999. The aquatic Oligochaeta (Annelida: Clitellata) of Montezuma Well, Arizona: A near-thermally constant limnocrene. Southwestern Naturalist 44:514–518.

White, P. S. 1979. Pattern, process, and natural disturbance in vegetation. Botanical Review 45:229–299.

White, R., D. Withers, N. Kanim, K. Stubbs, T. Klahr, and J. Hanlon. 1995. U.S. Fish and Wildlife Service, Region 1, report on conservation actions undertaken during 1993 for federally listed and candidate fishes and other aquatic species in California, Idaho, Nevada, and Oregon. Proceedings of the Desert Fishes Council 24 (1994): 5–9.

Whiting, P. J., and J. Stamm. 1995. The hydrology and form of spring-dominated channels. Geomorphology 12:235–240.

Williams, C. D. 1984. The decline of Ash Meadows, a unique desert ecosystem. Pages 717–719 in R. E. Warner and K. M. Hendrix, editors. California riparian systems: Ecology, conservation, and productive management. University of California Press, Berkeley.

Williams, D. D., and H. V. Danks. 1991a. Arthropods of springs: Introduction. Memoirs of the Entomological Society of Canada 155:3–4.

———, editors. 1991b. Arthropods of springs with particular reference to Canada. Memoirs of the Entomological Society of Canada 155:7–28.

Williams, J. E. 1991. Preserves and refuges for native western fishes: History and management. Pages 171–190 in W. L. Minckley and J. E. Deacon, editors. Battle against extinction: Native fish management in the American West. University of Arizona Press, Tucson.

Williams, J. E., D. B. Bowman, J. E. Brooks, A. A. Echelle, R. J. Edwards, D. A. Hendrickson, and J. J. Landye. 1985. Endangered aquatic ecosystems in North American deserts, with a list of vanishing fishes of the region. Journal of the Arizona-Nevada Academy of Science 20:1–62.

Williams, J. E., J. Johnson, D. A. Hendrickson, S. Contreras-Balderas, J. D. Williams, M. Navarro-Mendoza, D. E. McAllister, and J. E. Deacon. 1989. Fishes of North America endangered, threatened or of special concern: 1989. Fisheries 14 (6): 2–20.

Williams, J. E., G. C. Kobetich, and C. T. Benz. 1985. Management aspects of relict populations inhabiting the Amargosa Canyon ecosystem. Pages 706–715 in R. E. Warner and K. M. Hendrix, editors. California riparian systems: Ecology, conservation, and productive management. University of California Press, Berkeley.

Williams, J. E., and D. W. Sada. 1985. Status of two endangered species, *Cyprinodon nevadensis mionectes* and *Rhinichthys osculus nevadensis*, from two springs in Ash Meadows, Nevada. Southwestern Naturalist 30:475–484.

Wilson, E. 2000. Geologic framework and numerical groundwater models of the South Rim of the Grand Canyon, Arizona. M.S. thesis, Northern Arizona University, Flagstaff.

Wilson, K. P., and D. W. Blinn. 2007. Food web structure, energetics, and importance of allochthonous carbon in a desert cavernous limnocrene: Devils Hole, Nevada. Western North American Naturalist 67:185–198.

Wilson, W. E., and J. E. Moore, editors. 1998. Glossary of hydrology. American Geological Institute, Alexandria, Va.

Winograd, I. J., and F. J. Pearson. 1976. Major carbon-14 anomaly in a regional carbonate aquifer: Possible evidence for megascale channels. Water Resources Research 12:1125–1143.

Winograd, I. J., and W. Thordarson. 1975. Hydrogeologic and hydrochemical framework, south-central Great Basin, with special reference to the Nevada Test Site. U.S. Geological Survey Professional Paper 712-C. U.S. Government Printing Office, Washington, D.C.

Winsborough, B. M. 1990. Some ecological aspects of modern fresh-water stromatolites in lakes and streams of the Cuatro Cienegas basin, Coahuila, Mexico. Ph.D. dissertation, University of Texas, Austin.

Winsborough, B. M., J.-S. Seeler, S. Golubic, R. L. Folk, and B. Maguire Jr. 1994. Recent fresh-water lacustrine stromatolites, stomatolitic mats and oncoids from northeastern Mexico. Pages 71–100 in J. Bertrand-Sarfati and C. Monty, editors. Phanerozoic stromatolites II. Kluwer Academic Publishers, Netherlands.

Winter, T. C. 1988. A conceptual framework for assessing cumulative impacts on the hydrology of non-tidal wetlands. Environmental Management 12:605–620.

Winter, T. C., J. W. Harvey, O. M. Franke, and W. M. Alley. 1998. Ground water and surface water: A single resource. U.S. Geological Survey Circular 1139. U.S. Government Printing Office, Washington, D.C.

Withers, K., and J. L. Mead. 1993. Late Quaternary vegetation and climate in the Escalante River basin of the central Colorado Plateau. Great Basin Naturalist 53:145–161.

Witt, J. D. S., D. W. Blinn, and P. D. N. Hebert. 2003. The recent evolutionary origin of the ecologically novel amphipod *Hyalella montezuma* offers an ecological explanation for morphological stasis in a closely allied species complex. Molecular Ecology 12:405–423.

Wong, D. 1999. Community analysis of hanging gardens at Zion National Park, Utah, and the Colorado Plateau. M.S. thesis, Department of Biology, Northern Arizona University, Flagstaff.

Wood, W. R., and R. B. McMillan. 1976. Recent investigations at Rodger Shelter, Missouri. Archaeology 20:52–55.

Woodbury, A. M. 1933. Biotic relationships of Zion Canyon, Utah, with special reference to succession. Ecological Monographs 3:147–245.

Wormington, H. M., and D. Ellis, editors. 1967. Pleistocene studies in southern Nevada. Nevada State Museum Anthropology Papers 13. Carson City.

Zabriskie, J. G. 1979. Plants of Deep Canyon. Philip L. Boyd Deep Canyon Research Center, University of California, Riverside.

Zeidler, W., and W. F. Ponder, editors. 1989. Natural History of Dalhousie Springs. South Australia Museum, Adelaide, Australia.

Zervas, G. 1998. Quantifying and optimizing grazing regimes in Greek mountain systems. Journal of Applied Ecology 35:983–986.

About the Contributors

Diana E. Anderson is an associate professor of environmental sciences and education, and quaternary sciences at Northern Arizona University. She received her BS and MS from NAU and her PhD in geological sciences from the University of California, Riverside. Her research interests are in arid land geomorphology, meteorology, paleoclimatology, desertification, and watershed restoration.

Dean W. Blinn received his BA from Simpson College, his MA from the University of Montana in terrestrial plant ecology, and his PhD in aquatic ecology from the University of British Columbia in Vancouver, Canada. Dr. Blinn retired from Northern Arizona University as a Regents' Professor in 2001 and now lives in Bellingham, Washington. He has published more than 125 articles on aquatic ecology with students and colleagues during his career.

Adam E. Cohen received his master's degree from the University of Texas at Austin, studying fishes in Cuatro Ciénegas, Coahuila, Mexico. After working as an aquatic biologist with the Texas Natural Resources and Conservation Commission, he now works at the Texas Natural Science Center of the University of Texas at Austin, collaborating with Dean Hendrickson on Texas fish collection databases and a book project on the freshwater fishes of Texas.

James W. Cornett has lived and worked in the Sonoran Desert since 1972. For nearly thirty years he was with the Palm Springs Desert Museum as its director of natural sciences. He has written widely on desert subjects in his weekly newspaper column, popular magazines, scientific journals, and more than two dozen books. Currently, his formal research activities are sponsored by the Joshua Tree National Park Association.

Eric Dinger spent eight years studying the aquatic invertebrates and stromatolites of Cuatro Ciénegas while working on his MSc and PhD degrees from Northern Arizona University. He now works in colder climes in the mountains of northern Utah as a researcher at the National Aquatic Monitoring Center at Utah State University.

Stephen P. Flora is currently a project hydrogeologist working with Miller Brooks Environmental, Inc., in Phoenix. He graduated from Northern Arizona University in May 2004 with a master's degree in geology. His research while at NAU entailed characterizing, inventorying, and monitoring springs throughout the Verde River watershed in central Arizona.

Tim B. Graham grew up in New Mexico and earned a bachelor's degree at Evergreen State College in marine ecology and a PhD from Utah State University in biology/ecology. Graham has worked for the Department of the Interior for more than twenty years and is currently a research ecologist with the U.S. Geological Survey. His research focuses on the ecological relationships of terrestrial and aquatic invertebrates in deserts, and on amphibian and reptile population monitoring and dynamics.

C. Vance Haynes Jr. is a Regents' Professor Emeritus at the University of Arizona and a member of the National Academy of Sciences. He obtained his doctoral degree from the University of Arizona in 1965. For the past 50 years he has specialized in the geochronology of the peopling of the Americas. He was awarded the GSA Archaeological Geology Award in 1984 and the AMQUA Distinguished Career Award in 2002. Haynes has conducted field research at many of the important Paleoindian sites in North America, has performed Paleolithic research in East Africa and the eastern Sahara, and has published extensively on these topics.

Dean A. Hendrickson is curator of ichthyology at the Texas Natural Science Center of the University of Texas at Austin, where his research focuses on the evolution, ecology, and conservation of North American freshwater fishes, with a focus on those of deserts, predominantly in Mexico. Recently his conservation efforts broadened to include other aquatic organisms, the control of invasive species, and aquatic habitat restoration in Cuatro Ciénegas in Coahuila, Mexico. He has also opened a research station in Cuatro Ciénegas to facilitate research by others and to improve communication between researchers and the local community.

David K. Kreamer is a professor in the Department of Geoscience at the University of Nevada, Las Vegas, and served as director of the Water Resources Management Graduate Program for 12 years there. He is an internationally renowned authority on groundwater hydrology, water quality, contaminant transport, and aquifer remediation, and he has advised numerous government agencies on issues related to groundwater management. He is an avid whitewater kayaker and river runner.

Lisa A. Levin is a professor of biological oceanography at the Scripps Institution of Oceanography in La Jolla, California. She is an international expert on coastal wetlands and the marine biodiversity of continental margin cold springs.

Jane C. Marks is an aquatic ecologist at Northern Arizona University, where her research focuses on the conservation and restoration of freshwater ecosystems. Her research integrates natural history, stable isotopes of food web structure, and manipulative experiments to evaluate threats and restoration opportunities for freshwater species. In addition to Mexico, she is working at Fossil Creek in Arizona and in the Catalunya region of Spain.

Vicky J. Meretsky is an associate professor of conservation biology at the School of Public and Environmental Affairs at Indiana University. She is interested in conservation issues in managed landscapes, from the species to the ecosystem level. "Science helps us to find answers," she believes, "it doesn't help us to implement them. For that, you need a bigger tool box."

W. L. Minckley was a professor of biology at Arizona State University from 1963 until his passing in 2001. He published more than 200 scientific articles and books, and 45 students completed graduate degrees under his supervision. Dr. Minckley was especially interested in aquatic biology, with a strong emphasis on conservation and the responsible management of natural resources.

Angela B. Moline is a faculty member of the Audubon Expedition Institute at Lesley University in Massachusetts. Her research focuses on the riparian invasions and aquatic ecology of the western United States and Mexico. Specifically, she studies the effects of *Tamarix* on Colorado Plateau stream communities.

Gary Paul Nabhan is one of the best-known authors in the Southwest. He served as the director of the Arizona-Sonora Desert Museum in Tucson and the director of the Center for Sustainable Environments at Northern Arizona University in Flagstaff. He is now at the Southwest Center of the University of Arizona. He has conducted ethnoecological studies on desert springs, including Quitobaquito and Quitovac, for more than 30 years.

Nancy Nelson is a 1997 graduate of the University of Colorado Fleming School of Law. She served as a law clerk for a senior judge on the United States Tenth Circuit Court of Appeals. She lives in Flagstaff, Arizona.

Roderic A. Parnell is a professor of geology and environmental sciences at Northern Arizona University and chair and director of the Center for Environmental Sciences and Education. He received his PhD in geology from Dartmouth College in 1982 and has been a faculty member at St. Lawrence University, the University of Virginia, and Northern Arizona University. He has published extensively on the effects of acid rain, volcanic emissions, and sulfide mineral deposits on terrestrial and aquatic ecosystems, and on the biogeochemistry and geomorphology of southwestern U.S. rivers.

Duncan T. Patten recieved his PhD in plant ecology from Duke University. He is a research professor in the Department of Land Resources and Environmental Sciences at Montana State University and Professor Emeritus of plant biology at Arizona State University. His research includes understanding the processes, management, and long-term sustainability of mountain and aridland riparian and wetland ecosystems.

Bianca S. Perla is a doctoral student in ecosystems analysis at the University of Washington, Seattle. She obtained her bachelor's degree in earth systems biology from Stanford University and her MS in Biology from Northern Arizona University. As a conservation scientist for Grand Canyon Wildlands Council, Inc., Perla enjoyed working with Larry Stevens on an extensive inventory of desert springs in Arizona.

Amadeo M. Rea is an ethnobiologist and ornithologist who has worked with birds and southwestern indigenous peoples for the past four decades, and he teaches anthropology courses at the University of San Diego. His research focuses on the interrelationships between Piman peoples and the natural world, including cognition and naming, myth and metaphor, and traditional subsistence. His books include *At the Desert's Green Edge, Folk Mammalogy of the Northern Pimans, and Wings in the Desert,* all dealing with the ethnobiology of the Pimans of Arizona, Sonora, and Chihuahua.

Leigh Rouse is an ecologist with ERO Resources Corporation in Denver. She earned her MS degree at Arizona State University, studying avian habitats at springs in the Mojave Desert. She presently specializes in wetland issues, including delineation, mitigation, and monitoring, and provides natural resource assessments for public and private landowners.

John R. Spence received his BSc in botany and MSc in plant ecology from Utah State University, and his PhD from the University of British Columbia in 1986, where he studied the ecology of alpine mosses of the Pacific Northwest. He has worked on plants in many parts of the world, including East Africa, New Zealand, and Australia. His research interests include bryophyte taxonomy and ecology, floristics, and the evolution and reproductive ecology of endemic plants on the Colorado Plateau, Gentianaceae systematics, and bird distribution. He currently works for the National Park Service's Resource Management Division of Glen Canyon National Recreation Area in Page, Arizona.

Abraham E. Springer is an associate professor of geology in the Department of Geology at Northern Arizona University, and he is the NAU Water Coordinator for the Arizona Water Institute. He received his BS in geology from the College of Wooster and his MS and PhD in hydrogeology from Ohio State University. Dr. Springer and his students study local and regional groundwater flow systems and human impacts on them, apply principles of sustainability to aquifer management through models, quantify the hydrological function of groundwater-dominated ecosystems, the role of land-use change and disturbance on groundwater flow systems, and the restoration of riparian ecosystems.

Lawrence E. Stevens received his BA from Prescott College and his MS and PhD from Northern Arizona University. He served as the ecologist for Grand Canyon National Park, working on the downstream impacts of Glen Canyon Dam. He is the senior scientist for the Grand Canyon Wildlands Council, Inc., and the curator of ecology and conservation at the Museum of Northern Arizona in Flagstaff. He conducts basic and applied research on the biota and ecosystems of the Southwest.

Juliet Stromberg is an associate professor in the School of Life Sciences at Arizona State University. She and her graduate students have been studying desert rivers for more than two decades, seeking to understand the processes that shape riparian plant populations, communities, and landscapes. She serves on the editorial board of the journal *Wetlands.*

Peter J. Unmack was originally from Melbourne, Australia, and recently completed his PhD on the biogeography of Australian freshwater fishes at Arizona State University. His main interests are in the biogeography, systematics, ecology, and conservation of freshwater fishes and springs ecosystems.

Index